FUNDAMENTAL MATHEMATICS FOR ELEMENTARY TEACHERS

FUNDAMENTAL MATHEMATICS FOR ELEMENTARY TEACHERS:
A BEHAVIORAL OBJECTIVES APPROACH

A. RICHARD POLIS
Beaver College
Glenside, Pennsylvania

EARL M. L. BEARD
University of Maine
Orono, Maine

HARPER & ROW, Publishers
New York, Evanston, San Francisco, London

FUNDAMENTAL MATHEMATICS FOR ELEMENTARY TEACHERS:
A Behavioral Objectives Approach

Harper & Row, Publishers, Inc.
10 East 53rd Street, New York, N.Y. 10022

Standard Book Number: 06-045244-7
Library of Congress Catalog Card Number: 72-86370

In memory of Betty Lou Martin and Beatrice Polis

CONTENTS

Preface   XI
To the Instructor   XIII
To the Student   XVII
Acknowledgments   XIX

1. SETS   1

A.   Notation   1
B.   Set Relationships   6
C.   Set Equality and Equivalence Relations   19
D.   Operations on Sets   26
E.   Fundamental Properties of Operations   37
F.   The Cartesian Set, $A \times B$ 48
Capsule I: Structure in School Mathematics   51

2. NUMERATION SYSTEMS   54

A.   Characteristics of Numeration Systems   54
B.   Early Systems of Numeration   59
C.   Hindu-Arabic System   69
D.   Bases Other Than Ten   72

3. NATURAL NUMBERS AND WHOLE NUMBERS   93

A.   Operations on N and W 93
B.   Properties of Addition and Multiplication on N and W 107
C.   Properties of Order on N and W 128
Capsule II: Discovery — Its Role in the Teaching
of Elementary School Mathematics   133

4. NUMBER THEORY   138

A.   Factorization   138
B.   Divisibility   145
C.   Least Common Multiple and Greatest Common Divisor   153
D.   Special Numbers   161

5. INTEGERS   167
A.   Number Lines   167
B.   Equivalence Classes   170
C.   Addition   178
D.   Subtraction   191
E.   Multiplication   194
F.   Division   209
G.   Order   212
Capsule III: Discovery Sequences   215

6. STRUCTURES   219
A.   Groups   219
B.   Modular Systems   227
C.   Cyclic Groups   235
Capsule IV: Braids   242

7. RATIONAL NUMBERS   248
A.   Operations on the Rationals   248
B.   Properties of Addition and Multiplication   258
C.   Subtraction   272
D.   Division   282
E.   The Properties of Order on the Set of Rational Numbers   288

8. RATIONAL NUMBERS AS FRACTIONS AND DECIMALS   301
A.   Fractions   301
B.   Addition and Subtraction   308
C.   Fundamental Properties of Addition and Subtraction   314
D.   Multiplication and Division   315
E.   Fundamental Properties of Multiplication and Division   323
F.   Fractions and Decimals   324
G.   Operations with Decimals   328
Capsule V: Mathematics Laboratory   340

9. REAL NUMBERS   342
A.   Fundamental Properties of Operations on the Reals   342
B.   Solutions and Solution Sets   352

10. LOGIC   375

    A.   Statements   375

    B.   Conditional and Biconditional   386

    C.   Tautology   396

    D.   Disjunction, Syllogism, and Proof   399

    Capsule VI: Games   407

11. GEOMETRY   409

    A.   Points, Lines, and Planes   409

    B.   Angles   412

    C.   Curves and Polygons   415

    D.   Rigid Motions   420

    E.   Reflections   438

    F.   Congruence   443

Index   447

PREFACE

The purpose of using behavioral objectives has been clearly stated during the past few years by many writers in the field of education. It has been pointed out by these writers that the use of well-defined objectives has many advantages. For example, the student is aware of what the instructor wishes him to learn or what has priority. Once a student is aware of the goals of the instructor, he no longer has to search for hidden meaning in the course material. He simply goes on to the business at hand — learning.

The process of identifying objectives is, of itself, valuable. The instructor becomes sensitive to the explicit goals of the course. He no longer thinks about "teaching the rational numbers," but decides what his students should learn. His focus is, first, on the outcome of the course and, second, on the ways to achieve that outcome. Thus, he avoids the easy trap of thinking of what he will do (the means). His tests are not constructed on the basis of the question "What can I ask?" Rather, his early identification of explicit objectives leads naturally to the evaluation of achieving those objectives.

This approach has been field tested by the authors over a 3-year period with approximately 200 students. Each student was required to show mastery of the stated objectives of each unit. (Students were required to complete objectives they had initially missed.) As a result of student surveys, it was apparent that student learning was more efficient because the students were made aware of their accomplishments as well as their deficiencies.

The amount of work required to produce detailed lists of objectives and tests is overwhelming for most instructors. The iterative process used to refine the instructional material requires a great amount of out-of-class time. During the trial period we found that the number of objectives increased until each unit was well delineated.

We believe that a textbook that includes statements of objectives for each unit will provide instructors with a comprehensive syllabus for the course. An instructor may elect to delete particular units for his course or may develop new units for objectives not in the text or may modify the sequence of units to fit his own approach.

We suggest that the instructor pre-test his students to identify their understanding of the listed objectives and then teach those areas that require clarification. A post-test should be administered following the period of instruction or at a time when students feel ready to indicate their mastery of the objectives. Students who have mastered the objectives can then proceed to the next unit; however, students who have not mastered a particular set of objectives should be assisted until they are able to demonstrate their competence. The problem exercises following each subset of objectives allow the student to employ self-evaluation during the learning process and, in particular, to identify those areas where his efforts might best be directed.

In addition to the fundamental mathematics of the real number system, this text contains much of use to the prospective elementary teacher. A chapter on logic is included, as well as a development of Euclidean geometry based on transformations and including material suitable for the elementary classroom. Although the text was not specifically written to C.U.P.M. recommendations, the prospective instructor will find that the committee recommendations are generally satisfied.

Several examples accompany each mathematical concept in the text. In addition, there are exercise problems at the end of each section. The behavioral objectives around which the text has been organized are listed at the beginning of each section and are also keyed into the text. Thus, the student can pinpoint within the text itself the important concepts, ideas, and facts.

TO THE INSTRUCTOR

This book grew out of pre-service and in-service courses for elementary school teachers. Basically, it is a content text; however, we believe that it is essential for these students to be exposed to current thought and methods in mathematics education. To do this, we have included in the text five "Capsules." These are spaced throughout the text and can be used as the instructor desires. In addition to our ideas on what comprises and aids teaching in the elementary school, several of the Capsules include a suggested reading list that will add to the students' knowledge of the field.

Chapters 2, 4, 6, 10, and 11 are independent; therefore, a one-semester course can be organized around Chapters 1, 3, 5, 7, 8, and 9. If additional material is desired, it can be taken from the omitted chapters, or from the Capsules if a content — methods course is desired.

Although a one-semester course is possible, it would virtually eliminate topics such as numeration systems, number theory, and geometry. These topics, among others, are essential to the elementary school teacher and, therefore, along with many others we recommend a two-semester course. A two-semester course usually contains 28 to 30 weeks of classes. For instructors who would like an estimate of how this text could be used in such a course, we offer an approximate schedule below. We point out that this schedule is only a suggestion and that individual schedules might differ greatly from this one.

Chapter 1	Sets	2½–3½ weeks
Chapter 2	Numeration Systems	1½–2½ weeks
Chapter 3	Natural Numbers and Whole Numbers	2–2½ weeks
Chapter 4	Number Theory	2–2½ weeks
Chapter 5	Integers	2½–3½ weeks
Chapter 6	Structures	1½–2 weeks
Chapter 7	Rational Numbers	3–4 weeks
Chapter 8	Rational Numbers as Fractions and Decimals	2–2½ weeks
Chapter 9	Real Numbers	2–2½ weeks
Chapter 10	Logic	2½–3 weeks
Chapter 11	Geometry	2½–3 weeks

Since every instructor spends more or less time on a particular subject, the above should be viewed as a schedule of minimum suggested times. Many of the chapters contain enough material to considerably lengthen the actual time spent. This is particularly true of Chapters 1, 5, and 7.

We would like to give a brief résumé of the chapters to indicate their relationship to each other:

Chapter 1 is basic to the whole book. Very little can be overlooked or deleted. All the operations on the whole and natural numbers are built upon sets.

Chapter 2 deals with the history of number systems and includes work with better-known systems such as the Egyptian, Roman, and Babylonian. We have also included work on the Mayan system. Our development of number bases is rather standard, although few books discuss direct translation from base two to four or two to eight, as we do.

Chapter 3 develops the operation on whole and natural numbers. The properties developed in this chapter are used for proof of properties in the set of integers.

Chapter 4 deals with primes, composites, divisibility, G.C.D., and L.C.M. We include a discussion of polygonal numbers, Pythagorean triples, perfect numbers, etc.

Chapter 5 includes a development of the integers, and operations on the set. The integers are developed as sets of equivalence classes. We emphasize the distinction between elements that can represent the integer, and the integer or equivalence class itself. Our notation clearly delineates this difference, which is often misunderstood by students.

Chapter 6 deals with two structures, groups and fields. The primary emphasis is on the group structure. We develop modular systems, rotation groups, cyclic groups, and transformation groups. This topic can be dealt with nicely and applied to the teaching of children.

Chapter 7, the rational numbers, draws on the integers to develop the properties of operations. Again, we differentiate between objects in the equivalence class and the equivalence class itself. The development is quite rigorous and gives students a new insight into mathematics that seems familiar to them.

Chapter 8 develops the operations on rational fractions from the properties found in Chapter 7. Several good physical interpretations are used.

Chapter 9 continues with the real numbers. The development is more intuitive than the previous chapters, and main emphasis is on solving open sentences.

Chapter 10 deals with logical statements as connectives. Since simple proofs are developed and rules of logic employed, a great deal of the emphasis is on demonstration of proofs using truth tables.

Chapter 11 begins with an informal discussion of the basic objects that the elementary school student meets in his study of geometry. Then, a more formal approach based on transformations follows. This approach permits the prospective teacher to operate with Euclidean geometry from a transformational perspective.

The instructor will find an added benefit from our use of behavioral objectives when he begins to write test questions. Since students are cognizant of the explicit objectives of the course, they will tend to try to master these; and the instructor, knowing this, will tend to ask better questions and sample the material more adequately to evaluate the students' learning.

We hope you find this approach useful in your courses; we believe it benefits both the instructor and the student.

TO THE STUDENT

The intention of this book is to make mathematics easier for you to read and understand. To do this we have listed at the beginning of each section the behavioral objectives that we feel are most important for a course of this nature.

To find the text relating to a specific objective, look for the key number in the left margin. The block of text relating to that objective is marked at the beginnng and at the end by a dark square. If a second objective falls within the block of another objective, the start and finish of the *second* objective are marked by a dark triangle. A third and fourth objective falling within the same block are marked by a circle and a diamond, respectively. There may be more than one block of text relating to a specific objective.

Based on the experiences of students who have used this material in its prepublication form, we have several suggestions that may prove helpful.

If you are already familiar with some of the subject matter:

1. read the objectives before reading the section
2. check off the objectives that you know or can do
3. read the areas in the text corresponding to the unchecked objectives

Keep a pencil and paper handy when you read the text. There are a great number of examples, and you should work the problems as you read.

Finally, we suggest that you read everything in the text. If questions arise in your mind as you read, jot them down in the margin, and try to answer or get the answers to them.

Our hope is that you will find this text readable and that the formal statement of objectives, along with the examples and exercises, will help you learn and understand.

ACKNOWLEDGMENTS

The authors wish to thank the many students who used these materials and made suggestions for their improvement. We are especially indebted to those people who read, reviewed, and criticized this text in its various forms and to our many colleagues whose comments and suggestions were of great help. We thank Norm Miller for supporting our use of this material at Beaver College and at Lehigh University.

FUNDAMENTAL MATHEMATICS
FOR ELEMENTARY TEACHERS

CHAPTER I **SETS**

SECTION A **NOTATION**

BEHAVIORAL OBJECTIVES

A.1 Be able to identify well-defined sets and sets which are not well defined.

A.2 Be able to describe a given set in words.

A.3 Given a verbal description of a set, be able to tabulate the set.

A.4 Be able to distinguish between finite and infinite sets.

A.5 Be able to decide whether an element belongs to a specific well-defined set or not.

A.6 Given a set written in set builder notation, be able to tabulate the set.

A.7 Given a tabulated set, be able to symbolize it, using set builder notation.

SET DESCRIPTION

Man has used the notion of a *set* intuitively since he found a need to describe objects in terms of their common characteristics. We have in our language many words which describe sets. Cowboys round up herds of cattle. We describe sets of geese as flocks. A platoon of men is another type of grouping. Such words as swarm, covey, hive, and legion have all been used to describe special kinds of sets. Although we do not usually use a set of cattle when we have the word herd at our disposal, the word set denotes the characteristic of being a collection. In mathematics we use the word set to describe certain collections of objects such as numbers, points, or transformations. The word set is an example of what we call an *undefined term*. We will simply describe attributes for the term *set*.

Notice that all our examples of words which describe sets were very specific in nature. We would not discuss a platoon of geese or a flock of men. The one thing that herd, platoon, swarm, and covey have in common is the fact that they refer to specific kinds of sets. Some of these words have other connotations. For example, platoon refers not only to a grouping of men but also implies that the men are soldiers. Since it is also true that platoon is used in reference to football teams or police groups, it is often necessary to use the word in context to be sure of its meaning.

In mathematics we like to be precise. Sets are described in such a way that there is no question as to exactly what objects comprise the set. Thus, we require that our sets have the additional property that given any object, we can decide whether it belongs to our set or does not belong to our set. A set with this property is called *well-defined*.

WELL-DEFINED SETS

■ A.5　Let us consider some examples of sets. The set of flowers in a field is not a well-defined set since we do not know what field is meant nor what the boundaries of the field might be. That is, given a specific flower, it is impossible to decide whether or not the flower is in the set. On the other hand, if we specify the field, the boundaries, and the time, the set can be

▲ A.1　well defined. For example, the set of flowers in Malcolm Berry's field at 3 P.M. on August 8, 1973, is a well-defined set, because we are able to determine exactly which flowers comprise the set. Another example of a well-defined set is the set of letters in the English alphabet. The letters α and β are not in this set, but the letters a and b are. Given any letter, we

■ ▲　can determine whether or not it is in the set.

The set of people who will earn an A in this course is not well defined because we cannot now determine the elements of this set. The set of beautiful women is not well defined because the word "beautiful" is not universally defined in a consistent manner.

WRITING SETS IN WORDS AND TABULATING SETS

In each example above we have described a set in words. We can also define sets symbolically. You should be able to designate sets in both these ways.

Here are several sets described in two ways:

■ A.3　The first three letters

▲ A.2　　in the alphabet　　　　$F = \{a, b, c\}$

The days of the week　　$D = \{$Monday, Tuesday, Wednesday, Thursday, Friday, Saturday, Sunday$\}$

The New England states　$S = \{$Maine, New Hampshire, Vermont, Massachusetts, Connecticut, Rhode Island$\}$

In each example both the words used and the symbols represent the same collection of objects. When a set is described as those sets on the

■　right are, we call this the *tabulation* of the set. If a set is described using words and you are asked to tabulate the set, you simply list its *members* or *elements*. Notice that the members of each set are separated by commas

▲　and enclosed in braces. This is a standard way of writing sets. There is a certain ambiguity that now arises in tabulating sets. This ambiguity is not due to any fundamental difficulty with the concept of a set, but has to do with the fact that we use the English (or any other) language in our description. To illustrate this difficulty, the set $D = \{$Monday, Tuesday, Wednesday, Thursday, Friday, Saturday, Sunday$\}$ can have two interpretations: It might symbolize the set of *days* of the week or it might symbolize the set of *names* of days of the week. Thus, when we tabulate a set using English words, there is an ambiguity as to whether we are listing the words or listing the objects that the words represent. For now we must live with this ambiguity, but we are able to resolve it and will do so later in this chapter.

We often wish to discuss sets without writing a description of the set or tabulating the set. In these cases we use *capital letters* to represent the set as in the example above. We often use *lowercase letters* to represent elements of a set.

Many situations arise in mathematics in which we wish to tabulate a very large set but we do not wish to take the time to list each element, or it might be impossible to do so.

■ A.3 **Examples**
▲ A.4   1. A is in the set of letters in the alphabet $A = \{a, b, c, \ldots, z\}$
    2. F is the set of natural numbers less than 50 $F = \{1, 2, 3, \ldots, 49\}$
    3. N is the set of natural numbers $N = \{1, 2, 3, \ldots\}$
■   4. E is the set of even natural numbers $E = \{2, 4, 6, \ldots\}$

Note in Examples 3 and 4 above that we could never tabulate all of the elements in sets N and E.

FINITE AND INFINITE SETS .

Notice in the first two examples that the sets contained a *finite* number of elements. We listed the first few elements to indicate the pattern and the last element to show that the pattern terminated. In the last two examples the number of elements in the set is *infinite*; therefore, we cannot indicate the last element in the set. But we use "..." to indicate that the pattern of elements continues without terminating. We might look at the difference between the *finite sets* in Examples 1 and 2 above and the *infinite sets* in Examples 3 and 4. Notice that the sets are described in terms of having or not having a terminating element in the pattern. For the present, that is the main difference between infinite and finite sets. Thus, *infinite sets* are sets in which the tabulation does not terminate, whereas *finite sets*
▲ are sets in which the tabulation does terminate.

ELEMENTS OR SET MEMBERSHIP

So far in this section we have discussed ways of writing sets, special set notation, finite sets, and infinite sets. We discussed set membership but did not indicate the notation which states that an element is a member
■ A.5 of a set. In Example 3 above, the number 1 is an element of the set N, the set of natural numbers. We symbolize this by "$1 \in N$" and read the symbols as "1 is in N." If we wish to indicate that $\sqrt{2}$ is *not* an element of N, we use a slash through the "is in" symbol, that is, $\notin$.

Examples
  "$\sqrt{2} \notin N$" reads "$\sqrt{2}$ is not an element of N"
■   "$3 \in N$" reads "3 is an element of N," or "3 belongs to N"

The symbols $\in$ and $\notin$ are examples of a dichotomy that often arises in mathematics. For example, given a set N and some object a, it is obvious that either a belongs to N or a does not belong to N. In symbols, $a \in N$ or $a \notin N$. The dichotomy "belongs to" is often the way that people learn the characteristics of things in their environment. If you see an animal, your reaction may be that it belongs to the set of dogs, in which case it is a dog, or it does not belong to the set of dogs. It is either a vertebrate or not a vertebrate, etc.

SET BUILDER NOTATION

We often describe sets by using *set builder notation*. We may wish to specify a set in terms of a larger set, for example, the days of the week whose names begin with S. This could be done by tabulating the set as {Saturday, Sunday} or by using the convention below.

$A = \{x | x$ is a day of the week and the name for x begins with an $S\}$

Graphically,

$A | = | \{$ | x | | | x is a day of the week ...}
A | is | the set | whose elements, | have the | x is a day of the week ...
 | | | symbolized by x, | property that |

This set builder notation uses the braces to indicate a set; within the braces is a description of the set, which is divided into two parts by a vertical line. To the left of the line is placed the symbol that will represent a typical element of the set whose description is placed to the right of the vertical line. Thus, we can translate the vertical line by the words "such that" or "with the property that."

Example 1
■ A.6 $A = \{x | x$ is a natural number less than five$\}$
Tabulation: $A = \{1, 2, 3, 4\}$

Example 2
$B = \{x | x$ is the name of a geometric figure with four sides$\}$
Tabulation: $B = \{$quadrilateral, trapezoid, parallelogram, rectangle,
■ rhombus, square$\}$

Is it obvious that set builder notation allows us to resolve the ambiguity between the set of names of objects and the set of the objects themselves?
You may often wish to use set builder notation to write a given set. Several examples follow:

Example 1
■ A.7   Tabulation: $A = \{1, 2, 3, 4, \ldots\}$
Set builder notation: $A = \{x | x$ is a natural number$\}$

Example 2

Tabulation: $B = \{$Franklin Roosevelt, Harry Truman, Dwight Eisenhower, John Kennedy, Lyndon Johnson$\}$

Set builder notation: $B = \{x|x$ is a president of the United States between the years 1933 and 1968$\}$

EXERCISES

1. From the written description, tabulate the following sets.
   a. The set of letters used to write the word *dog*.
   b. The set of natural numbers greater than 2 and less than 10.
   c. The set of months of the year whose first letter is J.
   d. The set of three-digit numerals which can be represented using the numerals 1, 2, and 3 only.
   e. The set of even numbers less than 10.
   f. The set of natural numbers less than 100.
   g. The set of natural numbers greater than 100.

2. Some of the sets below are well defined. Tabulate these sets.
   a. The set of large natural numbers.
   b. The set of months of the year with common characteristics.
   c. The set of games which Americans enjoy.
   d. The set of vowels in the English language.
   e. The set of sports which are played by a team and use some kind of stick as an implement.
   f. The set of months of the year which have 31 days.

3. Use set builder notation to describe the sets below.
   a. $A = \{$Tuesday, Thursday$\}$
   b. $B = \{$Hawaii, Alaska$\}$
   c. $C = \{1, 4, 9, 16, 25\}$
   d. $D = \{$January, February, March$\}$
   e. $E = \{$blue, yellow, red$\}$
   f. $F = \{10, 11, 12, \ldots\}$
   g. $G = \{12, 14, 16, \ldots, 30\}$

4. Give a verbal description of each of the sets tabulated below.
   a. $A = \{1, 2, 3\}$
   b. $B = \{1, 2, 3, \ldots\}$
   c. $C = \{$summer, winter, spring, fall$\}$
   d. $D = \{$Monday, Wednesday, Friday$\}$
   e. $E = \{$March, May$\}$
   f. $F = \{3, 6, 9, \ldots\}$
   g. $G = \{1, 2, 3, 4, 6, 12\}$

5. Which of the following sets are infinite?
   a. The number of molecules in the head of a pin.
   b. Natural numbers larger than 1,000,000,000.
   c. The set of insects alive today on earth.
   d. The set of words in all of the books in all of the libraries of the world today.

6. If $S = \{$a, e, i, o, u$\}$, which of the following are true?

   a. $a \in S$ b. $c \in S$ c. $e \notin S$

   d. $o \in S$ e. $q \notin S$

7. Distinguish between the following pairs of sets.
  a. $\{y \mid y$ is a day of the week whose name begins with S$\}$
  $\{y \mid y$ is Saturday or y is Sunday$\}$
  b. $\{z \mid z$ is the integer between 4 and 6$\}$
  $\{z \mid z$ is 5$\}$

8. What is the difference between 7 and $\{7\}$?

SECTION B   SET RELATIONSHIPS

BEHAVIORAL OBJECTIVES

B.1 Be able to define the empty set, $\varnothing$.

B.2 Be able to define the universe, U.

B.3 Be able to define a subset $(A \subseteq B)$.

B.4 Be able to define a proper subset $(A \subset B)$.

B.5 Given a set, be able to tabulate its subsets.

B.6 Given a set, be able to tabulate its proper subsets.

B.7 Given two sets, decide if one is a subset of the other.

B.8 Given two sets, decide if one is a proper subset of the other.

B.9 Be able to define $n(A)$.

B.10 Be able to find $n(A)$ given A.

B.11 Be able to define power set $p(A)$.

B.12 Be able to tabulate $p(A)$ given A.

B.13 Be able to find $n[p(A)]$ given A.

B.14 Given U, A, B, etc., be able to draw Venn diagrams based on the given tabulation.

B.15 Be able to define the complement of a set A, $\bar{A}$.

B.16 Be able to tabulate $\bar{A}$ given A.

B.17 Be able to draw a Venn diagram of A.

B.18 Be able to define the relative complement of B in A, $(A - B)$.

B.19 Be able to tabulate $A - B$ given A and B.

B.20 Be able to draw a Venn diagram of $A - B$.

B.21 Be able to tabulate $p(A - B)$.

B.22 Be able to find $n(A)$ and $n(A - B)$.

B.23 Be able to relate $(A - B)$ and $(U - B)$ for $B \subset A$.

B.24 Be able to define disjoint sets.

SPECIAL SETS

There are two types of sets which have specific characteristics that make them very useful in mathematics. These are *empty sets* and *universal sets*, or universes. An empty set is, as you may have guessed, a set with no elements. This set may not appear to be very useful. How many times have you looked for a set which contained no elements? Probably not too often. However, you may have found that there was no food in your refrigerator or no money in your bank account. In each case you were dealing with an empty set. We might consider other examples of empty sets below.

■ B.1
■

Examples

1. The set of dogs which fly
2. The set of fish who sing
3. The set of states in the United States which are larger than Alaska
4. The set of presidents of the United States who were less than 35 years old when elected

All of the sets described are empty in that they contain no elements. We usually denote empty sets in one of two ways: by braces with no elements inside, { }, or as the symbol $\varnothing$. An empty set is often referred to as a *null set*. Note: We never use both in combination, since the set $\{\varnothing\}$ is *not* empty. $\{\varnothing\}$ is the set that has *one* element and that element is a set—namely, an empty set.

The second type of set which we will use quite often is called a universe or universal set. This set can be described as the set that contains all of the elements related to a specific discussion. In arithmetic our universe may be a set of numbers. In geometry our universe may be a set of points. By limiting the universe, we limit the elements that we include in the discussion. We symbolize the universe by U. As a child you may have been taught that you can not subtract 2 from 1. The reason for this is that the answer to the problem is not a number in the universe of numbers that is used in the early grades. Therefore, many problems have "no solution" simply because the universe in which the discussion is taking place is limited. Actually, you can subtract 2 from 1 in certain universes.

■ B.2
■

It is common practice to specify a universe for a given set of problems. By choosing a different universe we may produce different solutions to a given problem. It is often the case that the universe is specified by the context of the situation at hand. For example, if a club wishes to have a committee plan a party, the universe is usually assumed to be the set of members of that club. If a party wishes to run a candidate for President of the United States, they cannot run a man who is not a citizen, nor can they nominate someone who is under 35 years old. In this case, the Constitution designates the universe to be all natural-born citizens not younger than 35.

Definitions

■ B.1 ■   *An empty set or null set is a set which has no elements.*

▲ B.2 ▲   *A universal set or universe is the set of all elements which may be discussed.*

SUBSETS

In the previous section we discussed the universal set which contains all elements permitted in a discussion. The universal set also contains other sets. Suppose our universe in a given discussion is the set of all natural numbers, N. We indicate this by writing $U = N$. Then we know that $U = \{1, 2, 3, \ldots\}$. It follows that $1 \in U$, $2 \in U$, $3 \in U$, and so forth. We can, however, describe sets which contain 1 or 2 or 3 or combinations of elements of U. Suppose we name some sets as follows:

$$A = \{1\}$$
$$B = \{2\}$$
$$C = \{3\}$$
$$D = \{1, 2\}$$
$$E = \{1, 3\}$$

A, B, C, D, and E are sets. All of the elements in each of these sets are also contained in U. Thus, we can now state that "A is a subset of U," "B is a subset of U," and so forth. We call these sets subsets of the set U. The symbol $\subseteq$ is used for subset. We can state that $A \subseteq D$, $B \subseteq D$, and $C \subseteq E$ and read this "A is a subset of D," "B is a subset of D," and "C is a subset of E." If we wish to say that C *is not* a subset of D, we use the symbol for subset with a slash through it: $C \nsubseteq D$.

We should note that $A \subseteq U$ but that $A \notin U$, because A is not an element of U. In a similar fashion, if $a \in A$, then $a \nsubseteq A$ while $\{a\} \subseteq A$. This distinction between $\in$ and $\subseteq$ is important.

We should also note that all the elements of U are contained in U, so $U \subseteq U$. In general, any set is a subset of itself because all its members are contained in the set. We can make a similar remark about empty sets by noticing that almost every statement we make about the empty set is true. For example, every pink-eared flying elephant is a chair in this room. If you doubt this, try to find a pink-eared flying elephant that is *not* a chair in this room. Therefore, we can say, "The set of pink-eared flying elephants is a subset of the set of chairs in this room." More generally we say that an empty set is a subset of any set because the statement "every element of the empty set is contained in the given set" is true. We write this $\varnothing \subseteq A$ for every set A.

Example

■ B.5   What are the subsets of the set $A = \{1, 2, 3\}$?

SOLUTION $\varnothing$, A, $\{1\}$, $\{2\}$, $\{3\}$, $\{1, 2\}$, $\{1, 3\}$, $\{2, 3\}$.

■   Notice that all elements in the subsets of A are contained in A itself.

Let us consider the attributes of the relationship $\subseteq$.

1. Two sets are related by the symbol $\subseteq$ if every element of the first set is an element of the second set.
2. It is always the case that a set is a subset of itself.
3. An empty set is a subset of any set.

These requirements lead to the formal definition for a set B to be a *subset* of set A.

Definition

B.3 ■ $B \subseteq A$ *if every element of B is an element of A*.

PROPER SUBSET

Sometimes we wish to differentiate between all subsets of a given set and the set itself. To do this we consider only those subsets which are different from the given set. By different sets we mean sets that do not contain exactly the same elements. To designate a subset which is not the same as the given set, we use the symbol $\subset$ and read the symbol as "proper subset of." Thus, a set *cannot* be a proper subset of itself. A good way to remember this is to think of the bar in $\subseteq$ as an equal sign. The symbol $\subseteq$ includes the possibility that the subset may in fact be the same set as the given set, and the symbol $\subset$ without the bar excludes the given set as a subset.

Example

◀ B.6   What sets are proper subsets of $A = \{a, b\}$?

■   SOLUTION $\emptyset, \{a\}, \{b\}$. Note that $\{a, b\} \not\subset \{a, b\}$. Why?

We define proper subsets as follows.

Definition

B.4 ■   *A is a proper subset of B, written $A \subset B$, if $A \subseteq B$ and $A \neq B$*.

Many times we wish to see if some set B is a subset of a given set A. To do this we must see if every element in B is in A. Consider the examples below.

Example 1

◀ B.7   From the given sets, choose pairs of sets which are subsets of each other.

$$A = \{a, b, c, d\} \qquad B = \{b, a, d\}$$
$$C = \{b, o, y\} \qquad D = \{d, a, b\}$$

SOLUTION $B \subseteq A$, $D \subseteq A$, $D \subseteq B$, $B \subseteq D$. Notice that $C \not\subseteq A$, $C \not\subseteq B$,
■ $C \not\subseteq D$.

Example 2

◀ B.8   From the given sets in Example 1, which pairs of sets are proper subsets?

SOLUTION $B \subset A$ and $D \subset A$. Note that $B \not\subset D$ and $D \not\subset B$ since they contain *exactly* the same elements.

EXERCISES

1. Find the subsets of the set $A = \{2, 3, 5\}$.

2. Find the proper subsets of the set A from Exercise 1.

3. If $U = \{2, 4, 6, 8, 10\}$,
a. which of the following are subsets of U?

$A = \{2, 3, 5\}$	$B = \{2, 8, 10\}$
$C = \{1, 3, 5\}$	$D = \{2, 4, 6, 8, 10\}$

b. which of the following are proper subsets of U?

$A = \{2, 3, 5\}$	$B = \{2, 8, 10\}$
$C = \{1, 3, 5\}$	$D = \{2, 4, 6, 8, 10\}$

c. which of the following are true statements?

$2 \in U$	$\{2, 4\} \in U$	$\{2, 4\} \subseteq U$
$\{2\} \in U$	$2 \notin U$	$\{2\} \subseteq U$
$U \in U$	$2 \in U$	$U \subseteq \{2, 4\}$
$\{2, 4\} \subset \{U\}$		

4. If a set contains one element, how many subsets does it have?

5. If a set contains two elements, how many subsets does it have?

6. If a set contains three elements, how many subsets does it have?

7. See if you can find a pattern for the number of subsets of a given set. (*Hint*: How many subsets has a set with n elements?)

8. If $A \not\subseteq B$, is it possible that $B \subseteq A$?

9. If $A \subset B$, is it possible that $B \subset A$?

10. If $A \subseteq B$ and $B \subseteq A$, what can you say about A and B?

11. Is $\varnothing \subseteq \varnothing$?

12. What is the difference between the statement "$\{a\} \in A$" and the statement "$\{a\} \subseteq A$"?

13. If $A \subseteq B$ and $B \subseteq C$, what can you conclude about A and C?

14. If $A \subset B$ and $B \subseteq C$, what can you conclude about A and C?

15. If $A \subset B$ and $B \subset C$, what can you conclude about A and C?

16. If $A \subseteq B$ and $C \subseteq B$, what can you conclude about A and C?

$n\,(A)$, NUMBER PROPERTY OF SETS

We will be interested in the *number* of elements in a given set. Every set has a number associated with it that describes the number of elements in the set. The number of letters in our alphabet is 26. Our numeration system employs 10 symbols. When we wish to describe the number property of a set, we use the notation $n(A)$.

Consider the set of letters in our alphabet:

$A = \{a, b, c, \ldots, z\}$; in this case $n(A) = 26$

B.9

We read $n(A)$ as "n of A" or "the number of elements in A." Notice that $n(A)$ is a number and not a set; it is simply a number associated with a set. Consider the following examples.

Examples

B.10

Given set S_i	Find $n(S_i)$
$S_0 = \varnothing$	$n(S_0) = 0$
$S_1 = \{1\}$	$n(S_1) = 1$
$S_2 = \{1, 2\}$	$n(S_2) = 2$
$S_3 = \{1, 2, 3\}$	$n(S_3) = 3$

In the examples above we used *subscripts* to differentiate between the sets instead of different letters for each set. The use of subscripts does not imply anything about the number of elements in a given set unless we choose the subscripts in a special way. It is true in the examples above that the subscripts of the sets were in each case $n(S_i)$. The sets listed above are sometimes called *standard sets*. Notice that with standard sets we list the elements as natural numbers and the last element corresponds to $n(A)$. This is the way children often learn to count. Later in the text we will discuss this idea more fully. For the present we define a standard set by $S_i = \{1, 2, 3, \ldots, i\}$, where $i \in N$.

In the previous exercise set, Exercises 4, 5, 6, and 7 dealt with subsets of a given set. You were asked to find the number of subsets of sets which contained one, two, or three elements. Let us summarize the results below, using the standard sets in the previous example.

Given set	$n(S)$	Subsets	Number of subsets
$\varnothing$	0	$\varnothing$	1
$\{1\}$	1	$\varnothing, \{1\}$	2
$\{1, 2\}$	2	$\varnothing, \{1, 2\}, \{1\}, \{2\}$	4
$\{1, 2, 3\}$	3	$\varnothing, \{1, 2, 3\}, \{1\}, \{2\}, \{3\}, \{1, 2\}, \{1, 3\}, \{2, 3\}$	8

Notice that we might describe the pattern in the number of subsets by doubling the number of subsets each time we include a new element. When we pair the number of elements of a given set with the number of its subsets, we get the following:

(0, 1)
(1, 2)
(2, 4)
(3, 8)
(4, ?)

Is there some easy rule which we can use to find the number of subsets of a given set? Looking at the pairs listed, there are several rules that might come to mind. We notice that the number of subsets of a given set doubles when another element is included in the given set. That is, the number of subsets of a set of two elements is four whereas the number of subsets of a set of three elements is eight.

Thus, the above pairs of numbers could be written:

$(1, 2)$
$(2, 2 \cdot 2)$ instead of $(1, 2)$
$(3, 2 \cdot 2 \cdot 2)$ $(2, 4)$
$(4, 2 \cdot 2 \cdot 2 \cdot 2)$ $(3, 8)$
    $(4, 16)$

Now we may be able to see how the number of subsets of a given set is related to the number of elements in the given set. If $n(A) = 4$, then the number of subsets of A is $\underbrace{2 \cdot 2 \cdot 2 \cdot 2}_{4 \text{ times}}$. You may recall that $2 \cdot 2 \cdot 2 \cdot 2 = 2^4$.

Using this notation we have that the number of subsets of a set with four elements is 2^4, or 16.

See if you can use the rule in the example below.

Example

■ B.10   The set $B = \{x, y, z, q, r\}$ has how many subsets?

SOLUTION $n(B)$ is 5. Therefore, B has $2 \times 2 \times 2 \times 2 \times 2 = 32$ subsets,
■   or 2^5 subsets.

See if you can list all the subsets for a set A for which $n(A)$ is 4 or 5. When you do this, consider the pattern of subsets that is most efficient for you. Notice how it was done for sets in which $n(A)$ was 0, 1, 2, and 3. In each case the empty set and the given set were listed and then sets of elements taken one at a time, two at a time, and so on. This is one pattern to use. It does not matter what method you use to list the subsets.

POWER SETS

Up to this point we have not discussed sets in which the elements of the given set are also sets. In the previous paragraph we found all the subsets of a given set. These subsets are a collection of objects and this collection is certainly well defined. That is, a set is either a subset of the given set or it is not. Consider the set $A = \{a, b\}$, and notice that the subsets of A are $\emptyset$, $\{a\}$, $\{b\}$, and $\{a, b\}$.

■ B.12   The special set whose elements are the subsets of A is called the *power set* of A and is $\{\emptyset, \{a\}, \{b\}, \{a, b\}\}$. We will use the notation $p(A)$ for the power set of A. Thus, if $A = \{a, b\}$, then

■ $p(A) = \{\emptyset, \{a\}, \{b\}, \{a, b\}\}$

Whereas the set A is a set whose elements are letters, the set $p(A)$ is a set whose elements are sets, specifically, the subsets of A.

We have introduced three ideas in this section:

1. $n(A)$
2. the number of subsets in a set A
3. $p(A)$, the set of subsets of A

We can relate each of these ideas in a single statement. The number of elements in the power set of A, $n[p(A)]$, can be found by raising 2 to the power of the number of elements of A, $n(A)$. This can be stated more simply as follows: The number of elements in $p(A)$ is the same as the number of subsets of the set A. Symbolically, $n[p(A)] = 2^{n(A)}$.

Consider an example to clarify the rule shown above.

B.13   **Example 1**
Find $n[p(A)]$ where $A = \{1, 2\}$.

12 ▲   SOLUTIONS   1. $p(A) = \{A, \varnothing, \{1\}, \{2\}\}$
$$n[p(A)] = 4$$
2. Since $n(A) = 2$,
$$n[p(A)] = 2^{n(A)}$$
$$= 2^2$$
$$= 4$$

Example 2
Find $n[p(A)]$, where $A = \{x, y, z\}$.

SOLUTION   Since $n(A) = 3$,
$$n[p(A)] = 2^{n(A)}$$
$$= 2^3$$
$$= 8$$

Definitions

9 ■ $n(A)$ is the number of elements in a given set A.

11 ▲ $p(A)$ is the set of subsets of the set A.

EXERCISES

1. Find $n(A)$ for each of the following sets.
   a. $A = \{$red, blue, green$\}$
   b. $A = \{\varnothing, \{1, 2\}\}$
   c. $A = \{a, e, i, o, u\}$

2. How many elements does the power set of each of the sets in Exercise 1 contain?

3. Write the power set of each of the sets given below.
   a. $A = \{$red$\}$
   b. $B = \{$red, blue$\}$
   c. $C = \{3, 5, 7\}$

4. What is $n[p(A)]$ in each of the examples in Exercise 3? What is $p[p(A)]$?

5. Is $p(A)$ a set or a number?

6. Is $n[p(A)]$ a set or a number?

7. Is $n(A)$ a set or a number? $p[p(A)]$?

8. If $A = \{c, d, e\}$,
   a. $n(A) = ?$
   b. $\{n(A)\} = ?$
   c. $p[\{n(A)\}] = ?$
   d. Is $p[\{n(A)\}]$ a set or a number?
   e. Is $p[\{n(A)\}] = n[p(A)]$? Why or why not?

COMPLEMENT AND RELATIVE COMPLEMENT

Consider a club to which you may belong. Usually clubs have officers and non-officers. If we name the set of members of the club M and the set of officers A, how might we describe the set of people who are not officers? Simply designating that set as B does little to suggest its composition. We could suggest that the set of non-officers is the set M with A removed. To describe M with A removed, we could write $B = M - A$.

■ B.15   If the universe U consists of all people in the world, then the set of people who are not members of the club is symbolized as $U - M$. This particular set consisting of all members of the universe except those in a particular set M is usually called the *complement of M* and is symbolized
■   by $\bar{M}$.

Let us use this notation in the following example.

■ B.16   **Example**
▲ B.23
$$U = \{1, 2, 3, \ldots, 10\}$$
$$A = \{2, 4, 6, 8\}$$
$$B = \{2, 4\}$$

Tabulate: $\bar{A}, \bar{B}, U - A, U - B, A - B$

SOLUTIONS $\bar{A} = \{1, 3, 5, 7, 9, 10\} = U - A$
$\bar{B} = \{1, 3, 5, 6, 7, 8, 9, 10\} = U - B$
■ $A - B = \{6, 8\}$

Notice in the example above that $B \subseteq U$, $A \subseteq U$, and $B \subseteq A$. It is not necessary that $B \subseteq A$ in order to discuss $A - B$. In the case that $B \nsubseteq A$, by $A - B$ is meant the set of elements of A that are not elements of B.
● B.18 ● That is, $A - B = \{x \mid x \in A$ and $x \notin B\}$. For example, if M is the set of members of the club discussed earlier and B is the set of people who own a car, then $M - B$ is the set of club members who do not own a car. It should be obvious that $M - B$ could be empty. Can you see when $M - B$ would be an empty set? Notice that when $M = U$, the sets $M - B$ and $U - B$ are the same. Thus, if $A = U$, $A - B = \bar{B}$.

Since the set $\bar{A}$ was named the complement of A, it is natural to call $A - B$ the "relative complement of B with respect to A" or "*the relative complement of B in A*." Let us consider a few examples of relative com-
▲ plement.

Examples
■ B.19 $$U = \{1, 2, 3, \ldots, 10\}$$
$$A = \{1, 2, 3, 5, 7, 9\}$$
$$B = \{2, 4, 6, 8, 10\}$$
$$C = \{1, 2, 3\}$$
$$D = \{1, 3, 5, 7, 9\}$$

Tabulate the following sets: $A - C$; $A - D$; $A - \bar{B}$; $B - \bar{D}$; $(\overline{A - C})$;
$B - D$; $C - D$

SOLUTION
$$A - C = \{5, 7, 9\}$$
$$A - D = \{2\}$$

B.16
$$A - \bar{B} = \{2\}$$
$$B - \bar{D} = \varnothing$$

▲
$$(\overline{A - C}) = \{1, 2, 3, 4, 6, 8, 10\}$$

■
$$B - D = \{2, 4, 6, 8, 10\}$$
$$C - D = \{2\}$$

To summarize the attributes of the terms "the complement of a set" and "the relative complement of two sets," we see from the examples that:

1. The complement of a set A is itself a set.
2. The complement of a set A is a subset of the universe.
3. The complement of A contains all the elements outside A which are contained in the universe.
4. We can write the complement of A in two ways, $\bar{A}$ or $U - A$.
5. The relative complement of B with respect to A is a set.
6. The elements contained in the set $A - B$ are all of the elements which are members of A and not members of B.
7. $A - B$ is a subset of A.

Definitions

B.15
■ *The complement of A, written $\bar{A}$, is a set which contains all the elements in the universe U that are not in the set A.*

B.18
■ *The relative complement of B in A, written $A - B$, is the set which contains all of the elements in A that are not contained in B.*

It is often helpful to see a picture or diagram of a given mathematical situation. Diagrammed sets are called *Venn diagrams* or *Euler diagrams*. By convention the diagram consists of a rectangular region which represents the universe and one or more circular regions which represent sets in the universe. The relative size of the regions is of no importance; no attempt is made to convey size or number of elements within a given diagram. The diagrams that follow illustrate the sets discussed in this section. Shading is used to indicate specific sets. Figure 1.1 represents the set A and its complement $\bar{A}$.

B.17

■

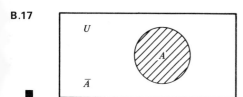

Figure 1.1

In Figure 1.2, the set A is cross-hatched to the left while the set B is cross-hatched to the right. Note that the set $A - B$ corresponds to the area in A which is cross-hatched to the left only, namely, the area in A which has no cross-hatching to the right. $B - A$ is cross-hatched only

■ B.20

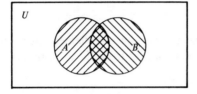

■

Figure 1.2

to the right. In Figure 1.3, since A and B contain no elements in common, the set $A - B = A$. If $B \subset A$ as shown in Figure 1.4, then $A - B$ would look much like a donut, since we remove a subset from A.

■ B.20

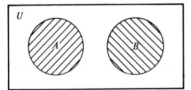

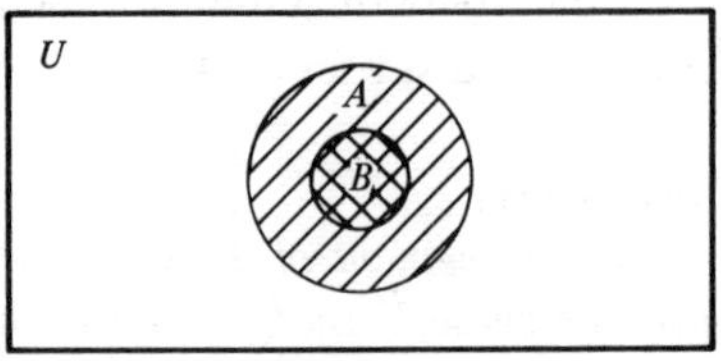

■

Figure 1.3　　　　　　　**Figure 1.4**

Notice that in Figures 1.2, 1.3, and 1.4, the set $A - B$ is cross-hatched in one direction only. In Figures 1.2 and 1.3 $B - A$ is cross-hatched in the other direction only. This is true whether the sets contain elements in common or not. However, in Figure 1.4, since $B - A = \emptyset$, there is no set which exhibits *right cross-hatching only*.

It might be of help to the student to consider an example to illustrate each of these diagrams.

Examples

■ B.19

For Figure 1.2.　If $A = \{1, 2, 3, 4\}$ and $B = \{3, 4, 5, 6\}$, then $A - B = \{1, 2\}$ and $B - A = \{5, 6\}$.

For Figure 1.3.　If $A = \{$red, blue, yellow$\}$ and $B = \{$green, brown$\}$, then $A - B = \{$red, blue, yellow$\}$ and $B - A = \{$green, brown$\}$.

For Figure 1.4.　If $A = \{$trees, flowers, dogs, cats$\}$ and $B = \{$trees, flowers$\}$, then $A - B = \{$dogs, cats$\}$ and $B - A = \emptyset$.

■

Notice in Figure 1.5 that the set A is cross-hatched to the left, the set B is cross-hatched to the right, and $\bar{A}$ is shaded in gray. When we remove

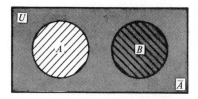

Figure 1.5 $\bar{A} - B = \boxed{}$

the set B from $\bar{A}$, we write $\bar{A} - B$ and note it as the set which has *gray shading only.*

In every diagram we have indicated a universe U by a rectangle. The form or shape of the figures used in Venn diagrams does not affect the relationships between the sets. With children it might be interesting to use shapes such as triangles, circles, squares, and so on.

Up to this point we have shown Venn diagrams for arbitrary sets. We did not indicate what elements were contained in the sets. If we have more information regarding the sets U, A, B, and so on, then the diagrams can be easily made to describe a specific situation. Consider an example below.

Example 1

B.14
Let $U = \{$Joan, Bill, Jon, Harry, David, Sara, Daniel, Sandy$\}$
Let $A = \{$Bill, Jon$\}$
Let $B = \{$David, Sara$\}$
Let $C = \{$Jon, Harry, David$\}$

Draw a Venn Diagram to describe this situation.

SOLUTION   There are several ways to do this. We can first draw U as a rectangle and then place A in the universe arbitrarily. We must then see if A shares elements with B, and we place B in the universe. Lastly we see if C shares elements with A or B, and place C in the universe. The result is shown in Figure 1.6. Notice that since Jon is an element in both A and C and David is an element in B and C, these sets overlap. However, A and B are disjoint and do not overlap.

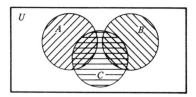

Figure 1.6

In this diagram the element Jon is represented by the overlap of horizontal and left cross-hatching and the element David is represented by the overlap of horizontal and right cross-hatching.

Example 2

B.17   From the given sets in Example 1, diagram the set $\bar{A}$.

SOLUTION   Notice in Figure 1.7 that the set $\bar{A}$ is represented by gray shading only, the universe is shaded in gray, and the set A is striped horizontally. Since A is in the universe, you notice that the stripes and gray shading overlap in the set A.

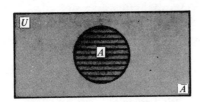

Figure 1.7

From the Venn diagrams in Figure 1.5 it should be obvious that some pairs of sets have no common elements, for example, (1) A and B, (2) A and $\bar{A}$, and (3) $\bar{A} - B$ and A. We name sets which have no elements in common *disjoint sets*.

Definition

B.24   *Two sets A and B are said to be disjoint if and only if there is no element which is a member of A and a member of B.*

In the previous section we defined $n(A)$ and $p(A)$. From what you know about complement and relative complement, you should be able to find $n(\bar{A})$, $n(A - B)$, $p(\bar{A})$, and $p(A - B)$. Consider the following example.

B.21   **Example**
▲ B.22
$$U = \{a, e, i, o, u\}$$
$$A = \{a, i, u\}$$
$$B = \{a, u\}$$
Find $n(\bar{A})$, $p(\bar{A})$, $n(A - B)$, and $p(A - B)$.

SOLUTION $n(\bar{A}) = 2$
$p(\bar{A}) = \{\varnothing, \bar{A}, \{e\}, \{o\}\}$
$n(A - B) = 1$
$p(A - B) = \{\varnothing, \{i\}\}$

EXERCISES

In Exercises 1–10 below, use the following as given sets.

$U = \{1, 2, 3, \ldots, 10\}$ $B = \{4, 5, 6\}$
$A = \{1, 2, 3\}$ $C = \{2, 4, 5, 6, 8, 10\}$

1. Tabulate $\bar{A}$, $\bar{B}$, and $\bar{C}$.

2. Draw a Venn diagram to illustrate $\bar{A}$, $\bar{B}$, and $\bar{C}$ as they relate to the given sets.

3. Find $n(\bar{A})$, $n(\bar{B})$, and $n(\bar{C})$.

4. Tabulate $p(\bar{A})$ and $p(\bar{C})$.

5. Find $n[p(\bar{A})]$, $n[p(\bar{B})]$, and $n[p(\bar{C})]$.

6. Tabulate $\bar{B} - A$ and $C - B$.

7. Draw a Venn diagram of $B - A$ and $C - B$.

8. a. Find $n(A)$, $n(\bar{A})$, and $n(U)$.
   b. Find a relationship between $n(A)$, $n(\bar{A})$, and $n(U)$.

9. Find a relationship between $n(C)$, $n(B)$, and $n(C - B)$.

10. Tabulate $p(C - B)$.

11. Suppose that $M \subseteq B$; what is $M - B$?

SECTION C   SET EQUALITY AND EQUIVALENCE RELATIONS

BEHAVIORAL OBJECTIVES

C.1 Be able to define equality.

C.2 Be able to define 1–1 correspondence.

C.3 Be able to define equivalent sets.

C.4 Given two sets, be able to decide whether they are equivalent or not.

C.5 Be able to define equivalence relations.

C.6 Given two sets, be able to decide whether they are equal or not.

C.7 Given equivalent sets, be able to place them in 1–1 correspondence.

C.8 Given equivalent sets, be able to decide whether they are equal or not.

C.9 Be able to show that set equality is an equivalence relation.

C.10 Be able to show that set equivalence is an equivalence relation.

C.11 Given a relation, be able to decide which properties of an equivalence relation the given relation possesses.

C.12 Be able to decide which properties of an equivalence relation the subset property satisfies.

C.13 Be able to state the properties of set equality.

EQUALITY OF SETS

While you are reading this, you may have some coins in your possession. The set of coins is the same set no matter what the location or position of the coins may be. If you move this set from one pocket to another or from one place to another, the set stays the same. The order in which you place the coins does not affect the fact that the set of coins contains exactly the same elements.

When two sets contain exactly the same elements, we call them *equal sets* and use the symbol $=$. Although it is not such a simple idea, the notion of equality is fundamental to the way we think. Two dollar bills are not equal because they are obviously not the same; however, their monetary measure or value is equal. Two chairs are not equal although they may have the same shape and as far as we can see they may have the same weight, color, and volume. Often two objects are called equal when we really mean that they have in some sense the same measure. We will, however, use "equal" to mean "the same." Thus, "the set A is equal to the set B" means that A and B are, in fact, the same set; that is, they are the same collection of objects.

For example, all the sets below are equal because they are the collection of the same objects.

$$A = \{1, 2, 3\} \qquad B = \{2, 3, 1\} \qquad C = \{3, 1, 2\}$$

From this example we can make the following statements:

$$A = B \qquad A = C \qquad B = C \qquad A = A \qquad C = C$$
$$B = A \qquad C = A \qquad C = B \qquad B = B$$

We might wish to consider pairs of sets to determine whether they are equal or not. The examples below should clarify this idea.

Example 1

■ C.6 From the given sets, determine pairs of sets which are equal.

$$A = \{x, y, z\}$$
$$B = \{b, a, d\}$$
$$C = \{a, y, z\}$$
$$D = \{a, b, d\}$$

SOLUTIONS $A = A, B = B, C = C, D = D, B = D,$ and $D = B$

Example 2
From the given sets in Example 1, which sets are not equal ($\neq$)?

SOLUTION $A \neq B,\ B \neq A,\ A \neq C,\ C \neq A,\ A \neq D,\ D \neq A,\ B \neq C,$
$C \neq B, C \neq D, D \neq C.$

C.13   Now we can list some of the properties of "set equality." Notice that

1. A set is equal to itself.
2. If a set A is equal to a set B, then the set B is also equal to the set A.
3. If a set A is equal to a set B and if B is equal to C, then it is also true that A is equal to C.

These statements may seem obvious, and we might expect them to be true for all relationships between objects.

Consider a relationship other than equality, for example, the relationship LOVES. We might accept as true that A LOVES A, that is, a person loves himself. If A LOVES B, does it follow that B LOVES A? If A LOVES B and B LOVES C, then does it follow that A LOVES C? The answer to these questions could all be No. Notice that this type of relationship does not work in the same manner as the relationship which we call equality.

The definition that we will formally give for equality between sets is a working definition. That is, it actually gives a procedure for testing for equality. If we have two sets, A and B, that we consider equal, that is, that are really the same set, then it is clear that $A \subseteq B$ and $B \subseteq A$. By the same token, if for sets A and B we have that $A \subseteq B$ and $B \subseteq A$, then every element of A is a member of B and every element of B is a member of A. Is it clear in this case that A and B are really the same set? Using this idea we say that the statement "$A = B$" is equivalent to or means the same as the statement "$A \subseteq B$ and $B \subseteq A$."

Definition
C.1   *For sets A and B, $A = B$ means that $A \subseteq B$ and $B \subseteq A$.*

EXERCISES

1. Can you show that there is only one empty set? (*Hint*: Assume that $\emptyset_1$ and $\emptyset_2$ are two empty sets; show they are equal.)

2. Use the definition to show that for any set A, $A = A$.

3. Use the definition to show that for arbitrary sets A and B, if $A = B$, then $B = A$.

4. If $A = B$ and $B = C$, show that $A = C$ using the definition.

ONE-TO-ONE CORRESPONDENCE

In the days of shepherds tending their flock, it was necessary for the shepherds to keep track of their sheep. Often a record was kept by making a pile of stones, one stone for each animal. At the end of the day the stones were discarded one at a time as the animals passed the shepherd. This was a primitive method of counting and was essentially a correspondence between the stones in the pile and the sheep in the flock. Children often do the same thing when distributing cookies or candy to their friends.

When two sets are matched in such a way that for every element in the first set there is an element in the second set and for every element in the second set there is an element in the first set, we say that the sets are in *one-to-one correspondence*. We abbreviate this notion by writing "the two sets are 1–1" or "the sets are in a 1–1 correspondence."

■ C.7

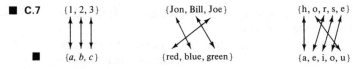

■

Figure 1.8

In the examples in Figure 1.8, the sets are in 1–1 correspondence. Notice that we can match their elements in more than one way.

Definition

■ C.2   *Two sets A and B are in 1–1 correspondence means (if and only if) for every element in A there is exactly one element in B and for every element*
■   *in B there is exactly one element in A.*

Notice the use of the word "exactly" in this definition. It is possible to have a correspondence between the elements of two sets that is not 1–1. For example, if the boys Adam, Bram, and Caleb were each wearing a red shirt, the correspondence between boys and colors of shirts would be as diagrammed in Figure 1.9. In this case every element of the set of boys corresponds to one element in the set of colors. However, the color red does not correspond to exactly one boy in the set of boys; it corresponds to three boys.

{Adam, Bram, Caleb}

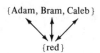

{red}

Figure 1.9

EQUIVALENCE

If two sets can be placed in 1–1 correspondence, we say that they are *equivalent sets*. Notice that sets which are equal are also equivalent. Are all equivalent sets equal? Consider the examples in Figure 1.8. All of the sets shown as having 1–1 correspondence are equivalent sets and yet none of the pairs of sets are equal sets. We indicate that two sets A and B are equivalent by the notation $A \sim B$.

Definition

■ C.3   *Two sets A and B are equivalent, written $A \sim B$, if and only if they can*
■   *be placed in a 1–1 correspondence.*

Before we continue, consider a few examples to illustrate pairs of sets which are equivalent or not.

Example 1

■ C.4   From the given sets, determine which sets are equivalent.

$A = \{*, @, 0\}$ $B = \{3, 2\}$
$C = \{1, 2, 3\}$ $D = \{a, b, c, d\}$
$E = \{d, b\}$

SOLUTION $A \sim C, B \sim E, A \sim A, B \sim B, C \sim C, D \sim D, E \sim E, C \sim A,$
           $E \sim B.$

Example 2

From the sets in Example 1, determine which pairs of sets are not equivalent.

SOLUTION $A \nsim B, A \nsim D, A \nsim E, D \nsim A, D \nsim B, B \nsim A, B \nsim C, B \nsim D,$
■ $D \nsim C, D \nsim E, C \nsim B, C \nsim D, C \nsim E, E \nsim D, E \nsim A, E \nsim C.$

It is often the case that we deal with sets which are equivalent, as shown above. Sometimes these sets are equal sets since the definition of equality implies 1–1 correspondence. Consider sets which are equivalent and see if you can determine which pairs of sets are equal.

Example 1

■ C.8   From the given sets, determine which equivalent sets are equal.

$A = \{1, 2, 3\}$ $B = \{2, 3, 4\}$
$C = \{3, 2, 1\}$ $D = \{b, a, d\}$
$E = \{d, b, c\}$ $F = \{b, c, d\}$

SOLUTION   Notice that all the given sets are equivalent, but only $A = C,$
           $C = A, E = F, F = E,$ and, of course, $A = A, B = B,$ and so
           on.

Example 2

Which of the given sets from Example 1 are not equal?

SOLUTION $A \neq B, B \neq A, A \neq D, D \neq A, A \neq E, E \neq A, A \neq F, F \neq A,$
           $B \neq C, C \neq B, B \neq D, D \neq B, B \neq E, E \neq B, B \neq F, F \neq B,$
           $C \neq D, D \neq C, C \neq E, E \neq C, C \neq F, F \neq C, D \neq E, E \neq D,$
■ $D \neq F, F \neq D.$

When we discussed the equality relation for sets, we pointed out certain properties of that relation. The equality relation satisfied three specific properties that many other relations also satisfy. Any relationship which possesses these three properties falls into the category of an *equivalence relation.*

Definition

■ **C.5**   *If a relation R between elements of a set A has the following properties, then the relation is called an equivalence relation on the set A.*

1. For $a \in A$, aRa. That is, an element is in this relationship to itself. (This is called the *reflexive* property.)
2. For $a, b \in A$, if aRb, then bRa. That is, if a is in this relationship to b, then b is in this relationship to a. (This is called the *symmetric* property.)
3. For a, b, and $c \in A$, if aRb and bRc, then aRc. That is, if a is in this relationship to b and b is in this relationship to c, then a is in this relationship to c. (This is called the *transitive* property.)

■ **C.10**   As an example of a particular equivalence relation, let us consider the relation symbolized by $\sim$. A set S is certainly in a 1–1 correspondence with itself, so

1. $S \sim S$

If S is in a 1–1 correspondence with T, then T is in a 1–1 correspondence with S. That is,

2. if $S \sim T$, then $T \sim S$

Finally, if S is in a 1–1 correspondence with T and T is in a 1–1 correspondence with R, then S is in a 1–1 correspondence with R. In symbols this is

3. if $S \sim T$ and $T \sim R$, then $S \sim R$

Since the relation symbolized by $\sim$ satisfies the properties 1, 2, and 3 of an equivalence relation, we can say that "set equivalence, $\sim$, is an equivalence relation." Thus, the relation of set equivalence is reflexive, symmetric, and transitive.

■ **C.9**   From the discussion of equivalence as an equivalence relation, one should be able to show that equality satisfies the properties of an equivalence relation. Why?

■ **C.11**   Consider a row of at least three chairs and the relation R, where R is the relation described by "sits in the same row as." Let us check to see if this is an equivalence relation. There are three chairs in the row (there could be more than three). Adam, Bram, and Caleb are sitting in the chairs. If a is a boy sitting in one of the chairs, does aRa hold for all a? Adam sits in the same row as himself, Bram sits in the same row as himself, and Caleb sits in the same row as himself. This relationship certainly is reflexive. Does the symmetric property hold for this relation? If Adam sits in the same row as Bram, aRb, then does Bram sit in the same row as Adam, bRa? You might repeat this same question for Adam and Caleb as well as for Bram and Caleb. This should satisfy you that the relation is symmetric. The transitive property requires that if Adam sits in the same row as Bram, aRb, and Bram sits in the same row as Caleb, bRc, then Adam sits in the same row as Caleb, aRc. We see that the property "sits in the same row as R" satisfies the reflexive, symmetric, and transitive properties and so is an equivalence relation.

Consider the relation described by "is the brother of." Since you can not be your own brother, this relation is not reflexive. We can describe this by $a\cancel{R}a$ or, verbally, by "it is not the case that aRa." If Jon is the brother of Bill, does that mean that Bill is the brother of Jon? Be careful here! We must consider our universe. If we include females in our universe, we find that this relation is not symmetric. Jon can be Mary's brother, but Mary cannot relate to Jon in the same way. Suppose Jon is the brother of Bill and Bill is the brother of Mary. Is Jon the brother of Mary? You see that this relationship is transitive, but not reflexive nor symmetric.

In the last example you were cautioned to carefully consider your universe. Notice that if we consider our universe to include males only, the relationship "is the brother of" is symmetric. Another caution to observe in making a decision regarding relations is that the relation be clearly defined. For example, we might consider the relation "is the neighbor of." This relation may be defined in many ways. The definition which you stipulate may change the properties of the relation. If by neighbor you mean "lives next door to," the properties will be different from the properties if you mean "lives in the same community as." Notice that you cannot live next door to yourself; however, you do live in the same community as yourself. The relation "lives next door to" is symmetric only, but the relation "lives in the same community as" is reflexive, symmetric, and transitive.

Remember, to decide whether a relation is an equivalence relation, be sure to consider the following questions.

1. Is your universe carefully defined?
2. Is the relation clearly defined?

EXERCISES

1. Which of the sets below can be placed in a 1–1 correspondence?

$A = \{2, 3, 4\}$ $B = \{3, 2, 4\}$ $C = \{a, b, c\}$

$D = \{b, a, d, c\}$ $E = \{h, o, r, s, e\}$ $F = \{a, b, c, d\}$

2. If there is a 1–1 correspondence between the sets A and B, what can you say about $n(A)$ and $n(B)$?

3. Which pairs of sets in Exercise 1 are equivalent sets?

4. Which pairs of sets in Exercise 1 are equal sets?

5. Write two sets which are equivalent but not equal.

6. Show at least three ways in which you can place sets *A* and *B* of Exercise 1 in a 1–1 correspondence.

7. Decide which properties of an equivalence relation are present in each of the following relations.
a. U = {all people in the world}; R = "is a first cousin of"
b. U = {cities and towns in your home state}; R = "is north of"
c. U = {rooms in a hotel}; R = "on the same floor with"
d. U = {sides of a triangle}; R = "has at least one point in common with"
e. U = a set of sets; R = "is a subset of" (R = $\subseteq$)

■ **C.12** **8.** a. Is $\subseteq$ an equivalence relation?
■　　　　 b. Can you find a set such that $\subseteq$ is an equivalence relation?

SECTION D **OPERATIONS ON SETS**

BEHAVIORAL OBJECTIVES

D.1 Be able to define the operation of union on sets.

D.2 Be able to define the set $A \cup B$.

D.3 Be able to tabulate $A \cup B$ for given sets.

D.4 Be able to give a verbal description of $A \cup B$ for given sets.

D.5 Be able to perform the operation of union, $A \cup B$, on given sets using Venn diagrams.

D.6 Be able to define the operation of intersection on sets.

D.7 Be able to tabulate $A \cap B$ for given sets.

D.8 Be able to define the set $A \cap B$.

D.9 Be able to give a verbal description of $A \cap B$ for given sets.

D.10 Be able to perform the operation of intersection, $A \cap B$, on given sets using Venn diagrams.

D.11 Given A and B, be able to find $n(A \cup B)$.

D.12 Given A and B, be able to find $n(A \cap B)$.

D.13 Given A and B, be able to find $p(A \cup B)$.

D.14 Given A and B, be able to find $p(A \cap B)$.

D.15 Given A and B, be able to relate $n(A \cup B)$ to $n(A)$ and $n(B)$.

D.16 Given A and B, be able to relate $n(A-B)$ to $n(A)$ and $n(B)$.

D.17 Given A and B, find $n[p(A \cup B)]$.

D.18 Given A and B, find $n[p(A \cap B)]$.

D.19 Generalize $\varnothing \cup A$ for any set A.

D.20 Generalize $U \cup A$ for any set A.

D.21 Generalize $\varnothing \cap A$ for any set A.

D.22 Generalize $U \cap A$ for any set A.

SET UNION

We can join two sets in a number of ways. The first operation which we will consider is called *union*. Union is a binary operation. This means that union operates on two sets to produce a single set. We can perform the operation of union on two sets that are disjoint (have no elements in common) or on sets that have elements in common (not disjoint). The set that results by unioning or "taking the union" of two sets is called—logically even if ambiguously—the *union* of the two sets. The union of two sets is the set which contains all the elements of both sets upon which the operation is performed.

The symbol for union is $\cup$. Do not confuse $\cup$ with the symbol for universe, U. The union of two sets is written $A \cup B$. Consider the following examples.

Examples

■ D.3   Let U = the set of counting numbers.

1. If $A = \{1, 2, 4\}$ and $B = \{3, 5, 6\}$,
   $A \cup B = \{1, 2, 3, 4, 5, 6\}$

2. If $A = \{1, 2, 4\}$ and $B = \{3, 4, 5\}$,
   $A \cup B = \{1, 2, 3, 4, 5\}$

3. If $A = \{1, 2, 3, 4\}$ and $B = \{2, 4\}$,
■ $A \cup B = \{1, 2, 3, 4\}$

Notice in the examples above that we wrote the union of A and B by tabulating the set; that is, we listed the elements in the union. There were three situations involved in the examples shown. The sets A and B were disjoint in the first example, had some elements in common in the second example, and one set was a subset of the other in the last example.

We can illustrate the union of two sets by a symbolic description, or by a Venn diagram, as well as by enumeration. Consider how we might describe the union of two sets by symbolic description, that is, without actually writing all of the elements of the two sets. In certain cases we can easily see what the result of unioning two sets is from the characteristics of the sets themselves.

◀ D.20 **Example**
▲ D.19   For any set A,

■   1. $A \cup U = U$
▲   2. $A \cup \varnothing = A$
     3. If $A \subseteq B$, then $A \cup B = B$

Notice in these examples we did not tabulate or enumerate the sets because we knew that the union of a set with the universe consists of the elements in the universe. Do you know why? If we union the set A with an empty set, we produce the set A. In the last example, since every element of A is a member of the set B, the union of A and B is the set B.

To describe the union of two sets using a verbal description only, that verbal description must be clearly stated. Some examples follow.

Examples

■ **D.4**
1. If A = the set of even natural numbers and B = the set of odd natural numbers, then $A \cup B$ = the set of all natural numbers.
2. If A = the set of people younger than 20 years of age and B = the set of newborn babies, then $A \cup B$ = the set of people younger than 20 years of age.
3. If A = the set of months of the year with 30 days and B = the set of months of the year with 31 days, then $A \cup B$ = the set of all the months
■ of the year but February.

Venn diagrams are often used to "picture" or "model" the unions. When we use Venn diagrams, we shade areas to indicate the sets we are describing. Since this can sometimes be confusing, we use the convention which most mapmakers use, namely, writing a key at the bottom of the map or diagram. In discussing union, we have pointed out several possible situations. Some of these are illustrated in Figures 1.10–1.13.

■ **D.5**

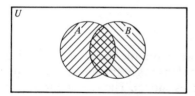

Figure 1.10 $A \cup B =$ ▨, ▨, and ▨.

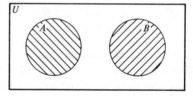

Figure 1.11 $A \cup B =$ ▨ and ▨.

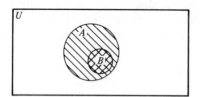

Figure 1.12 $A \cup B =$ ▨ and ▨.

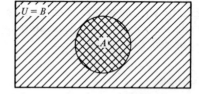

Figure 1.13 $A \cup B =$ ▨ and ▨.

Note in Figures 1.10, 1.12, and 1.13 that areas cross-hatched in both directions also belong to $A \cup B$. Notice in the diagrams above that it is not necessary to follow any particular convention for shading. It is often useful, as you will see later, to shade in more than one way. By using different shading we can see more clearly where sets overlap if they are not disjoint.

To summarize the attributes which define the operation of union, consider what has been said thus far.

1. Set union is an operation.

2. The operation is performed on two sets (union is a binary operation).

3. The operation produces a set which contains all of the elements in both sets and is called "the union of the sets."

4. Two sets can be unioned whether they have common elements or not.

5. We do not repeat common elements when we write the union of sets.

These attributes together produce the definition of union. We write the union of sets A and B as $A \cup B$.

Definition

■ **D.1** ■ *Union is a binary operation on sets that produces a set whose elements are all the elements of both sets joined by the operation.*

■ **D.2** *The set that results when we union the sets A and B is symbolized by $A \cup B$.*

We can use set builder notation to describe the set $A \cup B$. In doing so we also describe how the operation $\cup$ works to produce the set $A \cup B$.

Definition

■ $A \cup B = \{x \mid x \in A \text{ or } x \in B\}$

In an earlier section we showed how we could make Venn diagrams for specific sets, namely, $\bar{A}$ and $A - B$. We can find a diagram to describe set union for given sets, as shown in the example below.

Example

■ **D.5** Let $U = \{$Ethan, Mary, David, Adam, Daniel, Sara$\}$
$A = \{$Mary, Adam, Sara$\}$
$B = \{$Adam, David$\}$

Make a Venn diagram of $A \cup B$.

SOLUTION Since Adam is contained in both A and B, we would expect the sets to overlap as shown in Figure 1.14.

■

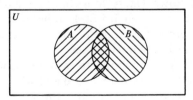

Figure 1.14 $A \cup B = $ ▨ . ▧ .
and ▩ .

EXERCISES

Use the sets listed for Exercises 1–3, where $U = N$.

$A = \{2, 4, 6, \ldots\}$ $\qquad\qquad\qquad$ $B = \{2, 4, 5, 6\}$
$C = \{3, 6, 7, 8\}$ $\qquad\qquad\qquad\qquad$ $D = \{1, 3, 5, \ldots\}$
$E = \{1, 2, 3, \ldots, 10\}$ $\qquad\qquad\quad$ $F = \{2, 4, 6, 8, 10\}$
$G = \{1, 3, 5, 7, 9\}$

1. Tabulate the following sets.
a. $A \cup C$ $\qquad\qquad\qquad\qquad\qquad$ b. $C \cup E$
c. $B \cup E$ $\qquad\qquad\qquad\qquad\qquad$ d. $U \cup D$

2. Give the name A, B, C, D, E, F, or G for each set below.
a. $A \cup D$ $\qquad\qquad$ b. $D \cup G$ $\qquad\qquad\qquad$ c. $F \cup G$

3. Make Venn diagrams to illustrate the union of sets below. Do not include the elements of the sets within the diagram.
a. $E \cup F$ $\qquad\qquad$ b. $F \cup G$ $\qquad\qquad\qquad$ c. $A \cup E$

4. If $U = \{x \mid x$ is any state in the United States$\}$, $A = \{$Maine, Massachusetts$\}$, $B = \{$New York, Pennsylvania$\}$, $C = \{$California, New Mexico$\}$, $D = \{$Tennessee, Mississippi$\}$, describe the following:
a. $A \cup B$ $\qquad\qquad\qquad\qquad\qquad$ b. $A \cup D$
c. $C \cup U$ $\qquad\qquad\qquad\qquad\qquad$ d. $B \cup \varnothing$

5. Using what you know about special sets, write a single symbol for each of the following unions.
a. $A \cup U$ $\qquad\qquad\qquad\qquad\qquad$ b. $A \cup \varnothing$
c. $A \cup \bar{A}$ $\qquad\qquad\qquad\qquad\qquad$ d. $U \cup \varnothing$

SET INTERSECTION

In the previous section we discussed the union operation on sets. There is another important binary operation on sets called *intersection*. The result of performing the operation of intersection is called "the intersection of the set." You should be quite accustomed to using the same word in more than one sense; for years you have probably used "take the sum of 3 and 5" or "sum the numbers 3 and 5" to mean perform the operation of addition on 3 and 5, while at the same time you would say that "8 is the sum of 3 and 5," meaning that the sum is the result of "summing" or adding 3 and 5.

One use of the word "intersection" relates to street crossings. The intersection of 3rd and Main St. in Sometown, U.S.A., looks like a rectangle which is on Main St. and at the same time is on 3rd St. In other words, both streets share some of the same ground or some of the same area. This is a good way to think of the operation of intersection.

If two streets such as 3rd St. and 4th St. run parallel to each other and someone asked how they might get to the intersection of 3rd and 4th Sts., they would be told that there is no intersection of 3rd and 4th Sts. This means that the intersection is an empty one. If two sets have no elements in common, we say that the intersection is an empty set. Thus, we can characterize *disjoint sets* as sets whose intersection is the empty set.

A third example to consider is the intersection of a street in a town with the town itself. This intersection includes all of the ground which makes up the street and which also belongs to the town. Such an intersection consists of the street itself. We might compare this to the intersection of a set with the universe. Since the set is a subset of the universe, the intersection is the set itself.

We symbolize intersection by using an upside down $\cup$. This is easy to remember since it is the opposite of the symbol for union. The intersection of two sets A and B is written $A \cap B$, and we read this statement as "A intersection B" or "the intersection of A and B." The operation of intersection produces a set just as the operation of union produces a set.

■ **D.8**   From the examples used above, one can see that the intersection of two sets produces a set containing elements which are common to both

■   sets. In symbols, $A \cap B = \{x \mid x \in A \text{ and } x \in B\}$.

In our discussion of union, we considered writing the result of the operation in four ways: (1) tabulating or enumerating the set, (2) using accepted names for the set produced, (3) describing the set by using words, and (4) making a Venn diagram of the set. We will consider each of these methods below.

Tabulation of the intersection of two sets involves finding in some systematic fashion the elements which are common to both sets. One way of doing this accurately is to circle or underline common elements as you see them and then check to see that you haven't missed any common elements. Consider the following examples.

Examples

■ **D.7**   1. If $A = \{2, 3, 5, 7, 9\}$ and $B = \{1, 2, 3, 4, 5\}$, then
$A \cap B = \{2, 3, 5\}$

2. If $A = \{\text{Jon, Bill, Mary}\}$ and $B = \{\text{Jon, Mary}\}$, then
$A \cap B = \{\text{Jon, Mary}\}$

3. If $A = \{\text{red, blue, green}\}$ and $B = \{\text{black, white}\}$, then
■ $A \cap B = \{\ \}$

Notice in the first example that the sets have a non-empty intersection which consists of the three elements common to both sets. In the second example, set B was a subset of set A and the intersection produced the set B. In the last example, there were no elements common to both sets. Therefore, the intersection was empty. Notice that we could have written $A \cap B = \varnothing$.

With certain special sets we can indicate intersections without actually tabulating the sets. Some examples of these are illustrated below. Notice that most of these examples involve the use of sets such as the universe, empty set, relative complement, or complement of a set.

Examples

■ D.21 ■ $A \cap \varnothing = \varnothing$
$(A - B) \cap A = A - B$
$\bar{A} \cap A = \varnothing$

■ D.22 ■ $A \cap U = A$
$(A - B) \cap B = \varnothing$

We might wish to perform intersection on sets which are described verbally without enumerating the elements. In this kind of situation, the verbal description of the set must be clear and the set must have characteristics which allow us to do this. Consider the next set of examples:

Examples

■ D.9   1. If A = the set of states in the United States and B = the set of southern states, then $A \cap B$ = the set of southern states.

2. If E = the set of even natural numbers and M = the set of multiples of three, then $E \cap M$ = the set of even multiples of three.

3. If B = the set of all brown-eyed people and G = the set of all blue-eyed
■   people, then $B \cap G$ = the empty set.

The first two examples listed above should be clear to the reader. You might, however, suggest that in the last example the intersection would include people who have one green eye and one brown eye, or people who have greenish-brown eyes. This would be true if we had such people in our original sets. The universe in this example was not specified. As we indicated earlier, a universe is implied by many problems, but many problems cannot be solved unless a carefully defined universe is stated in advance. Therefore, if we insist in the problem involving brown- or green-eyed people that our universe is a set of people whose eyes are either brown or green but not both, the intersection is empty.

Consider next the use of Venn diagrams to show the intersection of two sets. We will follow the convention of shading each set in different ways. If the first set is cross-hatched to the left and the second set is cross-hatched to the right, then the intersection set is the set that is cross-hatched in both directions, that is, the region which is cross-hatched both left and right. If there are no regions shaded in two directions, then the intersection is empty. Figures 1.15–1.17 represent $A \cap B$ when A and B overlap, when A and B are disjoint, and when B is a subset of A. Figure 1.18 represents $(A - B) \cap A$.

In Figure 1.15 you can see that the overlapping or shared region is cross-hatched to both the right and the left. This region represents the intersection. In Figure 1.16 it is clear that there is no region which is common to both A and B. In Figure 1.17, since B is contained in A, the common region of both sets is the set B. In Figure 1.18 the set $A - B$ is represented by left cross-hatching only, and the set A is represented by left cross-hatching as well as left and right cross-hatching. These

■ D.10

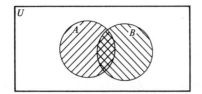

Figure 1.15 $A \cap B = $ ▓▓.

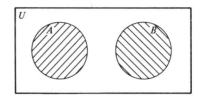

Figure 1.16 $A \cap B = \varnothing$.

■

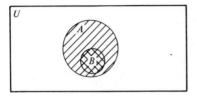

Figure 1.17 $A \cap B = $ ▓▓
or $A \cap B = B$.

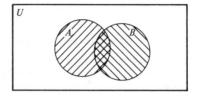

Figure 1.18 $A - B = $ ▨ ;
$(A - B) \cap A = $ ▨
or $(A - B) \cap A = A - B = $ ▨ .

two sets share the region of A which is not contained in B; in other words, they share $A - B$. As stated below Figure 1.18, $(A - B) \cap A = A - B$.

We have considered several examples of intersection thus far. To summarize the attributes of the operation we may consider the following:

1. Intersection is a binary operation.
2. The operation produces a set.
3. The set produced consists of those elements that are common to both sets.
4. The intersection of two sets is symbolized by $A \cap B$.

By considering all of these attributes together, we define the intersection.

Definition

■ D.6
▲ D.8
▲ ■

Intersecting the sets A and B is a binary operation that produces a set $A \cap B$ which contains all the elements common to both the set A and the set B. Symbolically, $A \cap B = \{x \mid x \in A \text{ and } x \in B\}$.

EXERCISES

Use the sets listed for Exercises 1–3, where $U = N$.
$A = \{2, 4, 6, \ldots\}$ $B = \{2, 4, 5, 6\}$
$C = \{3, 6, 7, 8\}$ $D = \{1, 3, 5, \ldots\}$
$E = \{1, 2, 3, \ldots, 10\}$ $F = \{2, 4, 6, 8, 10\}$
$G = \{1, 3, 5, 7, 9\}$

1. Tabulate the following sets:
   a. $A \cap C$
   b. $C \cap E$
   c. $B \cap E$
   d. $U \cap D$

2. Rename each of the following sets with a single symbol, for example, $A \cap F = F$
   a. $A \cap D$
   b. $D \cap G$
   c. $F \cap G$

3. Make Venn diagrams to illustrate the intersection of sets below. Do not include the elements of the sets within the diagram.
   a. $E \cap F$
   b. $F \cap G$
   c. $A \cap E$

4. Use symbols for the sets given and special sets with which you are familiar, for example, $A \cap \varnothing = \varnothing$, to symbolize the following.
   a. $A \cap B$ when $A \subset B$
   b. $A \cap B$ when A and B are disjoint
   c. $B \cap \varnothing$
   d. $B \cap U$
   e. $(A - B) \cap B$
   f. $(A - B) \cap \bar{B}$

5. Make Venn diagrams to illustrate Exercise 4.

6. Describe the intersection of the sets given below using words.
   a. The set of months of the year beginning with the letter J; the set of all the months of the year.
   b. The set of New England states of the United States; the set of southern states of the United States.
   c. The set of all people in the set of high school graduates with the set of all females.
   d. The set of natural numbers greater than 6 and less than 12; the set of natural numbers less than 10.

$n(A)$ AND UNION OF SETS

We designated $n(A)$ to be the number of elements in the set A. We want to consider the number property of the union of sets A and B, $n(A \cup B)$. If $n(A)$ is 3 and $n(B)$ is 2, what is $n(A \cup B)$? $n(A \cup B)$ is a number. Does this number depend on the specific sets A and B? That is, is $n(A \cup B)$ some fixed number regardless of the composition of A and B? To answer these questions, we consider the ways in which two sets can form a union. We know that two sets are either disjoint, or partially overlap, or one set is contained in the other.

■ D.11
▲ D.15
▲   If the sets A and B are disjoint, what is $n(A \cup B)$? $A \cup B$ will consist of all the elements of A and all the elements of B. These elements are distinct because A and B are disjoint, so $n(A \cup B) = n(A) + n(B)$.

What happens if one set is a subset of the other? We know that if B is a subset of A, then $(A \cup B)$ is the set A. This leads us to the conclusion that $n(A \cup B) = n(A)$. Now suppose that $A \cap B$ is not empty. That is, there are some elements common to both A and B. For example, if $A = \{a, b, c\}$ and $B = \{c, d\}$, then $A \cap B = \{c\}$. In this case, since $A \cup B =$
■ $\{a, b, c, d\}$, $n(A) = 3$, $n(B) = 2$, and $n(A \cup B) = 4$.

From the preceding remarks we see that the number of elements in the union of two sets cannot exceed the sum of the numbers of elements of each set, nor can it be less than the number of elements of either set. The key is that the number of elements in the intersection of A and B determines the size of $n(A \cup B)$.

D.16   In order to answer the question concerning the relative size of $n(A \cup B)$ in relation to $n(A)$ and $n(B)$, let us first note that $(A \cap B) \subseteq A$, so the set $A - (A \cap B)$ and the set B are disjoint. Can you see that $(A - (A \cap B)) \cup B = A \cup B$? Now use the number property that we know holds for the disjoint sets $A - (A \cap B)$ and B. That is, $n(A - (A \cap B)) + n(B) = n(A \cup B)$. Since $(A \cap B) \subseteq A$, we know that $n(A - (A \cap B)) = n(A) - n(A \cap B)$, and placing this expression in the previous equality, we have the relationship

$$n(A) - n(A \cap B) + n(B) = n(A \cup B)$$

Let us rewrite this as

■ $$n(A \cup B) = n(A) + n(B) - n(A \cap B)$$

D.15   It is of interest to one who teaches young children to note that the operation of addition is much like that of unioning disjoint sets. We considered the relationship $n(A)$, $n(B)$, and $n(A \cup B)$ in the last paragraph. Notice that when A and B are disjoint, $n(A) + n(B) = n(A \cup B)$.

For example, consider the sets $A = \{*, \text{\char"2A2F}, 0\}$ and $B = \{\text{☌}, \text{✐}\}$. Notice that $n(A) = 3$ and $n(B) = 2$. Since $A \cup B = \{*, \text{\char"2A2F}, 0, \text{☌}, \text{✐}\}$, $n(A \cup B) = 5$; that is, $n(A) + n(B) = n(A \cup B)$, or $3 + 2 = 5$. We will consider this
■   use of union later in Chapter 3.

EXERCISES

1. If $n(A) + n(B) = n(A \cup B)$, then $A \cap B = $?

2. If $n(A \cup B) = 7$ and $n(A \cap B = 3$, then $n(A) + n(B) = $?

3. If $n(A \cup B) = 7$ and $n(B) = 7$, then $n(A) = $?

4. If $n(B) = n(A \cap B)$, then $n(A) = $?

5. If $n(A - B) = n(A)$, then
   a. $n(B) = $?   b. $A \cup B = $?   c. $A \cap B = $?

6. Find two sets A and B such that $n(A \cup B) = n(A)$.

7. Find two sets A and B such that $n(A \cup B) = n(A) = n(B)$.

8. Find two sets A and B such that $n(A \cup B) = n(A) + n(B)$.

THE POWER SET OF UNION AND INTERSECTION OF SETS

Once we are able to produce union and intersection of given sets, it is relatively simple to find the power set of the union or intersection of the given sets. Consider the following examples of power sets for both the union and the intersection of sets A and B.

■ **D.13** **Examples**

▲ **D.14**

	1	2	3
A	$\{1, 2\}$	$\{1, 2, 3\}$	$\{1, 2\}$
B	$\{3\}$	$\{1, 3\}$	$\{2, 3\}$
$A \cup B$	$\{1, 2, 3\}$	$\{1, 2, 3\}$	$\{1, 2, 3\}$
$A \cap B$	$\varnothing$	$\{1, 3\}$	$\{2\}$
$p(A \cup B)$	$\{\varnothing, \{1\}, \{2\}, \{3\},$ $\{1, 2\}, \{1, 3\}, \{2, 3\},$ $\{1, 2, 3\}\}$	$\{\varnothing, \{1\}, \{2\}, \{3\},$ $\{1, 2\}, \{1, 3\}, \{2, 3\},$ $\{1, 2, 3\}\}$	$\{\varnothing, \{1\}, \{2\}, \{3\},$ $\{1, 2\}, \{1, 3\}, \{2, 3\},$ $\{1, 2, 3\}\}$
$p(A \cap B)$	$\{\varnothing\}$	$\{\varnothing, \{1\}, \{3\}, \{1, 3\}\}$	$\{\varnothing, \{2\}\}$

▲ ■

We notice in the illustration that $A \cup B$ is the same set in each example; therefore, its power set is the same in each example. The intersection of A with B is different in each case, which changes the power set of inter-section in each case. Notice also that $p(A \cap B)$ will always be a subset of

■ **D.18** $p(A \cup B)$ because $(A \cap B) \subseteq (A \cup B)$. It should be apparent that if the number of elements, $n[p(A \cap B)]$, in the set $p(A \cap B)$ is known, then $n(A \cap B)$ is known. For example, if there is only one element in the set $p(A \cap B)$, we know that 2^n is 1. This means that n is 0. If n is 0, then $n(A \cap B)$ is 0, which implies that the intersection is empty. In the second example, since $n[p(A \cap B)]$ is 4, 2^n is 4 and n must be 2. This means that there are 2 elements in the set $A \cap B$, or $n(A \cap B) = 2$. This concept is reversible. Suppose we know that there is only 1 element in $A \cap B$; then we know that $p(A \cap B)$ contains 2^1 elements, or simply 2 elements.

▲ **D.17** This relationship also holds for the number of elements in $p(A \cup B)$ and the number of elements in $A \cup B$. We can generalize these by stating the following:

▲ $$n[p(A \cup B)] = 2^{n(A \cup B)}$$

■ $$n[p(A \cap B)] = 2^{n(A \cap B)}$$

EXERCISES

1. Under what conditions are the following statements true?
 a. $n(A) + n(B) = n(A \cup B)$
 b. $n(A) + n(B) = n(A \cap B)$
 c. $n(A) + n(B) = n(A - B)$
 d. $n(A) - n(B) = n(A \cap B)$
 e. $n(A) - n(B) = n(A - B)$

2. Use the following sets to answer the questions stated below.
 $A = \{a, e, i, o, u\}$ $B = \{y, o, u\}$ $C = \{i, o, u\}$
 a. Find $n(A - C)$, $n(A \cup B)$, and $n(A \cap C)$.
 b. Find $p(A - C)$, $p(B \cup C)$, and $p(B \cap C)$.
 c. Find $n[p(A - C)$, $n[p(B \cup C)]$, and $n[p(B \cap C)]$.

3. Show using Venn diagrams that
 a. $A - (A \cap B)$ and B are disjoint
 b. $[A - (A \cap B)] \cup B = A \cup B$

SECTION E   **FUNDAMENTAL PROPERTIES OF OPERATIONS**

BEHAVIORAL OBJECTIVES

E.1   Be able to define closure.

E.2   Be able to state closure for ∪ of sets.

E.3   Be able to state closure for ∩ of sets.

E.4   Be able to state what is required for an operation to be commutative.

E.5   Be able to state the commutative property of ∪ on sets.

E.6   Be able to state the commutative property of ∩ on sets.

E.7   Be able to state the associative property of ∪ on sets.

E.8   Be able to state the associative property of ∩ on sets.

E.9   Be able to state what is required for an element to be an identity element.

E.10   Be able to identify the identity element of ∪ on sets.

E.11   Be able to identify the identity element of ∩ on sets.

E.12   Be able to list those fundamental properties that hold for $p(A)$ under union.

E.13   Be able to list those fundamental properties that hold for $p(A)$ under intersection.

E.14   Be able to perform mixed operations on sets.

E.15   Define a distributive property of ∪ over ∩.

E.16   Define a distributive property of ∩ over ∪.

CLOSURE

■ E.1   We recall from the sections dealing with union and intersection that each operation produced a set. In each case the set produced by the operation was a subset of the universe. When a binary operation takes two elements from a given set and produces a third element which is also in that set, we say that the *operation is closed in the given set*, or that the operation has the *closure property on the given set*, or simply that the

■   *operation is closed.*

In the case of set operations our elements are sets that belong to some larger collection of sets. This larger collection serves as the "given set" or universe.

To understand the idea of closure, it might help to consider an operation which is not closed. For example, consider the set of unmarried men and unmarried women as our given set or universe, and marriage as the operation on that set. This is a binary operation which produces an element called a married couple. If we consider our universe, we note that there are no married couples in the universe of unmarried men and women. We might symbolize this as follows: For U = (unmarried men and unmarried women} and an operation called marriage and written "∗", then if a and $b \in U$,

$$a * b = c \notin U$$

Therefore, U is not closed under ∗.

Consider another example of an operation which is not closed. The operation of chemical bonding is not closed if we let our universe be chemical elements. For example, we can take the chemical elements hydrogen and oxygen and combine them to produce water, which is not a chemical element but a chemical compound. This operation is not closed unless we wished to alter our universe to include both compounds and elements.

Finally, it might be helpful to consider a positive example of closure. The operation of mating in the animal kingdom is a closed operation. If we mate two horses from the universe of horses, we produce a horse as a result of the operation. The newly produced element is in the set of horses. It is not possible to produce a cow or a dog as the result of the operation of mating horses. This operation is closed in the set of horses. Even if we enlarge our universe to include all animals, it is the case that the mating of two animals produces an animal which is in the universe of animals.

Since our purpose was to consider the properties of union and intersection, let us summarize the attributes of the closure property for union and intersection.

1. Both union and intersection are binary operations.

2. The result of the operations performed is always a set.

3. The set produced is always a subset of the universe.

■ E.2
▲ E.3

▲ ■

We define the closure property for the operation union on the set U using symbols as follows: The operation union, $\cup$, is closed if $A \cup B \in U$ whenever $A \in U$ and $B \in U$. Similarly, the operation intersection, $\cap$, is closed if $A \cap B \in U$ whenever $A \in U$ and $B \in U$.

EXERCISES

1. Which of the following operations are closed? Why?
   a. The operation of making dessert from the set of ice cream, Jello, bananas, and apples.
   b. The operation of combining colors from the set of primary colors red, blue, and yellow.
   c. The operation of combining colors from the set of all colors.
   d. The operation of moving from floor to floor on an elevator for the set of all floors in a given building.

2. If $D = \{a, b, c\}$, is $p(D)$ closed under
   a. union?
   b. intersection?

3. Give an example of a set whose elements are sets for which the operation of union is
   a. closed
   b. not closed

4. Give an example of a set whose elements are sets for which the operation of intersection is
   a. closed
   b. not closed

5. If we consider p an operation on a set A that produces the power set $p(A)$, is p closed on U where U is the universe of all sets?

COMMUTATIVE PROPERTY

A commuter is a person who travels back and forth from home to work and from work to home. If we consider someone who uses the same route daily in both directions, we know that the distance traveled is the same whether he is going to or coming from work. We could translate this to mean that the order does not affect the distance traveled. We could shorten this idea by writing A for distance to home and B for distance to work, and the operation of commuting could be written "$*$". Then it would follow that

$$A * B = D \quad \text{and} \quad B * A = D$$

and we might conclude that

$$A * B = B * A$$

with respect to distance.

Certain operations have this property, which deals essentially with order. The order in which we do things may or may not affect the results. In the operation of making tea, one can add water to the tea or add tea to the water and the result will still be a cup of tea. No doubt some would argue that the taste would differ depending upon the order; however, one would still have tea. Therefore, making tea is a commutative operation.

When putting on one's shoes and socks, the order is quite important. If one tried to put one's shoes on first, the result would be somewhat different from the desired results. Therefore, the operation of "putting on" defined on the set of shoes and socks is a non-commutative operation.

Consider two examples which illustrate commutativity and non-commutativity.

Example 1

Consider the set $S = \{1, 2\}$ and the operation on S which produces the larger number, written "$\circ$". Since 2 is the larger, it is the case that

$$2 \circ 1 = 2 \quad \text{and} \quad 1 \circ 2 = 2$$

Therefore, $2 \circ 1 = 1 \circ 2$, and this operation is *commutative*.

Example 2

Consider again $S = \{1, 2\}$ and an operation called the *first number operation*. This operation will be symbolized as "$*$" and defined to produce the first number as its result. Therefore, $1 * 2 = 1$, but $2 * 1 = 2$. This means that $1 * 2 \neq 2 * 1$. Thus, the operation on S is *non-commutative*.

■ E.4   We say that a binary operation, $*$, is *commutative* on a set S if $a * b = b * a$
■   for any choice of elements from the set S.

Consider now the union and intersection operations. If we perform the union operation on the sets A and B in this order, we write all of the elements in A first and then list next those elements in B which are not contained in A. If we reverse the order of the operation, we list all of those elements in B first and then those elements in A which are not contained in B. The results are a set of elements which are in A only, elements in both A and B, and the set of elements contained in B only. The Venn diagram in Figure 1.19 may clarify this relationship; right cross-hatching was used first. Notice that in each of the diagrams the union of the two sets is different in direction of cross-hatching only. The intersection in both cases is the section which is cross-hatched in both directions. The order or position of cross-hatching did not affect the result. This is an intuitive notion and not really a proof. We say that the operations of union and intersection are commutative because

■ E.5 ■ $A \cup B = B \cup A$

and

■ E.6 ■ $A \cap B = B \cap A$

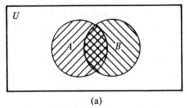 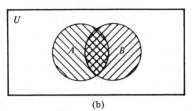

(a)   (b)

Figure 1.19   (a) $A \cup B = $ ▨ or ◺; $A \cap B = $ ▦. (b) $B \cup A = $ ▨ or ◺; $B \cap A$ ▦.

EXERCISES

1. Which of the following operations are commutative? Why?
     a. "Moving" from floor a to floor b and then from floor b to floor c, where the operation is defined in terms of distance traveled, the building has ten floors, and the elevator must go to the top floor before it will change direction. Note that the set is a set of distance between floors.
     b. The operation of "cleaning" on a set of household objects such as: $S = \{$dishes, glasses, floors, silverware$\}$. (*Hint:* Will the final result of cleaning the dishes and then the glasses be different from cleaning the glasses and then the dishes?)

c. The operation of "driving" between points A and B over the same road, where (1) distance traveled is the result and (2) arrival at one's destination is the result.

d. The operation of moving an object described as $\overline{AB}$ around a center to produce an object $\overline{BA}$. Let the set include two elements, namely, a single move which we call m_1, and a double move which we call m_2; thus, $S = \{m_1, m_2\}$. Note that the operation $m_1 \cdot m_1 = m_2$ and $m_1 \cdot m_2 = m_1$.

2. Use the definition of set equality to prove that for A and B sets
a. $A \cup B = B \cup A$
b. $A \cap B = B \cap A$

ASSOCIATIVE PROPERTY

We have not considered how the operations of union and intersection are performed when more than two sets are involved. Since we defined these operations as binary operations, you may get the impression that if you were given more than two sets to union, the task would be impossible. This is not true. Consider a few examples.

Example 1

Find the union of sets A, B, and C when $A = \{1, 2, 3\}$, $B = \{2, 3, 5, 7\}$, and $C = \{3, 5, 7, 9\}$.

The question arises as to what this problem involves. Should we union A and B and then union the result with the set C? Should we union A with the result of finding the union of B and C? To clarify what we mean in stating such a problem, we use parentheses to indicate which operation to perform first. Thus, $(A \cup B) \cup C$ means union C with the union of A and B, whereas $A \cup (B \cup C)$ means union A with the union of B and C. Notice that we did not change the order in which the elements are written! Let us consider what happens if we find the union in two ways for the sets listed above.

SOLUTION
1. $(A \cup B) \cup C = \{1, 2, 3, 5, 7\} \cup \{3, 5, 7, 9\} = \{1, 2, 3, 5, 7, 9\}$
2. $A \cup (B \cup C) = \{1, 2, 3\} \cup \{2, 3, 5, 7, 9\} = \{1, 2, 3, 5, 7, 9\}$

In the example above the change in grouping or association did not affect the result of the operation. With respect to an operation when the grouping of elements of a set does not affect the result, we say that the operation is *associative* on the set.

Example 2

Consider the intersection of those sets in the above example. Do you expect that operation is associative?

SOLUTION
1. $(A \cap B) \cap C = \{2, 3\} \cap \{3, 5, 7, 9\} = \{3\}$
2. $A \cap (B \cap C) = \{1, 2, 3\} \cap \{3, 5, 7\} = \{3\}$

Example 3

Find both the union and the intersection of the sets given. Show that grouping does not affect the final result.

$$A = \{s, x, r, e\} \qquad B = \{a, e, r, s, x\} \qquad C = \{x, y, s, a, e\}$$

SOLUTION
1. $(A \cup B) \cup C = \{s, x, r, e, a\} \cup \{x, y, s, a, e\}$
$= \{s, x, r, e, a, y\}$
2. $A \cup (B \cup C) = \{s, x, r, e\} \cup \{a, e, r, s, x, y\}$
$= \{s, x, r, e, a, y\}$

Thus,

■ E.7 ■
$$(A \cup B) \cup C = A \cup (B \cup C)$$
1. $(A \cap B) \cap C = \{s, x, r, e\} \cap \{x, y, s, a, e\}$
$= \{s, x, e\}$
2. $A \cap (B \cap C) = \{s, x, r, e\} \cap \{a, e, s, x\}$
$= \{s, x, e\}$

Thus,

■ E.8 ■
$$(A \cap B) \cap C = A \cap (B \cap C)$$

The associative property holds for the given examples for both union and intersection. Again, we will accept the property without proof; however, it will be instructive to look at Venn diagrams which illustrate this idea. In Figure 1.20, $(A \cup B) \cup C$, and in Figure 1.21, $A \cup (B \cup C)$. Thus,

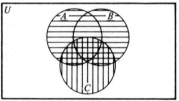

Figure 1.20   **Figure 1.21**

■ E.7 ■ $(A \cup B) \cup C = A \cup (B \cup C)$. Notice the use of shading here. In Figure 1.20, horizontal shading represents $(A \cup B)$. In Figure 1.21, cross-hatching to the left represents $(B \cup C)$.

Consider sets A, B, and C. In Figures 1.22–1.25 we have the following convention of shading; the first operation is performed with horizontal shading (see Figure 1.22). It should also be clear that $A \cap B$ is not the same set as $B \cap C$ (see Figure 1.24).

The intersection of the set $(A \cap B)$ with the set C in Figure 1.23 and the intersection of the set A with the set $(B \cap C)$ in Figure 1.25 contain the same region. Thus, $(A \cap B) \cap C = A \cap (B \cap C)$. It would be instructive for the student to make diagrams similar to these and to perform union and intersection on several sets.

■ E.7 *The associative property of union states that*

■ $(A \cup B) \cup C = A \cup (B \cup C)$

■ **E.8** *The associative property of intersection states that*

■ $(A \cap B) \cap C = A \cap (B \cap C).$

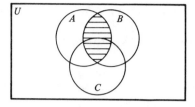

Figure 1.22 $(A \cap B) = \boxed{\equiv}.$

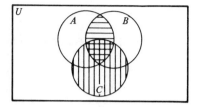

Figure 1.23 $A \cap B = \boxed{\equiv}$;
$(A \cap B) \cap C = \boxed{\#}.$

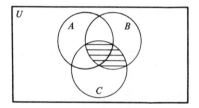

Figure 1.24 $B \cap C = \boxed{\equiv}.$

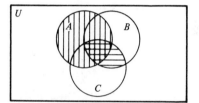

Figure 1.25 $B \cap C = \boxed{\equiv}$;
$A \cap (B \cap C) = \boxed{\#}.$

EXERCISES

1. Which of the following operations are associative?
   a. Refinishing furniture by sanding, using base coat of paint, and using final coat of paint. (*Hint*: Does the sequence of doing each thing affect the final outcome?)
   b. Cooking outdoors by gathering wood, making a fire, and heating the food. Consider resting between (AB) and C or resting between A and (BC).
   c. Eating a meal with salad, main course, and dessert.
   d. Consider C a subset of B and B a subset of A. Is $A - (B - C)$ associative, that is, is $A - (B - C) = (A - B) - C$?

2. State the associative property of addition and multiplication in the set of natural numbers.

3. Is division associative in the set of natural numbers?

IDENTITY ELEMENTS

When we start looking more deeply into the behavior of a binary operation on the elements of a set, we sometimes find particular elements that have a consistent behavior different from that of other elements under the operation. The behavior that we are interested in now is exhibited by a particular element having the property that whenever the binary operation operates on that particular element and any other element the result is the other element. It may be clearer to use symbolic notation

E.9

to describe the special behavior that we are looking at. Let $*$ symbolize a binary operation on a set S. If there is an element e in S with the property that for any element $a \in S$, $e * a = a$ and $a * e = a$, we say that e is an identity element for $*$ in S.

Notice we do not imply that every operation on a set S is such that there is an element in the set which is an identity. We are simply showing how an identity operates on other elements in S if the identity actually does exist.

To find the identity element with respect to set union, one must answer the question, Does a set exist such that its union with any set A will produce the set A? The same question should be asked with respect to the intersection operation. Is there a set such that its intersection with the set A will produce the set A? Consider the special sets U and $\emptyset$. It is left as an exercise to show that

■ E.10   1. The empty set is the identity for the union operation; that is,

■ $$A \cup \emptyset = A = \emptyset \cup A$$

■ E.11   2. The set U is the identity element for the intersection operation; that is,

■ $$A \cap U = A = U \cap A$$

Notice that we are stipulating that an identity element must produce the same result on the right as it does on the left. In this case, this follows automatically since the operation union and intersection are commutative. However, our definition specifically states that an identity element must commute, if it exists.

EXERCISES

1. In the mixing of colors, is there an identity color?

■ E.12   **2.** In $p(A)$ for $A = \{1, 2, 3\}$, and the operation union,
   a. is $p(A)$ closed?
   b. is $\cup$ commutative in $p(A)$?
■   c. is there an identity element in $p(A)$? If so, what is it?

■ E.13   **3.** In $p(A)$ for $A = \{1, 2, 3, 4\}$, and the operation intersection,
   a. is $p(A)$ closed?
   b. is $\cap$ an associative operation?
■   c. is there an identity?

   4. Suppose that e and i are identity elements for $*$ in S.
   a. Since e is an identity, $e * i = ?$
   b. Since i is an identity, $e * i = ?$
   c. Since $e * i = e * i$, what can you say about e and i?
   d. How many identity elements are there for $*$ in S?

MIXED OPERATIONS ON SETS

In the last sections we indicated how union and intersection might be performed with more than two sets. Consider the question of combining two sets by union and then intersecting that set with a third set, as opposed to finding the union of the first set with the intersection of the last two. We might symbolize this as follows:

Is $(A \cup B) \cap C$ the same as $A \cup (B \cap C)$?

A hasty reply to the question would be that both operations are associative; therefore, grouping does not affect the result. Let us consider three sets which have a common intersection which is not empty to see if this is true for a single example. If it is not true, we have supplied a counterexample, and, thus, we can conclude that the answer to the question, Does $(A \cup B) \cap C = A \cup (B \cap C)$? is not always Yes. Remember that for a statement to be true it must always be true. If it is not always true, we call it false.

E.14   Let
$A = \{1, 2, 3, \ldots, 10\}$
$B = \{1, 3, 5, 6, 7\}$
$C = \{6, 7, 8, 9\}$

Then $(A \cup B) \cap C$ is $\{1, 2, 3, \ldots, 10\} \cap C = \{6, 7, 8, 9\}$ and $A \cup (B \cap C)$
■ is $A \cup \{6, 7\} = \{1, 2, 3, \ldots, 10\}$. Obviously, $\{6, 7, 8, 9\} \neq \{1, 2, 3, \ldots, 10\}$.

Notice that we can perform the operations, but we produce two different sets. Therefore, to perform mixed operations, we must know how the operations are ordered. It is not sufficient to write $A \cup B \cap C$ because it is not clear in what order the operations are to be performed. This is true of any binary operation unless priority rules for operating are established.

EXERCISES

For Exercises 1 and 2, assume $U = \{1, 2, 3, \ldots, 15\}$, $A = \{1, 2, 5\}$, $B = \{6, 8, 10\}$, and $C = \{1, 3, 5, 7, 9, 11, 13, 15\}$, and tabulate the results.

1. a. $(A \cup B) \cap C = ?$
  b. $(A \cup B) \cap (A \cup C) = ?$
  c. $(A \cap C) \cup (B \cap C) = ?$
  d. $A \cup (B \cap C) = ?$

2. a. $(\bar{A} \cap \bar{B}) \cup B = ?$
  b. $(\varnothing \cap A) \cup (B \cap A) = ?$
  c. $(\varnothing \cup B) \cap A = ?$

3. Use Venn diagrams to decide which of the following are true statements. Give reasons based on properties, or use examples based on the given sets.
  a. $(A \cup B) \cap C = (B \cup A) \cap C$
  b. $(A \cup B) \cap C = (A \cup C) \cap (B \cup C)$
  c. $C \cap (A \cup B) = (C \cap A) \cup (C \cap B)$
  d. $A \cup (B \cap C) = (A \cup B) \cap (A \cup C)$
  e. $(B \cap C) \cup A = A \cup (B \cap C)$
  f. $(A \cup B) \cap (A \cup C) = (A \cup C) \cap (A \cup B)$

4. Make a Venn diagram for each of the following sets.
  a. $A \cup (B \cap C)$ b. $A \cap (B \cup C)$
  c. $(A \cap B) \cup (A \cap C)$ d. $(A \cup B) \cap (A \cup C)$

DISTRIBUTIVE PROPERTIES

In the last exercise set you performed several mixed operations. Consider Exercise 4 in that set. You were asked to diagram and compare the sets $A \cup (B \cap C)$, $A \cap (B \cup C)$, $(A \cap B) \cup (A \cap C)$, and $(A \cup B) \cap (A \cup C)$.

You may have noticed that the Venn diagrams for parts a and d in Exercise 4 were the same, so $A \cup (B \cap C) = (A \cup B) \cap (A \cup C)$. The results of b and c were the same, so $A \cap (B \cup C) = (A \cap B) \cup (A \cap C)$.

■ E.15 These equations are statements of a distributive property. $A \cup (B \cap C)$
▲ E.16 $= (A \cup B) \cap (A \cup C)$ is the distributive property of union over intersection, while $A \cap (B \cup C) = (A \cap B) \cup (A \cap C)$ is the distributive property
▲ ■ of intersection over union.

Consider the examples below for clarification of these properties.

Examples

■ E.14 For sets $A = \{2, 4, 6, 8\}$, $B = \{1, 2, 3, 4\}$, and $C = \{4, 5, 6\}$:

1. Tabulate $A \cap (B \cup C)$ and $(A \cap B) \cup (A \cap C)$ and compare the results.

SOLUTION $A \cap (B \cup C) = \{2, 4, 6, 7\} \cap \{1, 2, 3, 4, 5, 6\} = \{2, 4, 6\}$
$(A \cap B) \cup (A \cap C) = \{2, 4\} \cup \{4, 6\} = \{2, 4, 6\}$

2. Tabulate $A \cup (B \cap C)$ and $(A \cup B) \cap (A \cup C)$. Are the same results produced?

SOLUTION $A \cup (B \cap C) = \{2, 4, 6, 8\} \cup \{4\} = \{2, 4, 6, 8\}$
$(A \cup B) \cap (A \cup C) = \{1, 2, 3, 4, 6, 8\} \cap \{2, 4, 5, 6, 8\}$
■ $= \{2, 4, 6, 8\}$

As in Example 1, the results are the same.

3. Make Venn diagrams to show that $A \cap (B \cup C) = (A \cap B) \cup (A \cap C)$ for A, B, and C, where $A \cap B \neq \varnothing$, $A \cap C \neq \varnothing$, and $B \cap C \neq \varnothing$.

SOLUTION In Figure 1.26 the set $A \cap (B \cup C)$ is shaded ▦ and bounded

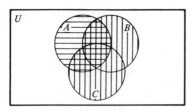

Figure 1.26 $B \cup C = $ ▥;
$A = $ ▤; $A \cap (B \cup C) = $ ▦.

by a dark outline. In Figure 1.27 $A \cap B$ is shaded ▨ as well as ▨; $A \cap C$ is shaded ▨ and ▨, and the union of these two sets, $(A \cap B) \cup (A \cap C)$, is shaded ▨ or ▨ or ▨.

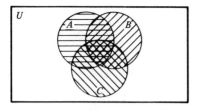

Figure 1.27 $A \cap B =$ ▨ and ▧;
$A \cap C =$ ▨ and ▧;
$(A \cap B) \cup (A \cap C) =$ ▨ or ▨ or ▧

Since $(A \cap B) \cup (A \cap C)$ and $A \cap (B \cup C)$ are represented by the same set, we say that they are equal.

The distributive properties permit us to simplify an expression for sets involving mixed operations or aid in finding equivalent expressions for sets, as the examples below will illustrate.

Example 1

$(A \cup B) \cap (C \cup A) = (A \cup B) \cap (A \cup C)$ (by commutative property of union)

$\qquad\qquad\qquad\quad = A \cup (B \cap C)$ (by distributive property of union over intersection)

Example 2

$(A \cap B) \cup [(B \cup C) \cup (A \cap C)]$
$= (A \cap B) \cup [(A \cap C) \cup (B \cup C)]$ (commutative property of union)
$= [(A \cap B) \cup (A \cap C)] \cup (B \cup C)$ (associative property of union)
$= [(A \cap (B \cup C)] \cup (B \cup C)$ (distributive property of intersection over union)

$= B \cup C$ Why?

E.15 *The distributive property of union over intersection is*
■ $A \cup (B \cap C) = (A \cup B) \cap (A \cup C)$

E.16 *The distributive property of intersection over union is*
■ $A \cap (B \cup C) = (A \cap B) \cup (A \cap C)$

EXERCISES

1. Use the sets $A = \{1, 2, 5\}$, $B = \{6, 8, 10\}$, and $C = \{1, 3, 5, 7, 9, 11, 13, 15\}$ to illustrate the distributive property of intersection over union, and the distributive property of union over intersection.

2. Simplify the following sets.
   a. $(A \cup B) \cap (A \cup B)$
   b. $(A \cap B) \cup (A \cap B) \cup (B \cap C)$
   c. $(A \cap C) \cap (C \cap A)$

3. Which statements below seem to be true? Why?
a. $A - (B \cup C) = (A - B) \cup (A - C)$
b. $A - (B \cap C) = (A - B) \cup (A - C)$
c. $A - (B \cup C) = (A - B) \cap (A - C)$
d. $A - (B \cap C) = (A - B) \cap (A - C)$
(*Hint:* Use the sets $A = \{1, 2, 3\}, B = \{2, 4, 6\}$, and $C = \{1, 3, 4, 5\}$ to illustrate your answer.)

SECTION F THE CARTESIAN SET, A × B

BEHAVIORAL OBJECTIVES

F.1 Define ordered pair.

F.2 Define $A \times B$.

F.3 For given A and B, be able to tabulate $A \times B$.

F.4 For given A and B, be able to find $n(A \times B)$.

F.5 Relate $n(A \times B)$ to $n(A)$ and $n(B)$.

F.6 For $C \subset B$, find a relation between $A \times C$ and $A \times B$.

F.7 Relate $A \times B$ to $B \times A$.

F.8 Relate $n(A \times B)$ to $n(B \times A)$.

F.9 For given A and B, graph $A \times B$.

CARTESIAN SETS

■ **F.1**   An ordered pair is simply a pair of objects in a set of parentheses and separated by a comma. (a, b) is an ordered pair where the position occupied
■ by a is called first and the position occupied by b is called second.

We will combine two sets, A and B, in such a way as to produce a set of ordered pairs of elements from A and B. The set produced is called the *cartesian set* or *cartesian product* of A and B. Before defining this new set, let us consider a simple example. If A is the set $\{1, 2\}$ and B is the set $\{2, 3, 4\}$, then the cartesian product, symbolized by $A \times B$, is the set of pairs $\{(1, 2), (1, 3), (2, 2), (2, 3), (1, 4), (2, 4)\}$. Notice that the set does not contain any pairs with leading element 3 or 4, nor does it contain any pairs with second element 1.

Consider the cartesian sets $A \times B$ and $B \times A$. The set that describes $A \times B$ above will not describe $B \times A$. The set $B \times A$ contains the pairs $\{(2, 1), (2, 2), (3, 1), (3, 2), (4, 1), (4, 2)\}$. Notice that $A \times B$ and $B \times A$ have only one ordered pair in their intersection. Why? If we look at both examples, we can see that the first element in a given ordered pair is an element of the first set forming the cartesian product and the second element in each pair is always an element from the second set forming the product.

Before we define cartesian product, we will illustrate the concept with several examples.

Example 1

F.3   Tabulate: $A \times B$ and $B \times A$
Given: $A = \{1, 2, 3, 4\}$ and $B = \{2, 5\}$

SOLUTION $A \times B = \{(1, 2), (1, 5), (2, 2), (2, 5), (3, 2), (3, 5), (4, 2), (4, 5)\}$
$B \times A = \{(2, 1), (2, 2), (2, 3), (2, 4), (5, 1), (5, 2), (5, 3), (5, 4)\}$

Example 2

Tabulate $B \times B$, $B \times C$, and $C \times B$
Given: $B = \{2, 5\}$ and $C = \{1\}$

SOLUTION $B \times B = \{(2, 2), (2, 5), (5, 2), (5, 5)\}$
$B \times C = \{(2, 1), (5, 1)\}$
$C \times B = \{(1, 2), (1, 5)\}$

Definition

F.2   *The cartesian set written $A \times B$, is the set of all ordered pairs whose first elements are elements from the set A and whose second elements are from the set B. Symbolically,*
$A \times B = \{(a, b) \mid a \in A \text{ and } b \in B\}$

EXERCISES

1. Tabulate the cartesian products indicated below.
$A = \{1, 2, 3\}$ $B = \{2, 3\}$ $C = \{5\}$ $D = \{4, 5\}$
a. $A \times B = ?$ b. $A \times C = ?$
c. $B \times C = ?$ d. $A \times A = ?$
e. $C \times C = ?$ f. $B \times D = ?$

F.4, 5   **2.** Find the number property of each set in Exercise 1.
7, 8
3. Find a rule to relate $n(A)$, $n(B)$, and $n(A \times B)$ for Exercise 1.

4. How would you describe the relationship between the following?
a. $A \times B$ and $B \times A$ b. $n(A \times B)$ and $n(B \times A)$

5. Under what conditions is $A \times B = B \times A$?

F.6   **6.** Tabulate the cartesian sets from the given sets below.
$A = \{1, 2, 3\}$ $B = \{2, 3, 4\}$ $C = \{3, 4\}$ $U = \{1, 2, 3, 4\}$
a. $A \times C$ b. $A \times B$

7. In Exercise 6, $C \subset B$. How does $A \times C$ relate to $A \times B$? Will this relationship always hold?

A GEOMETRIC MODEL OF $A \times B$

For this discussion we assume that sets A and B forming the set $A \times B$ are subsets of the natural numbers, that is, that $A \subseteq N$ and $B \subseteq N$. It should be ovbious that the set $A \times B$ is a subset of $N \times N$. Figure 1.28 indicates what a cartesian set might look like if we numbered the columns and rows of points by the natural numbers. If we let $A = \{1, 2, 3, 4, 5\}$ and $B = \{2, 4\}$, then it follows that $A \times B$ is a subset of $N \times N$. The points in the diagram represent ordered pairs of $N \times N$ and are often called *lattice points*. The set A is marked in a horizontal direction and the set B is marked in a vertical direction. In the figure, the lattice points of $A \times A$ are indicated by circled points. The lattice points of $A \times B$ are marked by a circled letter "x".

■ F.9

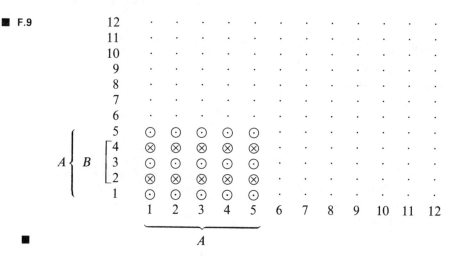

Figure 1.28

EXERCISES

Use the following sets in the problems below.
$$U = \{1, 2, 3, 4, 5\} \qquad A = \{1, 2, 3, 4\} \qquad B = \{1, 2\} \qquad C = \{4, 5\}$$

1. Tabulate
   a. $U \times U$
   b. $A \times B$
   c. $(A - B) \times B$
   d. $(A - B) \times A$
   e. $\bar{C} \times B$
   f. $\bar{B} \times A$

2. Make a lattice diagram showing the following:
   a. $U \times U$
   b. $U \times U$ and $B \times C$
   c. $U \times U$ and $(A - B) \times B$

3. Tabulate the following:
   a. $(A \times B) \cap (B \times A)$
   b. $[(A - B) \times B] \cap [(A - B) \times A]$
   c. $(A \times B) \cup (B \times C)$
   d. $U \times A$
   e. $U \times \bar{A}$

4. How does $U \times A$ relate to $U \times \bar{A}$?

5. Show that if $A \subseteq N$ and $B \subseteq N$, then $A \times B \subseteq N \times N$.

CAPSULE I   STRUCTURE IN SCHOOL MATHEMATICS

The past two decades have seen a revolution in the teaching of mathematics. This revolution started with the end of World War II and spiraled upward with the news of Sputnik. People started to show great concern with regard to the purpose and end product of the school programs in mathematics. Initially, the changes in mathematics training began at the secondary level as a result of national programs initiated by groups such as the School Mathematics Study Group (SMSG) and the University of Illinois Committee on School Mathematics (UICSM). Funding from the federal government and the influence of the National Science Foundation, the National Council of Teachers of Mathematics, the College Entrance Examination Board, and others put mathematics training in the limelight and forced changes to take place.

By the early 1960s this change or revolution in school mathematics began to affect mathematics programs in the elementary school. The effect was a change in the elementary school mathematics curriculum from a skills-based curriculum which emphasized only arithmetic to a many-faceted curriculum which includes new topics, such as sets, number theory, algebra, and geometry.

These changes came about because the emphasis was now placed on the structure of mathematics. What do we mean by structure? Why should we teach structure? How does structure help children to learn mathematics? These are a few of the questions which we will attempt to answer. It is obvious that such questions cannot be answered adequately in a few paragraphs or pages. We will simply summarize our point of view, which is based on the research and writings of others and our experience with children. There are some suggested readings at the close of this capsule where you may look for further information. The role of structure in the mathematics curriculum is of primary importance, and you should find exploration of this question to be of interest and help in your work with children.

What do we mean by structure? Most children are able to identify patterns in mathematics if they are asked to seek them. Certain patterns continually emerge throughout the study of mathematics and are often characterized as "properties" or "principles." In this chapter we have shown that a binary operation such as union or intersection has some specific properties, and we have developed the following ones—closure, commutativity, associativity, identity, and also distributive properties. These properties are of particular importance since they form threads which weave their way through the tapestry of mathematics. Later, you will see that certain sets under binary operations have some, none, or all of these properties.

Grossnickle, Brueckner, and Reckneh state that the many patterns in mathematics taken together form the structure of mathematics. They imply that discovering patterns is important to the learner and suggest relating mathematical ideas such as addition, subtraction, zero facts, and commutativity of operations. The properties we mentioned earlier are considered by most authorities to be of primary importance. It is important to teach these properties from use rather than from memory so as to foster understanding rather than verbalization.

Why should we teach structure? Bruner has pointed out in general that most disciplines have a structure which, if discovered and applied by the learner, will make the learning more permanent and meaningful. Memorization of isolated facts tends to lead to forgetting, whereas understanding the structure of mathematics permits students to memorize less and learn more. Some of the early research of Brownell, Thiele, Swenson, and others indicated that the learning of isolated facts inhibits the learner, whereas using certain basic mathematics principles increases the learning and retention. Compare some of Bruner's conclusions to the earlier conclusions of Brownell. (See *Thirty-second Yearbook*, NCTM, pp. 136–137.)

In the paragraph above we discussed the question of "Why should we teach structure?" It is often difficult to separate this question from the question "How does structure help children learn mathematics?" The structure of mathematics is basically similar patterns in different contexts. It is important for children to recognize and seek these properties and patterns, particularly if they can find the patterns with very little help from the teacher. It is just as important for children to notice that the same patterns do not always exist for all operations on all sets of objects, nor do all patterns thread their way through every part and portion of mathematics. Too often children are drilled into learning that everything is commutative or associative. The result of such teaching could be disastrous. Therefore, children should be encouraged to seek pattern, properties, and structure; at the same time, they should be encouraged to check for the lack of certain properties. In this way they will learn to recognize the properties of certain mathematical structures and will question the validity of applying a set of rules to all situations. If this is done in more classrooms, we may reach some of the goals set by groups which have instituted change. If the changes instituted simply substitute a new rote for the old one, then we will continue to produce children and adults who are mathematically crippled.

SUGGESTED READINGS

Adler, Irving. *Mathematics and Mental Growth*. New York: Day, 1968.

Brownell, William A. "When Is Arithmetic Meaningful?" *Journal of Educational Research*, March, 1945.

Bruner, Jerome S. *The Process of Education*. Cambridge, Mass.: Harvard Univ. Press, 1961.

Copeland, Richard W. *How Children Learn Mathematics*. London: Macmillan, 1970.

Diehl, Digby. "A Sense of Order in Mathematics," *The Arithmetic Teacher*, November, 1964.

Flourney, Frances. "Applying Basic Mathematical Ideas in Arithmetic," *The Arithmetic Teacher*, February, 1964.

Grossnickle, Foster E., Bruekner, Leo J., and Reckneh, John. *Discovering Meaning in Elementary School Mathematics*. New York: Holt, Rinehart and Winston, 1968.

National Council of Teachers of Mathematics. *Instruction in Arithmetic. Twenty-fifth Yearbook*. Washington, D.C.: NCTM, 1959.

———. *The Growth of Mathematical Ideas: Grades K–12. Twenty-fourth Yearbook*. Washington, D.C.: NCTM, 1959.

———. *A History of Mathematics Education in the United States and Canada. Thirty-second Yearbook*. Washington, D.C.: NCTM, 1970.

Shipp, Donald. *Developing Arithmetic Concepts and Skills*. Englewood Cliffs, N.J.: Prentice-Hall, 1964.

Spitzer, Herbert F. *Teaching Elementary School Mathematics*. Boston: Houghton Mifflin, 1967.

Spencer, Peter L., and Brydegaard, Marguerite. *Building Mathematical Competence in the Elementary School*. New York: Holt, Rinehart and Winston, 1966.

CHAPTER 2 NUMERATION SYSTEMS

SECTION A CHARACTERISTICS OF NUMERATION SYSTEMS

BEHAVIORAL OBJECTIVES

A.1 Be able to describe and give an example of an additive system.

A.2 Be able to describe and give examples of a system which is additive and subtractive.

A.3 Be able to explain how a multiplicative symbol is used in a numeration system.

A.4 Be able to give examples of a system in which order is not needed for symbolization.

A.5 Be able to give examples of a system in which order is needed for symbolization.

A.6 Be able to describe a system in which place value is a characteristic.

A.7 Be able to identify the base for a given system.

COUNTING

No one knows exactly when man started to count. We are sure that man always had a need for number ideas. For example, in hunting, the concepts of both size and distance were necessary for the hunter to track and kill his prey. Shepherds needed a way of keeping count of their sheep. They often used notched sticks, knotted ropes, or bags of stones as counting devices. The development of the abacus is a more sophisticated invention used by the Orientals as well as the Romans and Greeks. It has taken man thousands of years to develop the symbols and notation that we use in counting or naming numbers.

There are tribes today in South America, Australia, and Africa who still use primitive counting ideas. Their need for numbers extends to the use of one, two, and many. Some tribes have developed number names to four but no further. Some tribes count with their fingers, or fingers and toes, and use these names to convey number ideas.

Although they were cumbersome or inefficient, many of the early counting systems were used for thousands of years without much change. Roman numerals were used in Europe as late as the sixteenth century,, even though the more efficient Hindu-Arabic system was well known by that time. In fact, the Roman numeral system used today is only slightly changed from the system used by the Romans.

PRINCIPLES USED IN NUMERATION SYSTEMS

In this section consider a dawning civilization. As its economic and social needs became greater, its need to count became more apparent, and, as a result, efforts were made to improve its numeral system. Several principles used in counting and naming numbers will be illustrated to help you understand the sections which follow.

At first the members of this new civilization counted simply by using strokes. In this system, the only symbols were slashes, which stood for ones and were repeated as many times as was necessary. The numeral, or name, for three was /// or simply the repetition of one, three times. Thus, when looking at a large numeral like 15 it appeared as ///////////////. It is not possible to know what this means at a glance; thus, people had to count the strokes. This way of writing numerals employs two principles, repetition and addition. As more of these numerals appeared on records, the scribes decided that there must be a better way of recording numbers.

■ A.1

■

One very bright scribe decided to group every six strokes and named this grouping the *head*, since the head is made up of two ears, two eyes, a nose, and a mouth. Initially, he simply drew a line through the groups of six, and the numeral appeared as ʻʻʻʻʻ ʻʻʻʻʻ ///. Later he decided to use a head to represent six, and he wrote 15 as ☺ ☺ ///, which was read as two heads and three strokes for 15. The civilization grew and larger numerals were needed. When the scribe wrote 45 he needed seven heads and three strokes, ☺☺☺☺☺☺☺///. This scribe was too bright and too lazy to make all those characters, so he decided that since most households included at least six people, he would make six heads into a household, symbolized as ☐. Thus, the number 45 became ☐ ☺ ///, or 36 + 6 + 3.

A.7

This principle of grouping a number of symbols into a single symbol is found in most civilized numeration systems. We call this type of grouping a *base system*. The one used here has started to evolve with six as a base. Notice that the first set of characters are unit characters. The second set represents multiples of 6, or multiples of the base to the first power. The third character represents multiples of 36, or the base squared.

■

◄ A.2

In order to lessen the number of symbols used to name a number, the scribe decided that if he wanted to write the numeral /////, he would place a slash through the fact to represent one less than a face, or ⊘. He also decided that instead of five faces he would place a face in the household to represent one household minus one person. In this way the numeral for 30 became ⊡ instead of five faces. This system now used subtraction as a principle.

■

There were still some problems in the system. To represent a number like 275, the scribe needed to write more than six houses, since six houses meant only 216. Hence, a new symbol was developed to represent 216, or 6 houses. Since his village was shaped like a triangle, he decided to let a triangle represent six houses. Now he had a symbol for 216. The scribe felt that he would never need more than six △'s, or villages, in his numeration system for his peoples' needs.

As you probably can guess, he soon found a need to make even larger numerals. So all the scribes had a meeting and the most intelligent scribe

■ A.3   indicated that he was tired of always inventing new symbols. He suggested that they could multiply each symbol by six villages (6 × 216), or 1296, by simply drawing a vertical line under it. Thus ⌐ would then represent 1296; ☺ would represent 6 × 1296; ⊡ would represent 36 × 1296; and, lastly, △ would represent 216 × 1296. This multiplicative principle was adopted by the wise scribes, and there were no numeral problems for this civilization for thousands of years, since they could write numerals

■   well into the millions with the four characters and this modification.

We can summarize the principles used as follows:

1. a base six with four characters
2. principles of repetition, addition, subtraction, and multiplication
3. numerical value in the system unaffected by symbol order

■ A.2   Some examples of this system are written below.

$$56 = \square \, \diamondsuit\diamondsuit\diamondsuit \, // \qquad 1 \times 36 + (3 \times 6) + 2$$
$$33 = \boxdot \, /// \qquad (36 - 6) + 3$$
$$545 = \triangle\triangle\square\square\square \, \varnothing \qquad (2 \times 216) + (3 \times 36) + (6 - 1)$$
■ $$1695 = \ulcorner \triangle \triangle \, /// \qquad 1296 + 216 + (216 - 36) + 3$$

■ A.4   The system developed above did not use order as a principle. That is, the symbols used in writing a numeral had the same value regardless of the position in which they appeared in the numeral. For example:

■
$$\square \, \diamondsuit\diamondsuit\diamondsuit \, // \;\; = \;\; \diamondsuit \, \square \, / \, \diamondsuit \, / \, \diamondsuit$$

It should be noted that this system also lacks a symbol for zero.

If this system had developed as a base-six system which had a zero and used positional notation, then the names for numbers would have had a completely different composition. Let us consider the growing civilization across the river, which developed such a system. This group had started copying their neighbor's system, but one of their scribes soon

■ A.6   devised a new method. He represented the first five numerals as /, Z,

▲ A.5   E, #, and *. This scribe suggested that they use the position of the numerals to represent groups of 6 and its powers. In this way, /Z would represent the number eight (one group of 6 and 2) while Z/ would represent the number thirteen (two groups of 6 and 1). In this way, order became important to the system. Since most people did not quite understand his meaning, he made a table for the other scribes. Part of what the table looked like

▲ ■   is seen in Figure 2.1.

You will notice that one, six (one group), and thirty-six (six groups) are all represented by a slash. This means that confusion could occur. Because of this, a new symbol was developed which they called *no ones*, written ∅, and later shortened to read *nones*. Now they wrote / to read

Number Used in English	Numeral Used in Culture	Translation
one	/	one
two	Z	two
three	E	three
four	#	four
five	*	five
six	/	one group
seven	//	group + 1
eight	/Z	group + 2
nine	/E	group + 3
ten	/#	group + 4
eleven	/*	group + 5
twelve	Z	two groups
thirteen	Z/	two groups + 1
.		
.		
.		
twenty-four	#	four groups
twenty-five	#/	four groups + 1
twenty-six	#Z	four groups + 2
twenty-seven	#E	four groups + 3
.		
.		
.		
thirty	*	five groups
thirty-one	*/	five groups + 1
thirty-two	*Z	five groups + 2
thirty-three	*E	five groups + 3
thirty-four	*#	five groups + 4
thirty-five	**	five groups + 5
thirty-six	/	six groups
thirty-seven	//	six groups + 1
thirty-eight	/E	six groups + 2

Figure 2.1

one; $/\emptyset$ to read six, or one group plus nones, and $/\emptyset\emptyset$ to read 36, or one groupgroup plus nonesgroup and nones. (Notice that a groupgroup means six group of six.) Some examples of their system are written below. Compare these to the examples of the system of the people across the river. Below we use groupgroupgroup to mean six groupgroups or six groups of thirty-six, and so on.

Examples

A.5 $56 = /EZ$ (one groupgroup + three groups + two ones)
(1×36) $+ (3 \times 6)$ $+ (2 \times 1)$

$545 = ZE\emptyset*$ (two groupgroupgroups + three groupgroups
(2×216) $+ (3 \times 36)$
$+$ nones group + five ones)
$+ (0 \times 6)$ $+ (5 \times 1)$

$1695 = //* \varnothing E$ (one groupgroupgroupgroup + one groupgroup-
(1×1296) $+ (1 \times 216)$
group + five groupgroups + zero group + three
ones)
$+ (5 \times 36)$ $+ (0 \times 6)$ $+ (3 \times 1)$

Note: that the position where a symbol occurs in a name is very important. For example, $/Z$ is a name for 8, whereas $Z/$ is a name for 13. Systems using the position of a symbol to indicate its value are said to be *positional systems*.

A.7   The system just described is much like the Hindu-Arabic system which we use today. The base is six and uses six symbols instead of our base of ten, which uses ten symbols, namely, 0, 1, 2, 3, 4, 5, 6, 7, 8, 9. In later sections we will describe actual systems which have been used by the Egyptians and Romans as an illustration of simple non-positional grouping systems. Later we will consider the Babylonian and Mayan systems as examples of positional systems such as the one just developed.

The principles used in this numerical system are:

1. a base-six system with six symbols
2. positional notation
3. a symbol for zero

EXERCISES

1. Translate the following numerals from the first system discussed into the system we use today.
   a.   △△ ☐☐ ◉ ////
   b.   △ ☐☐ ◇◇∅ /
   c.   ⌐⌐ △△ ☐ ◇◇ //

2. Translate the following numerals from the second system discussed into our system.
   a. EZ/   b. ZE∅*   c. *∅/#

3. Using the base five, devise a system like the first one explained.

4. Using the base five, devise a positional system.

5. Write the numerals from 1 to 40 in the systems explained in the text.

6. Write the numerals from 1 to 40 in the systems you devised in Exercises 3 and 4.

7. Explain how the positional systems in bases five and six compare to each other.

8. How would you compare and contrast a positional system to a non-positional system? Consider the advantages and disadvantages of each.

SECTION B EARLY SYSTEMS OF NUMERATION

BEHAVIORAL OBJECTIVES

B.1 Be able to list the characteristics of the Egyptian system.

B.2 Be able to translate from Hindu-Arabic numerals to Egyptian numerals.

B.3 Be able to translate to Hindu-Arabic numerals from Egyptian numerals.

B.4 Be able to list the characteristics of the Roman system.

B.5 Be able to translate from Hindu-Arabic numerals to Roman numerals.

B.6 Be able to translate to Arabic numerals from Roman numerals.

B.7 Be able to list the characteristics of the Babylonian system.

B.8 Be able to translate from Hindu-Arabic numerals to Babylonian numerals.

B.9 Be able to translate to Arabic numerals from the Babylonian numerals.

B.10 Be able to list the characteristics of the Mayan system.

B.11 Be able to translate from Hindu-Arabic numerals to Mayan numerals.

B.12 Be able to translate to Arabic numerals from Mayan numerals.

The next few sections will deal with some of the systems used for recording numbers. These systems are referred to as *numeration systems* and are used for recording as well as for computation.

EGYPTIAN SYSTEM

The early hieroglyphics of the Egyptians were fully developed by 3000 B.C. or earlier. This system used the seven different numerals shown in the table below. The system had the following characteristics:

■ B.1 1. It was a base-ten system.
2. It used the addition principle.
3. It used a repetition principle.
■ 4. Order did not affect the result (a system with no place value).

𓀁 astonished man	million	(10^6)
𓆓 burbot fish	hundred thousand	(10^5)
𓂭 pointing finger	ten thousand	(10^4)
𓆼 lotus flower	thousand	(10^3)
𓏢 scroll	hundred	(10^2)
𓂪 heel bone	ten	(10)
/ staff	one	1

Numerals in this system were written by following the principles listed above. Some examples of writing numerals in this system are shown below. Notice that we are translating from the Arabic to the Egyptian system.

Examples

■ B.2 346 = 

3,452 = 

26,751 = 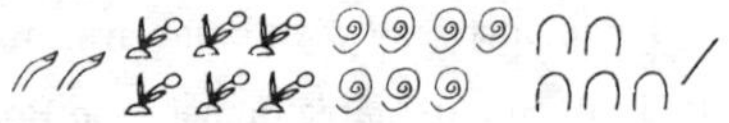 ■

Given numerals in the Egyptian system, we can translate to the Arabic as follows:

Examples

■ B.3  = 232

 = 2,255

 = 33,641 ■

In the examples above, you will notice that the characters are grouped in different ways (and might also have varied in size to facilitate grouping), which is done more for style than anything else. If a scribe made an error in writing, another character could be added without concern for order or placement. For example, a scribe might have tallied grain and have written the numeral  or 1,355. Later more grain may have arrived. It would not cause a problem in this system if 125 loads arrived and were added to the tally. The numeral would simply read 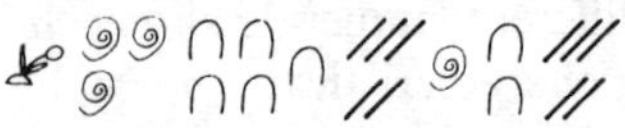. This should indicate to the reader that order did not affect the utility of this system. We might write the same numeral , since the ten strokes can be written as ∩; however, it is easily read in both ways. You can also see that the writing of large numbers is not very economical.

ROMAN SYSTEM

The Roman system should be quite familiar to you. It is still used today for numbering chapters of books, on clock faces, in building dates, and quite often to represent volume numbers of journals. This system dates to about 250 B.C. It is very much like the Egyptian system in principle, although the symbols are different. The Roman system changed over a period of years, and many of the conventions used by the Romans have been modified. For example, we are quite accustomed to reading IV as four, whereas the Romans actually used IIII. The introduction of IV for four, IX for nine, and IC for ninety-nine changed the system to a great extent. No longer was the `order of the system of little importance; in a subtractive system, order is important. In the early development of the system, both IV and VI would have been interpreted as six.

One of the important reasons for the continued use of this system well into the sixteenth century was the fact that computations involving addition and subtraction were easily done. The system used had characteristics which will be listed and illustrated below.

■ B.4   1. It was base-ten system.
2. It used the principles of addition, subtraction, multiplication, and repetition.
3. It was not a place-value system; however, order affected the meaning,
■   due to the subtractive principle.

The numerals are illustrated in the table below. In the paragraphs that follow, you will see examples that illustrate each principle. The numerals shown may have resulted from the language as well as economical changes. The numeral V may have changed from IIIII to IIII to ⅋ to V and the numeral X from VV to X. The numerals C and M probably resulted from the language, since the words for 100 and 1,000 were centum and mille.

Roman Symbol	Our Symbol
M	1,000
D	500
C	100
L	50
X	10
V	5
I	1

Notice that this system has a base of ten but actually there is a duality in the base. For example, the relation between I and V is one to five, between V and X is five to ten (two to one), between X and L is ten to fifty (again one to five), and so on. However, we consider this a base-ten system. Why do you think base ten was used by both the Romans and Egyptians?

$I = 1$, or 10^0
$X = 10$, or 10^1
$C = 10 \times 10$, or 10^2
$M = 10 \times 10 \times 10$, or 10^3
$V = 5 \times 1$
$L = 5 \times 10^1$
$D = 5 \times 10^2$

The use of the symbols for 5, 50, and 500 may be somewhat confusing;

however, it is explained in terms of historical development. These symbols were probably developed as an economy measure. It does not seem reasonable to write ///////// for nine as the Egyptians did. The Roman system may have improved on this by first writing IIIIIIIII, then VIIII, then VIV, then IX. This is just conjecture, but it seems reasonable to assume that such changes took place over the centuries.

Given numerals in the Roman system, we can translate to the Arabic as follows.

Examples

■ B.6   1. CCXXIX = 229
2. MCMXLVII = 1,947
3. $\overline{\text{XIX}}$DCXXI = 19,621

The use of repetition and addition is shown below:

332 = CCCXXXII, where
        100 + 100 + 100 + 10 + 10 + 10 + 1 + 1
        represents 300 + 30 + 2, or 332.

2772 = MMDCCLXXII, where
        1,000 + 1,000 + 500 + 100 + 100 + 50 + 10 + 10 + 1 + 1
        represents 2,000 + 500 + 200 + 50 + 20 + 2, or 2,000 + 700
■   + 70 + 2 = 2,772.

The subtractive principle was illustrated earlier; however, several more examples are written below. The subtractive principle was used primarily with four, nine, and multiples of four and nine. However, this practice was not strictly followed.

■ B.5   40 = XL, or 50 − 10
          90 = XC, or 100 − 10
         400 = CD, or 500 − 100
         900 = CM, or 1,000 − 100
         118 = CXIIX, or 100 + 10 + 10 − 2 (this was seldom done)

The multiplicative principle was accomplished by using a bar above the numeral, which multiplied the numeral by one thousand.

10,236 = $\overline{\text{X}}$CCXXXVI, or (10 × 1,000) + 236
■ 25,210 = $\overline{\text{XXV}}$CCX, or (25 × 1,000) + 210

If we were to compare the Roman and Egyptian systems, we would find few improvements from Egyptian to Roman, even though the time lapse was great. Consider the numerals written in both systems (Figure 2.2). Notice that the only difference is in the characters. Considering the time difference, one might expect great improvement in the systems, since one culture could learn from the other's mistakes. The fact that the civilizations were able to function at a very high level with the existing systems probably accounts for the lack of change. It is believed by some

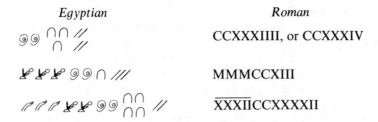

Figure 2.2 Comparison of Egyptian and Roman systems.

historians of mathematics that the development and use of our system of notation marked a milestone in the evolution of mathematics. The suppression of the Hindu-Arabic system during early centuries of its use in Europe may have prevented the growth of mathematical development for hundreds of years.

EXERCISES

1. Translate the Egyptian numerals below into the Hindu-Arabic system of numeration.

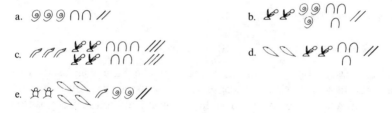

2. Translate the Roman numerals below into the decimal system (Hindu-Arabic).
   a. XXXIX
   b. CCLXIV
   c. C̄X̄VCCCXIX
   d. X̄IX̄CXLIX
   e. C̄V̄IDCXIII
   f. MMDCXVII

3. Translate the Hindu-Arabic numerals below into both Egyptian and Roman numerals.
   a. 42   b. 357   c. 483   d. 1,785   e. 25,843

4. Use the following symbols and characteristics to develop a system of numeration: A = 1, B = 10, Q = 100, and 1,000 = ⊗. Use the principle of repetition and addition. Write the following numerals.
   a. 236   b. 357   c. 8,543   d. 24,356

5. Using the symbols from Exercise 4, develop a system that uses addition, repetition, and multiplication as a principle. Illustrate how your system works by writing the numerals for the following.
   a. 34,256   b. 256,842   c. 1,987

6. Develop a system using any principles you wish. As a basis for your system, let the following symbols represent the numerals shown: X = 1, ⌂ = 5, ⊠ = 25, and ⊞ = 125. Explain how you can represent numerals over 1,000, 10,000, or 100,000. You may introduce more symbols if necessary.

7. Is order important in the Roman numeral system?

BABYLONIAN SYSTEM

The early Babylonian system was a simple grouping system which used base ten and both addition and subtraction as principles which governed the system. Later, about 2000 B.C., as the need for larger numerals arose, the Babylonians developed a positional system which may have been the forerunner of the Hindu-Arabic system we now use. The system employed essentially two symbols: ◄ for 10 and for ▼ for 1. A combination of symbols was used to show subtraction. The symbol ▼➤ was used to mean "subtract what follows". The base sixty was employed in the system, and we still maintain this base today for time measure (hours, minutes, and seconds) as well as angle measure (degrees, minutes, seconds). The characteristics of the Babylonian system are listed below.

■ B.7
1. It was a base-sixty system.
2. It used the principles of repetition, addition, and subtraction.
■
3. It was a place-value system.

The examples below illustrate the addition and subtraction principles used to represent numbers.

Examples

■ B.8 $45 =$ ◄◄◄◄▼▼▼▼▼ $10 + 10 + 10 + 10 + 1 + 1 + 1 + 1 + 1$, or $40 + 5$

■ $38 =$ ◄◄◄◄▼➤▼▼ $10 + 10 + 10 + 10 - 2$, or $40 - 2$

The examples above do not indicate how the base was used. Two examples follow to illustrate the base idea. The use of base sixty in conjunction with place value meant that the numeral 1 written in first position meant one. If the numeral was written in the second position, it was interpreted to mean 1×60. In the third position, it was interpreted to mean 1×60^2, or 1×360. The problem with the system was that gaps or spaces were used to indicate what position the numerals were in. It was usually necessary to use contextual clues to understand the numerals written, since the use of gaps alone was not sufficient to explain the numbers being represented. From the examples given below, you can see how confusion might occur in this system. The first example, as well as the second, can only be interpreted if the context of the writing is known. We will interpret the first one in only one way. The second one will be considered in more detail to point out the flaw in the system.

Example 1

■ B.9
■
◄◄◄ ▼▼ ◄◄ ▼▼▼▼ ◄◄◄◄ ▼▼▼▼▼

is to be understood as $(32 \times 60^2) + (24 \times 60) + (45 \times 1)$ or 116,685

We assumed in the example above that the first position on the right represented units, that the middle position represented groups of sixty, and that the left position represented groups of 60^2. It could have been the case that there were no units represented. The context of the writing would reveal this. In the next example, there is a gap between the first set of symbols and the second set. Again, we do not know whether the first set represents units, sixties, or some higher power of sixty.

Example 2

■ B.9

In the example above, it is clear that the two sets of symbols represent 25 and 53, respectively. There is a gap between the sets which indicates a place holder. The problem here is that we are not able to tell whether the gap is to mean 0×60 or 0 times some power of 60. If the 53 means 53×1, then the 25 represents 25×60^2 or 25×3600, in which case the number represented by the numeral is 90,053. If, however, the gap represents 0×60^2, then the 53 is taken as 53×60 and the 25 represents 25×60^3. If this is the case, then the numeral above means 5,403,180. In either case, it is apparent that a system such as this needs some improvement. The Hindu-Arabic system we now use has solved this problem by using a symbol for zero to fill these gaps. This place holder, or zero as we call it, is another development which has improved arithmetic notation over the ages.

Consider how we might translate from Hindu-Arabic to Babylonian.

■ B.8 1. $264 = (4 \times 60) + 24$ or $\textbf{VVVV} \textbf{<<} \textbf{VVVV}$

2. $29 = (30 - 1)$ or $\textbf{<<<} \textbf{V>V}$

■ 3. $425 = (7 \times 60) + 25$ or $\textbf{VVVVVVV} \textbf{<<} \textbf{VVVVV}$

EXERCISES

1. Translate the following Babylonian numerals into our system. Consider the first numeral on the right to be a multiple of one.

   a. $\textbf{<<<} \textbf{VV}$

   b. $\textbf{<<} \textbf{V>} \textbf{VVV}$

   c. $\textbf{<VV} \quad \textbf{<<<VV}$

   d. $\textbf{<<<} \textbf{VV} \textbf{VV} \quad \textbf{<<<VV}$

2. Translate the following from our system to the Babylonian.
   a. 25   b. 48   c. 84   d. 275

3. Are there any advantages or disadvantages in the Babylonian system which are not present in the Roman system of numeration? Egyptian? Hindu-Arabic?

4. How might we improve the Babylonian system so that confusion does not occur? Use examples to illustrate how you would do this.

5. What are the advantages or disadvantages of a system which employes a large base such as sixty?

MAYAN SYSTEM

The Mayan Indians of South America had developed a very sophisticated system of numeration. The system, dating to about 300 A.D., is closer to our present system than any of the systems discussed thus far. The fact that the Mayans had developed a zero makes their system somewhat superior to that of the Babylonians.

The Mayan system probably started as a repetitive additive system, although this is only conjecture. The system as it evolved was a positional system, with a zero, which used twenty digits. In other words, it developed a base-twenty positional system. The characters used employ the following symbols in combination.

ⓝ represents 0
. represents 1
—— represents 5

We suggest that in the beginning this system was an additive repetitive system since the numeral 4 was written •••• and the 9 was written ——————. It must be reported, however, that with the acquisition of the zero and the positional idea, this system actually must be viewed as having twenty symbols which always represent the same numeral or multiple of that numeral. The numerals from 0 to 19 are shown in Figure 2.3.

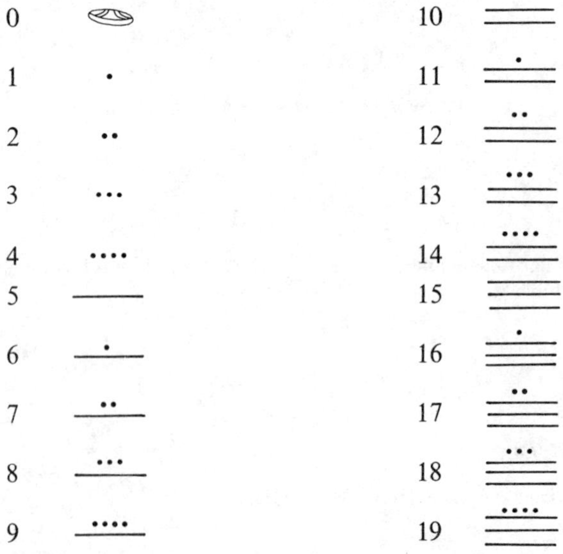

Figure 2.3

The reason we are considering the Mayan system is that it was developed rather late in history and seems to have been developed independently of the other systems mentioned. The Mayans employed base twenty; however, they used 20×18 instead of 20×20. This makes the system difficult to deal with. The most important aspects the systems are the development of a zero and the concept of the astronomical measure which seems implicit in the use of 20×18 (360), which roughly represents one year.

The use of 18 as a part of the base is only employed in place of the base squared. In general, the Mayans used $18 \cdot 20$ instead of 20^2, $18 \cdot 20^2$ instead of 20^3, or $18 \cdot 20^{n-1}$ instead of 20^n. The examples below illustrate this idea. Notice that the numerals are written vertically. The first example simply represents the powers of twenty. One can see how zero is used to hold a place.

Examples

B.12 20^0, or 1 20^1, or 20 20×18, or 360 $20^2 \times 18$, or 7,200

In the illustration above, the use of $\cdot$ always represents one group of the base raised to a power. Notice that the power is represented by the number of zeros employed. The examples below will use the same principle.

Examples

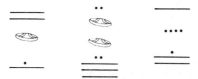

In the first example, from top to bottom the numerals are 10, 0, and 6, respectively. This translates as $(10 \times 20 \cdot 18) + (0 \times 20^1) + (6 \times 1)$ or 3,606. In the middle, the numerals are 2, 0, 0, and 17, respectively. This translates as $(2 \times 20^2 \cdot 18) + (0 \times 20 \cdot 18) + (0 \times 20^1) + (17 \times 1)$ or 14,417. In the last example, no zeros appear. The numerals are 5, 4, and 11. This numeral is translated to mean $(5 \times 20 \cdot 18) + (4 \times 20) + (11 \times 1)$ or 1,891.

Consider translation from the decimal system to the Mayan system.

Example 1

B.11   Translate 37 to the Mayan system.

SOLUTION   Since $37 = 20 + 17$, and
$20 = $ ⌾ and $17 = $ ⊜

therefore we write

$$
\begin{array}{rl}
\cdot & = 20 \\
\underline{\overset{\cdots}{}} & = 17 \\
\hline
& 37
\end{array}
$$

Example 2

Translate 395 to the Mayan system.

Since $395 = 360 + 20 + 15$,

$360 = \overset{\cdot}{\text{⟨◎⟩}}$ and $20 = \overset{\cdot}{\text{⟨◎⟩}}$ and $15 = \equiv$

We write

$$
\begin{array}{rl}
\cdot & = 360 \\
\cdot & = 20 \\
\equiv & = 15 \\
\hline
& 395
\end{array}
$$

Example 3

Translate 165 to the Mayan system.

SOLUTION Since $165 = (8 \times 20) + 5$,
we write

$$\frac{\cdots}{\text{⟨◎⟩}} = 160 \quad \text{and} \quad 5 = \underline{},$$

which is

$$
\begin{array}{rl}
\underline{\overset{\cdots}{}} & = 160 \\
\underline{} & = 5 \\
\hline
& 165
\end{array}
$$

The Mayan system was introduced here to indicate that the problem of the place holder was solved by people in two completely different cultures which were thousands of miles apart, namely, the Mayans and the Hindus. It has often been the case in history that discoveries in science and mathematics are made by two or more independent groups of people. Unfortunately, seldom have these joint discoveries spelled great progress, due mainly to a lack of communication and understanding between nations and people.

EXERCISES

1. Translate these Mayan numerals into our system.

2. Translate the following into Mayan numerals.

   a. 25   b. 48   c. 84   d. 275

3. Why do you think the Mayans might have used the base 20?

4. How would you compare the Mayan system to the Babylonian system?

5. Consider what would happen if the Mayans deleted the 18 × 20 and substituted 20^2 for it. After doing this, do Exercises 1 and 2 in terms of this new system.

SECTION C HINDU-ARABIC SYSTEM

BEHAVIORAL OBJECTIVES

C.1 Be able to list the characteristics of the Hindu-Arabic system.

C.2 Be able to write Hindu-Arabic numerals in expanded form.

C.3 Be able to write a numeral in Hindu-Arabic notation given the numeral in expanded form.

HISTORY OF THE HINDU-ARABIC SYSTEM

The system of numeration which we use today is called the Hindu-Arabic system. This system developed as a positional system sometime between A.D. 600 and 800. The system developed by the Hindus was introduced to Europe through trade with the Arabian tribes. There is some indication that the development of this system might have been influenced by the positional notation used by the Babylonians. This is purely conjecture, and no one has proven or disproven this theory.

The introduction of the system to Europe was aided greatly by the writings of the mathematician Leonardo of Pisa, better known as Fibonacci, around A.D. 1200. The way in which the symbols are written has never been completely standardized. For example, in Europe one often sees 1 as a symbol for 1, and 7 as a script symbol for 7.

As we mentioned earlier, this system was not completely adopted in Europe until the sixteenth or seventeenth century. The use of Roman numerals was continued for many reasons, not the least of which is that adding with Roman numerals is quite easy. There is also some suggestion that the Church withheld permission to use numerals which were developed by non-Christians.

CHARACTERISTICS AND USE
OF HINDU-ARABIC NUMERALS

All positional base systems which have place holders require as many digits or symbols as the base itself. Therefore, our system, which is a base-ten system, has ten symbols or digits, 0, 1, 2, 3, 4, 5, 6, 7, 8, 9. Every numeral can be written using these digits. The numbers zero through nine are each symbolized by one of the digits. Numbers greater than nine are symbolized by a numeral of more than one digit. Such a numeral implies that two operations are being performed. The first operation is multiplication based on the position of the digit. The first digit on the right is multiplied by 10^0, or 1, the digit in the second position from the right is multiplied by 10^1, or 10; the digit in the third position from the right is multiplied by 10^2, or 100; etc. Thus, starting with the first digit on the right, each succeeding position implies multiplication of the digit in that position by a power of ten.

The second operation is addition. The products formed by multiplying the digits by their corresponding powers of ten are then added. For example, the numeral 23 represents the number $(2 \times 10) + (3 \times 1)$. That is, the first digit on the right, 3, is multiplied by 1, producing 3. The second digit from the right is multiplied by 10, producing 2×10, or 20. These numbers are then added, producing $(2 \times 10) + (3 \times 1)$.

Each of these positions or places has a named value, as indicated in Figure 2.4.

6th place	10^5, or 100,000	hundreds of thousands
5th place	10^4, or 10,000	tens of thousands
4th place	10^3, or 1,000	thousands
3rd place	10^2, or 100	hundreds
2nd place	10^1, or 10	tens
1st place	10^0, or 1	units

Figure 2.4

Notice that the places are named in groups of three. The first three places are units, tens, and hundreds. The next three are thousands (which is short for units of thousands), tens of thousands, and hundreds of thousands. The same procedure is followed for the next three places, that is, millions, tens of millions, and hundreds of millions. Commas are often used to separate these groups of three digits. The numeral 2,364 represents $(2 \times 1,000) + (3 \times 100) + (6 \times 10) + (4 \times 1)$. When a numeral is written in such a way that both the multiplication and addition are shown, we call this the *expanded form* of the number, or *expanded notation*. The expanded notation in base ten can be written as follows: If the numeral in base ten is $a_n a_{n-1} \times \cdots \times a_2 a_1 a_0$, where each of the a_i's is in the set $\{0, 1, 2, \ldots, 9\}$, we mean that

$$a_n a_{n-1} \times \cdots \times a_2 a_1 a_0 = (a_n \times 10^n) + (a_{n-1} \times 10^{n-1}) + \cdots + (a_2 \times 10^2) + (a_1 \times 10^2) + (a_0 \times 1)$$

Consider some examples of expanded notation in base ten as follows.

Example 1

■ C.2 $3,642 = (3 \times 10^3) + (6 \times 10^2) + (4 \times 10) + (2 \times 10^0)$
$= (3 \times 1,000) + (6 \times 100) + (4 \times 10) + (2 \times 1)$
$= 3,000 + 600 + 40 + 2$

Example 2

$20,835 = (2 \times 10^4) + (0 \times 10^3) + (8 \times 10^2) + (3 \times 10) + (5 \times 10^0)$
$= (2 \times 10,000) + 0 + (8 \times 100) + (3 \times 10) + 5$
■ $= 20,000 + 0 + 800 + 30 + 5$

Consider some expanded notation in base ten and the numeral as written in base ten.

Example 1

■ C.3 $(2 \times 10^3) + (5 \times 10^2) + (6 \times 10^0) = 2,000 + 500 + 6$
$= 2,506$

Example 2

$(4 \times 10^5) + (6 \times 10^3) + (7 \times 10^2) + (4 \times 10)$
$= 400,000 + 6,000 + 700 + 40$
$= 406,740$

Example 3

$(2 \times 10^4) + (3 \times 10^3) + (2 \times 10) + (5 \times 10^0)$
$= 20,000 + 3,000 + 20 + 5$
■ $= 23,025$

From what we have said about bases, including the base-six system developed earlier in the chapter, you should have very little difficulty working with any base. Several bases other than base ten will be discussed in this chapter. While you are looking at each base system, notice the similar pattern of place value.

Before we consider other systems, we should look at the characteristics of the Arabic system, they can be summarized as follows:

◄ C.1   1. It is a base-ten system.
2. It uses ten digits or numerals.
■   3. It is a place-value system (which employs multiplication and addition).

EXERCISES

1. Write the following numerals in expanded notation.
a. 32,645   b. 243,644
c. 206,050   d. 2,000,234

2. In the numeral 23,465, what numbers do each of the following digits represent?
a. 4   b. 2   c. 6

3. In the numeral 21,356, which number is larger?
   a. the number represented by 1 or the number represented by 5
   b. the number represented by 3 or the number represented by 6

4. What number is represented by 6 in each of the following numerals?
   a. 236,402   b. 56,987   c. 6,834
   d. 264,920   e. 324,567

5. In what place do you find the digit 4 in each of the numerals below?
   a. 234,532   b. 425,678   c. 123,456
   d. 80,432   e. 456

6. Write the following in Hindu-Arabic numerals.
   a. $(2 \times 10^4) + (5 \times 10^3) + (3 \times 10^2)$
   b. $(5 \times 10,000) + (4 \times 1,000) + (6 \times 100) + (4 \times 10)$
   c. $(6 \times 10^5) + (4 \times 10^4) + (8 \times 10^3) + (9 \times 10^2)$
   d. $(6 \times 10^4) + (3 \times 10^2) + (2 \times 10^0)$

SECTION D BASES OTHER THAN TEN

BEHAVIORAL OBJECTIVES

D.1 Be able to list the characteristics of positional systems.

D.2 Be able to count in base five.

D.3 Be able to translate from base five to base ten.

D.4 Be able to translate from base ten to base five.

D.5 Be able to construct an addition table for base five.

D.6 Be able to add numbers written in base five.

D.7 Be able to construct a multiplication table in base five.

D.8 Be able to multiply numbers written in base five.

D.9 Be able to count in base two.

D.10 Be able to translate from base two to base ten.

D.11 Be able to translate from base ten to base two.

D.12 Be able to construct an addition table for base two.

D.13 Be able to add numbers written in base two.

D.14 Be able to construct a multiplication table in base two.

D.15 Be able to multiply numbers written in base two.

D.16 Be able to translate from any base to base ten.

D.17 Be able to translate from base ten to any base.

D.18 Be able to construct an addition table for any base.

D.19 Be able to add numbers written in any base.

D.20 Be able to construct a multiplication table for any base.

D.21 Be able to multiply numbers written in any base.

D.22 Be able to translate directly from base two to base four.

D.23 Be able to translate directly from base four to base two.

D.24 Be able to translate directly from base two to base eight.

D.25 Be able to translate directly from base eight to base two.

D.26 Be able to count in base twelve.

D.27 Be able to translate from base ten to base twelve.

D.28 Be able to translate from base twelve to base ten.

D.29 Be able to construct an addition table for base twelve.

D.30 Be able to add numbers in base twelve.

D.31 Be able to construct a multiplication table for base twelve.

D.32 Be able to multiply numbers in base twelve.

A SUMMARY OF POSITIONAL SYSTEMS

In this section we will develop several positional systems other than base ten. The purpose of this chapter and particularly this section is to give the reader an insight into the structure of our numeration system. We will consider base five in great detail and briefly discuss bases two and twelve.

As we saw in the first section of this chapter and in the section about the Hindu-Arabic system, there are several principles which characterize a positional system. These principles are a requisite for understanding our own system. Consider each one as we review them briefly.

■ D.1 1. There are as many symbols as the number of the base, including a symbol for zero.

In base ten this characteristic requires that we have ten symbols, $0, 1, 2, \ldots, 9$. In a base-five system, the symbols would be $0, 1, 2, 3, 4$. Notice here that the symbol for the base, in this case 5, is not included in the system.

We represent ten in our system as 10. Whenever 10 is written in a base system, it represents one group of the base and zero units.

2. The second position in any numeral represents groups of the base, and each of the other positions represents a group of increasing powers of the base.

In the first section we characterized the numeral for six, the base, as *group*. When we produced six groups of six, we called that a groupgroup. This indicates that the base squared or multiplied by itself produces the second grouping. In our base the numeral 100 represents one group of 10^2, or 10×10. This convention will be followed in any base. The base number of units is in second position, the base squared number of units is in third position, and so forth. This enables us to tell by position what value each digit represents.

3. Each digit has a specific value which is based on the digit's value as well as its position.

In our system the numeral 246 uses three digits, 2, 4, 6. Each character has its own value as a number. It also has a positional value which indicates by what power of the base the digit is multiplied to find its value in terms of units. Therefore, the 2 in base ten in the third position of 246 represents twice the base squared, that is, 2×10^2. The 4 is in second position, so it represents 4×10 units. The 6 is in first position, and therefore it represents 6×1 units. The only time a numeral represents its own unit value is in the first position.

4. Once the position and unit values are multiplied, we add them to produce the numeral's value.

Therefore, the numeral 246 in base ten represents $200 + 40 + 6$. Each of these principles will operate in every base. You should review these principles while you are working in various bases.

EXPRESSING NUMBERS IN BASE FIVE

In base five, we will use the symbols, 0, 1, 2, 3, and 4 only. Since we are using characters from base ten, we must distinguish the numbers in base five from the numbers in base ten. The convention which we will use is to write the name of the base as a subscript to the numeral. Thus, 23_{five} is a symbol for the number $(2 \times 5^1) + (3 \times 5^0)$, while 23_{seven} means $(2 \times 7^1) + (2 \times 7^0)$, and, of course, 23 represents $(2 \times 10) + (3 \times 10^0)$.

In base-five notation the first symbol on the right indicates how many groups of 5^0 there are, the second symbol from the right indicates how many groups of 5^1, the third indicates how many groups of 5^2, etc. For example, the number $abcde_{five} = [(a \times 5^4) + (b \times 5^3) + (c \times 5^2) (d \times 5) + (e \times 1)]_{ten}$. That is, the number $abcde_{five}$ means a number of 625's plus b number of 125's plus c number of 25's plus d number of 5's plus e number of units, all in base ten.

■ D.3   Let us now find what number in base ten the number 234_{five} is.

$$234_{five} = [(2 \times 25) + (3 \times 5) + 4]_{ten}$$
$$= [50 + 15 + 4]_{ten}$$
$$= 69_{ten}$$

Now how do we write 234_{ten} in base five?

■ D.4 $$234_{ten} = [(1 \times 125) + 109]_{ten}$$
$$= [(1 \times 125) + (4 \times 25) + 9]_{ten}$$
$$= [(1 \times 125) + (4 \times 24) + (1 \times 5) + 4]_{ten}$$
$$= 1414_{five}$$

Can you justify the last step? If so, you understand positional notation.

As a last exercise, we will count in base ten (our old friend) and base five at the same time.

Base Ten	Base Five		Base Ten	Base Five
1	1		11	21
2	2		12	22
3	3		.	.
4	4		.	.
5	10		.	.
6	11		99	344
7	12		100	400
8	13		101	401
9	14		.	.
10	20		.	.

■ D.2

EXERCISES

1. Count from 34_{five} to 201_{five}.

2. Translate the following to base ten.

   a. 2332_{five} b. 2030_{five}

   c. 4444_{five} d. 10001_{five}

3. Translate the following to base five.
   a. 247_{ten} b. 567_{ten} c. 435_{ten}
   d. 300_{ten} e. 624_{ten}

ADDITION IN BASE FIVE

Before we consider addition in base five, let us first look more closely at addition in base ten. How do we add the numbers 345_{ten} and 789_{ten}? First, we add units, 5_{ten} and 9_{ten} for a total of 14_{ten}, that is, 1 group of 10 plus 4. Next we add the ten's digits 4 and 8. But are we really adding 4 and 8? No! We are really adding 4 groups of 10 and 8 groups of 10, or 40 and 80, so that we have 12 groups of 10, or 120, that is 1 group of one hundred plus 2 groups of ten. Next we add the digits 3 and 7. Again, we are really adding 3 groups of 100 and 7 groups of 100 for a sum of 10 groups of 100 or 1 group of 1,000.

Adding these we have:

	1,000's digit	100's digit	10's digit	1's digit	
			3	4	5	
			7	8	9	
adding units digits				1	4	1 group of 10 + 4
adding ten's digits			1	2		1 group of 100 + 2 groups of 10
adding hundred's digit		1	0			1 group of 1,000 + 0 groups of 100
Total		1	1	3	4	

Thus, we have as the sum one group of 1,000 plus one group of 100 plus three groups of 10 plus 4. Can you see why when we add in our usual manner, we "carry the 1" when we add the 5 and 9 digits?

Another way of considering the addition problem is to consider each sum separately as follows:

$$5 + 9 \quad = \quad 14 \quad \text{or} \quad 10 + 4$$
$$40 + 80 = \quad 120 \quad \text{or} \quad 100 + 20 + 0$$
$$300 + 700 = 1,000 \quad \text{or} \quad 1,000 + 0 + 0 + 0$$

Combining the sums we have $1,000 + (100 + 20) + (10 + 4)$, or $1,000 + 100 + (20 + 10) + 4 = 1,000 + 100 + 30 + 4$, or 1,134.

Now let us carry this over into base five by performing some easy examples.

■ **D.6 Examples**

1. $3_{five} + 1_{five} = 4_{five}$
2. $3_{five} + 2_{five} = 10_{five}$
3. $3_{five} + 4_{five} = 7_{ten}$ Do you know why?

■ $= (1 \times 5) + 2_{ten}$
         $= 12_{five}$ because 7 is one group of 5 plus 2.

It is obvious that the symbols used in base five for addition sums do not have the same combination that we know from base ten. To facilitate addition, we include an addition table for the numbers 1_{five} to 10_{five} (Figure 2.5). Notice the blank spaces. How would you fill them in?

■ **D.5**

+	0	1	2	3	4	10
0	0	1	2	3	4	10
1	1	2	3	4	10	11
2	2	3	–	10	11	–
3	–	4	–	11	–	13
4	–	10	–	12	–	–
10	10	11	12	–	–	20

Figure 2.5 Addition table for base$_{five}$.

Now we are ready to add in base five. By this time all that should be necessary is an example written out in the same fashion as the example in base ten.

■ **D.6 Examples**

234_{five}		342_{five}		423_{five}	
$+ \ 401_{five}$		$+ \ 213_{five}$		$+ \ 303_{five}$	
10	1's sum	10	1's sum	11	1's sum
3	5's sum	10	5's sum	2	5's sum
11	25's sum	10	25's sum	12	25's sum
1140_{five} TOTAL		1110_{five} TOTAL		1231_{five} TOTAL	

EXERCISES

1. Find the following sums in base five.
   a. 234_{five} b. 214_{five} c. 3212_{five} d. 2302_{five}
   $+312_{five}$ $+32_{five}$ $+4232_{five}$ 4243_{five}

2. Add the following columns.
   a. 23_{five} b. 2002_{five}
       12_{five} 123_{five}
   $+24_{five}$ $+\ 411_{five}$

3. Use the addition table (Figure 2.5) to help answer these questions. Substantiate your answers.
   a. Is addition in base five a commutative operation?
   b. Is there an identity in base five? If so, what is it?
   c. Do you think that addition in base five is associative?

MULTIPLICATION IN BASE FIVE

Next we will consider the operation of multiplication in base five. A table of basic multiplication facts for base five is given in Figure 2.6. Notice again that the entries in the table are given as they were in the addition table. This standard format can be used for tables in any base. Fill in the blank spaces in the table.

D.7

×	0	1	2	3	4	10
0	0	0	0	0	0	0
1	–	1	2	3	4	10
2	–	–	4	11	13	20
3	–	–	–	14	22	30
4	–	–	–	–	31	40
10	–	–	–	–	–	100

Figure 2.6   Multiplication table for base$_{five}$.

In Figure 2.6 there is a row and a column of zeros, which represents the fact that 0 times any number in this base is zero. This is a particular property of zero that makes multiplication by 0 an easy operation. We can see that 1 acts as an identity element for multiplication, since $1 \times a = a = a \times 1$ for all a. There is symmetry in the table, which can be seen by drawing a diagonal from upper left to lower right through the symbols ×, 0, 1, 4, 14, 31, and 100. You notice that on either side of this diagonal the numerals that have corresponding positions with respect to the diagonal are equal. In Figure 2.7 we indicate numerals that correspond with respect to the diagonal.

■ D.7

×	0	1	2	3	4	10
0	0	0	0	0	0	0
1	0	1	2	3	4	10
2	0	2	4	11	13	20
3	0	3	11	14	22	30
4	0	4	10	22	31	40
10	0	10	20	30	40	100

■

Figure 27

Since these corresponding numerals are really 3×1 and 1×3 and 4×3 and 3×4, or, in general, $a \times b$ and $b \times a$, we see that if the elements corresponding to each other with respect to this diagonal are equal, then the operation is commutative. That is, since the multiplication table exhibits symmetry about the diagonal, the operation is commutative on the set of numbers

$$\{ 0_{\text{five}}, 1_{\text{five}}, 2_{\text{five}}, 3_{\text{five}}, 4_{\text{five}}, 10_{\text{five}}, 11_{\text{five}}, \ldots \}$$

We must at this time note that it is the set of *numbers* (in base ten the set $\{0, 1, 2, 3, \ldots\}$) that is commutative under multiplication and that our table, which was written in base-five numerals, indicates this property. Do not be confused between the name for a number (called a numeral) and the number itself. As we observed with sets, the name of an object and the object are two different things. As an example of this, let us observe that the numeral 11_{five} and the numeral 6 are two different objects but that they are both names for the number 6. Thus, we can say $11_{\text{five}} = 6$, meaning that the 11_{five} and 6 are the same number. On the other hand, the numeral 11_{five} is not equal to the numeral 6.

In order to develop an algorithm to multiply numbers expressed in base five, we will assume that the set of numbers $\{0, 1, 2, 3, \ldots\}$ has certain properties under the operations of multiplication and addition. In a later chapter we will justify these assumptions.

We will now list the properties of addition and multiplication that we will need in this section.

	Addition	Multiplication
Commutative property	$a + b = b + a$	$a \times b = b \times a$
Associative property	$(a + b) + c = a + (b + c)$	$(a \times b) \times c = a \times (b \times c)$
Identity element	$0 + a = a = a + 0$	$1 \times a = a = a \times 1$
Distributive property of multiplication over addition	$a \times (b + c) = (a \times b) + (a \times c)$ and $(b + c) \times a = (b \times a) + (c \times a)$	

Notice that the distributive property is listed as a property relating both operations, not as a property of each operation.

We will consider a few simple multiplication problems below as a basis for problems which follow. Suppose we wish to multiply 24_{five} by 10_{five}, 100_{five} or $1,000_{\text{five}}$. We know that 10×10 is 100, 10×100 is $1,000$, and $10 \times 1,000$ is $10,000$ from previous discussion of the meaning of base. Consider how this relates to the question

■ D.8
$$24_{\text{five}} \times 10_{\text{five}} = \{[(2 \times 10) + (4 \times 1)] \times 10\}_{\text{five}}$$

Using the distributive property we have

$$\{[(2 \times 10) + (4 \times 1)] \times 10\}_{\text{five}} = \{[(2 \times 10) \times 10] + [(4 \times 1) \times 10]\}_{\text{five}}$$

Using the associative property for multiplication twice,

$$= \{[2 \times (10 \times 10)] + [4 \times (1 \times 10)]\}_{\text{five}}$$

Now use the facts mentioned above to get

$$\{[(2 \times 10) + (4 \times 1)] \times 10\}_{\text{five}} = \{(2 \times 100) + (4 \times 10)\}_{\text{five}}$$
$$= \{200 + 40\}_{\text{five}}$$
$$= 240_{\text{five}}$$

Notice that $24 \times 10 = 240$ is true in any base. You should be able to see that 24×100 is $2,400$ and $24 \times 1,000$ is $24,000$ in any such base. Of course, this can be true only in bases which have 2 and 4 as digits.

It should follow from the example above that if $10_{\text{five}} \times 24_{\text{five}} = 240_{\text{five}}$, then $20_{\text{five}} \times 24_{\text{five}}$ can be found as follows:

$$20_{\text{five}} \times 24_{\text{five}} = (2_{\text{five}} \times 10_{\text{five}}) \times 24_{\text{five}}$$

By the associative property,

$$= 2_{\text{five}} \times (10_{\text{five}} \times 24_{\text{five}})$$

Now using the fact that $10_{\text{five}} \times 24_{\text{five}} = 240_{\text{five}}$

$$= 2_{\text{five}} \times 240_{\text{five}}$$

Rewriting 240_{five} we get

$$= 2_{\text{five}} \times (200_{\text{five}} + 40_{\text{five}})$$

Using the distributive property,

$$= 2_{\text{five}} \times 200_{\text{five}} + 2_{\text{five}} \times 40_{\text{five}}$$
$$= [2_{\text{five}} \times (2_{\text{five}} \times 10_{\text{five}}^2)] + [2_{\text{five}} \times (4_{\text{five}} \times 10_{\text{five}})]$$

Now use the distributive property to get

$$20_{\text{five}} \times 24_{\text{five}} = [(2_{\text{five}} \times 2_{\text{five}}) \times 10_{\text{five}}^2] + [(2_{\text{five}} \times 4_{\text{five}}) \times 10_{\text{five}}]$$
$$= 4_{\text{five}} \times 10_{\text{five}}^2 + (13_{\text{five}} \times 10_{\text{five}})$$
$$= 400_{\text{five}} + 130_{\text{five}}$$
$$= 1030_{\text{five}}$$

Consider the example in short form:

$$\begin{array}{r} 24_{\text{five}} \\ \times\ 20_{\text{five}} \\ \hline 1030_{\text{five}} \end{array} \quad \text{or} \quad \begin{array}{r} 20_{\text{five}} \\ \times\ 24_{\text{five}} \\ \hline 130 \\ 400 \\ \hline 1030_{\text{five}} \end{array}$$

Several more examples of multiplication given are below. You may wish to refer back to Figure 2.6, as you look at the algorithm.

Example 1	Example 2	Example 3
312_{five}	214_{five}	423_{five}
$\times\ 23_{\text{five}}$	$\times\ 24_{\text{five}}$	$\times\ 31_{\text{five}}$
1441	1421	423
11240	4330	23240
13231_{five}	11301_{five}	24213_{five}

Any calculation which is done in a base other than ten can be checked by translating the numerals into base ten, doing the calculation in base ten, and translating back to the original base. This process is quite simple and will be illustrated below.

■ D.6
▲ D.3 **Example 1**

$$\begin{array}{l} 3234_{\text{five}} = 3 \times 125 + 2 \times 25 + 3 \times 5 + 4 = 444_{\text{ten}} \\ + 1324_{\text{five}} = 1 \times 125 + 3 \times 25 + 2 \times 5 + 4 = 214_{\text{ten}} \\ \hline \end{array}$$

▲ ■ $\qquad 10113_{\text{five}} \qquad\qquad\qquad\qquad\qquad\qquad 658_{\text{ten}}$

■ D.4 Now $658_{\text{ten}} = 1 \times 625 + 0 \times 125 + 1 \times 25 + 1 \times 5 + 3 \times 1$, or
■ 10113_{five}.

■ D.8
▲ D.3 **Example 2**

$$\begin{array}{l} 312_{\text{five}} = 3 \times 25 + 1 \times 5 + 2 \times 1 = \quad 82_{\text{ten}} \\ \times\ 23_{\text{five}} = \qquad\qquad 2 \times 5 + 3 \times 1 = \quad 13_{\text{ten}} \\ \hline \quad 1441 \qquad\qquad\qquad\qquad\qquad\qquad 246 \\ \quad 11240 \qquad\qquad\qquad\qquad\qquad\quad\ 82 \\ \hline \end{array}$$

▲ ■ $\qquad 13231_{\text{five}} \qquad\qquad\qquad\qquad\qquad\quad 1066_{\text{ten}}$

■ D.4 Now $1066_{\text{ten}} = 1 \times 625 + 3 \times 125 + 2 \times 25 + 3 \times 5 + 1 \times 1$, or
■ 13231_{five}.

EXERCISES

1. Find the following products in base five.

a.	b.	c.	d.
324_{five}	212_{five}	341_{five}	213_{five}
$\times\ 31_{\text{five}}$	$\times\ 22_{\text{five}}$	$\times\ 42_{\text{five}}$	$\times\ 32_{\text{five}}$

2. Perform the indicated operations and check in base ten.

a. 312_{five}
$\underline{\times\ \ 12_{\text{five}}}$

b. $12_{\text{five}} \times 2_{\text{five}} \times 3_{\text{five}}$

BINARY SYSTEM

The *binary system* or the *base-two-numeration* system is the numeration system used by most computers. If base ten were used, each switch or relay that represented a position in some numeral would have to have 10 positions or states. In base two the switch or relay needs only 2 positions or states. From an engineering standpoint, such a switch is easier to construct; that is, the switch is either on or off, the relay is either activated or not.

Since counting, regardless of base, follows the principle of simply adding 1, we will illustrate counting in base two in Figure 2.8. This table includes bases ten, five, and two for comparison purposes.

◀ D.9

Base Ten	Base Five	Base Two	Base Ten	Base Five	Base Two
1	1	1	11	21	1011
2	2	10	12	22	1100
3	3	11	13	23	1101
4	4	100	14	24	1110
5	10	101	15	30	1111
6	11	110	16	31	10000
7	12	111	17	32	10001
8	13	1000	18	33	10010
9	14	1001	19	34	10011
10	20	1010	20	40	10100

■

Figure 2.8

If you look at the base-two column, you will notice that the numerals 1, 11, 111, 1111 always precede a power of 2. The powers of 2 which are shown in base ten numerals translate as follows:

$$2 = 10_{\text{two}}, 4 = 100_{\text{two}}, 8 = 1000_{\text{two}}, \text{ and } 16 = 10000_{\text{two}}$$

Translation to base two from base ten, and to base ten from base two, is done as illustrated in the previous section. Several examples are given below.

Example 1

D.10

Translate 1101_{two} to base ten.

$$1101_{\text{two}} = (1 \times 2^3 + 1 \times 2^2 + 0 \times 2 + 1 \times 1)_{\text{ten}}$$
$$= (1 \times 8 + 1 \times 4 + 0 \times 2 + 1 \times 1)_{\text{ten}}$$
$$= 13_{\text{ten}}$$

Example 2

Translate 10101_{two} to base ten.

$$10101_{two} = (1 \times 16 + 0 \times 8 + 1 \times 4 + 0 \times 2 + 1 \times 1)_{ten}$$
$$= 21_{ten}$$

Example 3

Translate 110011_{two} to base ten.

$$110011_{two} = (32 + 16 + 0 + 0 + 2 + 1)_{ten}$$
$$= 51_{ten}$$

Notice that in each succeeding example the process was shortened. In the beginning it is more accurate to use the complete process. As you understand better, you will begin to shorten the process; for the sake of accuracy, do not drop the zeros. There is a tendency to make errors if the zeros are deleted.

Consider the following examples, which translate from base ten to base two.

■ **D.11 Example 1**

Translate 51_{ten} to base two.

$$2\overline{\smash{)}51} \quad \text{25 rem 1} \qquad 51 = 2 \cdot 25 + 1$$

$$2\overline{\smash{)}25} \quad \text{12 rem 1} \qquad 51 = 2 \cdot (2 \cdot 12 + 1) + 1$$

$$2\overline{\smash{)}12} \quad \text{6 rem 0} \qquad 51 = 2 \cdot [2 \cdot (2 \cdot (6 + 0)) + 1] + 1$$

$$2\overline{\smash{)}6} \quad \text{3 rem 0} \qquad 51 = 2 \cdot \{2 \cdot [2 \cdot (2 \cdot 3 + 0)] + 1\} + 1$$

$$2\overline{\smash{)}3} \quad \text{1 rem 1} \qquad 51 = 2 \cdot [\![2 \cdot \{2 \cdot [2 \cdot (2 \cdot 1 + 1)]\} + 1]\!] + 1$$

Now use the distributive property on the last expression above. Notice that the first remainder (the 1 on the far right) is not multiplied by 2. Thus, the "unit" digit in base two is 1. Notice next that each of the succeeding remainders are multiplied by increasing powers of 2 and are thus the "2" digit, the "4" digit, the "8" digit, etc. Thus, $51_{ten} = 110011_{two}$.

Example 2

Translate 28_{ten} to base two.

$$2\overline{\smash{)}28} \quad \text{14 rem 0} \qquad 28 = 2 \cdot 14 + \underset{\text{unit's digit}}{0}$$

$$\frac{7 \text{ rem } 0}{2\overline{\big|\,14}} \qquad 28 = \underset{\substack{\text{2's digit} \qquad \text{unit's digit}}}{(2(2 \cdot 7 + 0) + 0}$$

$$\frac{3 \text{ rem } 1}{2\overline{\big|\,7}} \qquad 28 = \underset{\substack{\text{4's digit} \quad \text{2's digit} \quad \text{unit's digit}}}{2(2(2 \cdot 3 + 1 + 0) + 0}$$

$$\frac{1 \text{ rem } 1}{2\overline{\big|\,3}} \qquad 28 = \underset{\substack{\text{16's digit} \quad \text{8's digit} \quad \text{4's digit} \quad \text{2's digit} \quad \text{unit's digit}}}{2(2(2(2 \cdot 1 + 1) + 1) + 0) + 0}$$

Therefore, $28_{\text{ten}} = 11100_{\text{two}}$.

Example 3

Translate 19_{ten} to base two.

In this case the algorithm is combined into a simpler form:

$$\frac{0 \text{ rem } 1}{2\overline{\big|\,1}} \qquad (\text{two}^4)$$

$$\frac{1 \text{ rem } 0}{2\overline{\big|\,2}} \qquad (\text{two}^3)$$

$$\frac{2 \text{ rem } 0}{2\overline{\big|\,4}} \qquad (\text{two}^2)$$

$$\frac{4 \text{ rem } 1}{2\overline{\big|\,9}} \qquad (\text{two})$$

$$\frac{9 \text{ rem } 1}{2\overline{\big|\,19}} \qquad (\text{unit})$$

■ Therefore, $19_{\text{ten}} = 10011_{\text{two}}$.

The operations in base two are simpler to perform than in any other base even though it takes many digits to represent large numbers. The tables in Figure 2.9 are for addition and multiplication. Notice that there are only four basic facts in each table.

◄ D.12
▲ D.14

+	0	1		×	0	1
0	0	1		0	0	1
1	1	10		1	0	1

▲ ■

Figure 2.9

There are several examples below to illustrate addition and multiplication. These examples are all in base two, and we drop the subscript since there is no possibility of misunderstanding.

■ D.13 **Examples**

$$
\begin{array}{r}
101 \\
+\,110 \\
\hline
1011
\end{array}
\qquad
\begin{array}{r}
1101 \\
+\,1011 \\
\hline
11000
\end{array}
\qquad
\begin{array}{r}
1011 \\
+\,1001 \\
\hline
10100
\end{array}
$$

■

■ D.15

$$
\begin{array}{r}
101 \\
\times\,11 \\
\hline
101 \\
101 \\
\hline
1111
\end{array}
\qquad
\begin{array}{r}
1001 \\
\times\;\;101 \\
\hline
1001 \\
0000 \\
1001 \\
\hline
101101
\end{array}
\qquad
\begin{array}{r}
1101 \\
\times\,111 \\
\hline
1101 \\
1101 \\
1101 \\
\hline
1011011
\end{array}
$$

■

EXERCISES

1. Translate these base-two numerals to base ten.
   a. 10001_{two} b. 10101_{two} c. 11011_{two}
2. Translate these base-ten numerals to base two.
   a. 29_{ten} b. 47_{ten} c. 127_{ten}
3. Count in base two from 1101_{two} to 10111_{two}.

4. Find the following sums in base two, and check your answer by translating and adding in base ten.

$$
\begin{array}{ll}
\text{a.} & \begin{array}{r} 101 \\ +\,110 \\ \hline \end{array} \\
\end{array}
\qquad
\begin{array}{ll}
\text{b.} & \begin{array}{r} 1101 \\ +\,111 \\ \hline \end{array} \\
\end{array}
\qquad
\begin{array}{ll}
\text{c.} & \begin{array}{r} 1100 \\ +\,1010 \\ \hline \end{array} \\
\end{array}
$$

5. Multiply in base two and translate to base ten as a check.

$$
\begin{array}{ll}
\text{a.} & \begin{array}{r} 101 \\ \times\,11 \\ \hline \end{array} \\
\end{array}
\qquad
\begin{array}{ll}
\text{b.} & \begin{array}{r} 1101 \\ \times\,101 \\ \hline \end{array} \\
\end{array}
\qquad
\begin{array}{ll}
\text{c.} & \begin{array}{r} 1001 \\ \times\,111 \\ \hline \end{array} \\
\end{array}
$$

BASE SYSTEMS AND POWERS OF BASE SYSTEMS

Many base systems relate directly to other base systems due to the fact that the base in one system is a power of the base in another system. The base-four system relates to the base-two system in this manner. It is also the case that base eight relates to base two, since $2^3 = 8$, and base nine relates to base three, since $3^2 = 9$. To see the relation, let us first compare numerals in base two and base four by listing them in the table in Figure 2.10. To emphasize the relation that we are looking for, we use commas in base two.

Do you see that the first digit (on the right) in base four corresponds to the number that the first pair of digits in base two represents, and that the second digit in base four corresponds to the number that the second pair of digits represents in base two? Let us now try to find an explanation for this.

The first pair of numbers in base two represents the numbers 1_{two}, 10_{two}, and 11_{two}, that is, 1_{ten}, 2_{ten}, 3_{ten}. These are exactly the numbers represented by the first digit in a base-four representation. The second pair

Base Two	Base Four		Base Two	Base Four
1	1		11,01	31
10	2		11,10	32
11	3		11,11	33
1,00	10		1,00,00	100
1,01	11		1,00,01	101
1,10	12		1,00,10	102
1,11	13		1,00,11	103
10,00	20		1,01,00	110
10,01	21		1,01,01	111
10,10	22		1,01,10	112
10,11	23		1,01,11	113
11,00	30		1,10,00	120

Figure 2.10

of numbers in base two represents the number 100_{two}, 1000_{two}, and 1100_{two}, that is, 4_{ten}, 8_{ten}, and 12_{ten}. Again, these are exactly the numbers that the second digit in base four represents, namely, 10_{four}, 20_{four}, and 30_{four}. Check to see that the third pair of digits in base two represents the same numbers that the third digit in base four does.

Let us now illustrate this method of translating base two to base four with a few examples.

Example 1

D.22   Find $10,01_{two}$ in base four.
Since $01_{two} = 1_{four}$ and $10_{two} = 2_{four}$, then
$10,01_{two} = 21_{four}$

Example 2

Find $11,01_{two}$ in base four.
Since 01_{two} is 1_{four} and 11_{two} is 3_{four}, then
$11,01_{two} = 31_{four}$

Example 3

Find $10,11,00_{two}$ in base four.
Since 00_{two} is 0_{four}, 11_{two} is 3_{four}, and 10_{two} is 2_{four}, then
$101100_{two} = 230_{four}$

Now consider how base two relates to base eight. We will write base two numerals set off in groups of three this time. See Figure 2.11.

Can you see that the pattern is similar to that between base two and base four before looking at the examples?

Base Two	Base Eight		Base Two	Base Eight
1	1		1,101	15
10	2		1,110	16
11	3		1,111	17
100	4		10,000	20
101	5		10,001	21
110	6		10,010	22
111	7		10,011	23
1,000	10		10,100	24
1,001	11		10,101	25
1,010	12		10,110	26
1,011	13		10,111	27
1,100	14		11,000	30

Figure 2.11

Example 1

■ D.24   Find $1,001_{two}$ in base eight.

Since 001_{two} is 1_{eight} and 1_{two} is the same as 1_{eight}, then $1,001_{two} = 11_{eight}$

Example 2

Find $1,101_{two}$ in base eight.

Since 101_{two} is 5_{eight} and 1_{two} is 1_{eight}, then $1,101_{two} = 15_{eight}$

Example 3

Find $101,111_{two}$ in base eight.

Since 111_{two} is 7_{eight} and 101_{two} is 5_{eight}, then

■ $101,111_{two} = 57_{eight}$

Notice that we translated simply by grouping the binary numerals and translating them as we usually do. For example, we translated 11_{two} to 3_{four}. Actually $11_{two} = 3$ in any base larger than three. We translated 101_{two} to 5_{eight}. Again, $101_{two} = 5$ in any base larger than five.

Since you can see that translation simply involves using what you already know about translation from base two to base ten, we can simplify the translations as follows:

Example 1

■ D.22   Base two to base four $11, 01_{two}$

                                        $3 \quad 1_{four}$

▲ D.24   Base two to base eight $1,101_{two}$

                                        $1 \quad 5_{eight}$

Example 2

Base two to base four $10,11,01_{two}$

                                    $2\ \ 3\ \ 1_{four}$

Base two to base eight $101,101_{two}$

                                     $5\ \ \ 5_{eight}$

Example 3

■   Base two to base four $11,00,10_{two}$

                                    $3\ \ 0\ \ 2_{four}$

▲   Base two to base eight $110,010_{two}$

                                     $6\ \ \ \ 2_{eight}$

As a result of the previous work, we can see that translating from base eight or base four to base two simply involves translating each digit to base-two notation. Some examples follow:

D.23 **Example 1**

D.25 Base four to base two $3\ \ 2_{four}$

                                    $11,10_{two}$

D.24 Base eight to base two $1\ \ 6_{eight}$

                                    $1,110_{two}$

Example 2

Base four to base two $2\ \ 1\ \ 3_{four}$

                                    $10,01,11_{two}$

Base eight to base two $4\ \ 7_{eight}$

                                    $100,111_{two}$

Example 3

■   Base four to base two $1\ \ 0\ \ 2\ \ 1_{four}$

                                    $1,00,10,01_{two}$

●▲ Base eight to base two $1\ \ 1\ \ 1_{eight}$

                                    $1,001,001_{two}$

EXERCISES

1. Translate the following base-two numerals directly to base four and base eight.

  a. 11101_{two} b. 101011_{two} c. 110011_{two}

2. Translate the following base-four numerals directly to base two.

  a. 213_{four} b. 302_{four} c. 2013_{four}

3. Translate the following base-eight numerals directly to base two.

  a. 43_{eight} b. 64_{eight} c. 275_{eight}

4. Make a chart with the base-three and base-nine numerals from 1 to 25_{ten}.

5. By grouping base-three numerals in pairs, find a method for translating directly to base nine.

DUODECIMAL SYSTEM: BASE TWELVE

We will discuss the duodecimal system of numeration as an example of a system in which the base is greater than ten. In the systems introduced thus far, we were able to use symbols from the base ten system. However, in base twelve we need twelve symbols, which forces us to introduce two new symbols. We will use the following symbols for this system: 0, 1, 2, 3, 4, 5, 6, 7, 8, 9, T, E. Thus, we will represent 10_{ten} as T_{twelve} and 11_{ten} as E_{twelve}. The reason for introducing new digits should be obvious. In base twelve we must be able to write all the numbers up to 12 as one-digit numerals, since 12 is to be the first two-digit number, that is, $12_{ten} = 10_{twelve}$. This would mean that a base-fifteen system would require fifteen symbols.

Again, since counting in any base follows the same principles, we will illustrate counting in base twelve in the table in Figure 2.12. We introduced bases three and nine in the last section. We include them here for clarification and comparison purposes.

The powers of twelve which we will consider are: $12_{ten}^1 = 10_{twelve}$, $12_{ten}^2 = 144 = 100_{twelve}$, and $12_{ten}^3 = 1728 = 1000_{twelve}$. Using these facts, translation to base twelve from base ten and from base ten to base twelve is executed as it was shown in previous sections. Several examples follow below.

Example 1

D.28 Translate 135_{twelve} to base ten.

$$135_{twelve} = (1 \times 12^2 + 3 \times 12 + 5 \times 1)_{ten}$$
$$= (1 \times 144 + 3 \times 12 + 5 \times 1)_{ten}$$
$$= 185_{ten}$$

Example 2

Translate $15TE_{twelve}$ to base ten.

$$15TE_{twelve} = (1 \times 12^3 + 5 \times 10^2 + 10 \times 12 + 11 \times 1)_{ten}$$
$$= (1 \times 1728 + 5 \times 144 + 10 \times 12 + 11 \times 1)_{ten}$$
$$= (1728 + 720 + 120 + 11)_{ten}$$
$$= 2579_{ten}$$

Example 3

Translate $1T34_{twelve}$ to base ten.

$$1T34_{twelve} = (1728 + 1440 + 36 + 4)_{ten}$$
$$= 3208_{ten}$$

In the third example, we shortened the process. We do not suggest that you do this until you are quite sure that you understand the process. You will also find that using a short cut method may allow you to become careless in your work.

Consider next the examples of translation from base ten to base twelve.

■ D.26

Base Ten	Base Three	Base Nine	Base Twelve
1	1	1	1
2	2	2	2
3	10	3	3
4	11	4	4
5	12	5	5
6	20	6	6
7	21	7	7
8	22	8	8
9	100	10	9
10	101	11	T
11	102	12	E
12	110	13	10
13	111	14	11
.	.	.	.
.	.	.	.
.	.	.	.
22	211	24	1 T
23	212	25	1 E
24	220	26	20
25	331	27	21
.	.	.	.
.	.	.	.
.	.	.	.
117	11100	140	99
118	11101	141	9T
119	11102	142	9E
120	11110	143	T0
121	11111	144	T1
.	.	.	.
.	.	.	.
.	.	.	.
129	11210	153	T9
130	11211	154	TT
131	11212	155	TE
132	11220	156	E0
.	.	.	.
.	.	.	.
.	.	.	.
141	12020	166	E9
142	12021	167	ET
143	12022	168	EE
144	12100	170	100

■

Figure 2.12

Example 1

■ D.27   Translate 125_{ten} to base twelve.

$$10 \text{ rem } 5$$
$$12\overline{)125} \qquad 125 = 12 \times 10 + 5$$

Thus, $125_{ten} = T5_{twelve}$.

Example 2

Translate 162_{ten} to base twelve.

$$13 \text{ rem } 6$$
$$12\overline{)162} \qquad 162 = 12 \times 13 + 6$$

$$1 \text{ rem } 1$$
$$12\overline{)13} \qquad\qquad = 12 \times (12 \times 1 + 1) + 6$$

Using the distributive property on the last expression, we have

$$162_{ten} = 12 \times 12 + 12 \times 1 + 6$$
$$= 1 \times 12^2 + 1 \times 12 + 6$$

■ Thus, $162_{ten} = 116_{twelve}$.

EXERCISES

1. Count in base twelve from 25_{twelve} to 53_{twelve}.

2. How many characters are needed for a base-fourteen system? For base twenty-three? For base eleven?

3. Count in base eleven from 9_{eleven} to 32_{eleven}.

4. Develop a set of symbols for a base-fourteen system.

5. Translate the following base-twelve numerals to base ten.
   a. 346_{twelve} b. $14T_{twelve}$ c. $2TE_{twelve}$ d. $100T_{twelve}$

6. Translate the following base-ten numerals to base twelve.
   a. 2456_{ten} b. 1866_{ten} c. 3060_{ten} d. 2436_{ten}

The operations in base twelve are more difficult to perform in the sense that there are more addition and multiplication facts. The table in Figure 2.13 is an addition table in base twelve. When you use the table, look for patterns as well as properties which have previously been discussed.

The examples which follow should help you to add in base twelve. Refer to Figure 2.13 as you read the examples. We exclude the notation that indicates the base.

D.29

+	0	1	2	3	4	5	6	7	8	9	T	E
0	0	1	2	3	4	5	6	7	8	9	T	E
1	1	2	3	4	5	6	7	8	9	T	E	10
2	2	3	4	5	6	7	8	9	T	E	10	11
3	3	4	5	6	7	8	9	T	E	10	11	12
4	4	5	6	7	8	9	T	E	10	11	12	13
5	5	6	7	8	9	T	E	10	11	12	13	14
6	6	7	8	9	T	E	10	11	12	13	14	15
7	7	8	9	T	E	10	11	12	13	14	15	16
8	8	9	T	E	10	11	12	13	14	15	16	17
9	9	T	E	10	11	12	13	14	15	16	17	18
T	T	E	10	11	12	13	14	15	16	17	18	19
E	E	10	11	12	13	14	15	16	17	18	19	1T

Figure 2.13 Addition table for base twelve.

D.30 Examples

```
  243        347        578        86E
+ 115      + 273      + 246      + T73
-----      -----      -----      -----
  358        5ET        802       1722
```

Consider next the operation of multiplication in base twelve. a multiplication table is shown in Figure 2.14 to help you with the operation.

D.31

×	0	1	2	3	4	5	6	7	8	9	T	E
0	0	0	0	0	0	0	0	0	0	0	0	0
1	0	1	2	3	4	5	6	7	8	9	T	E
2	0	2	4	6	8	T	10	12	14	16	18	1T
3	0	3	6	9	10	13	16	19	20	23	26	29
4	0	4	8	10	14	18	20	24	28	30	34	38
5	0	5	T	13	18	21	26	2E	34	39	42	47
6	0	6	10	16	20	26	30	36	40	46	50	56
7	0	7	12	19	24	2E	36	41	48	53	5T	65
8	0	8	14	20	28	34	40	48	54	60	68	74
9	0	9	16	23	30	39	46	53	60	69	76	83
T	0	T	18	26	34	42	50	5T	68	76	84	92
E	0	T	1T	29	38	47	56	65	74	83	92	T1

Figure 2.14 Multiplication table for base twelve.

Since we have explained the operation in other bases, no explanation of the algorithm will be given. It will help if you use the table while looking at the examples.

Consider the examples below for illustration of the multiplication process in base twelve.

■ D.32 **Examples**

125	231	2E4	74T
× 12	× 23	× 36	× 8E
24T	693	1580	6952
1250	4620	8T00	4E280
■ 149T	50E3	T380	56012

EXERCISES

1. Find the following sums in base twelve. Check your answers to a and b by translating the addends and the sum to base ten.

a.   213   b.   348   c.   120E   d.   10 TE
   +314   +234   + 3E2   + 213

2. Find the following products in base twelve. Check your answers to a and b by translating to base ten.

a.   234   b.   2E6   c.   147T   d.   20E0
   × 23   × 100   × 201   × 36

■ D.18 ■   **3.** Construct a table for addition in base four.

■ D.19   **4.** Using the table which you made for base four, perform the additions below. The given numerals are written in base four.

a.   231   b.   123   c.   233   d.   123
   +102   +103   +132   12
                                    + 23

■ D.20 ■   **5.** Construct a table for multiplication in base three.

■ D.21   **6.** Using the table which you made for base three, perform the operations below. The given numerals are in base three.

a.   101   b.   1212   c.   1221   d.   212
   × 11   × 100   × 212   × 22

■ D.16   **7.** Translate the base-six numerals given below to base ten.
     a. 135   b. 430   c. 1254

■ D.17   **8.** Translate the base-ten numerals given below to base seven.
     a. 156   b. 278   c. 572

CHAPTER 3 NATURAL NUMBERS AND WHOLE NUMBERS

SECTION A

OPERATIONS ON *N* AND *W*

BEHAVIORAL OBJECTIVES

A.1 Be able to describe the counting characteristics of the natural numbers.

A.2 Be able to symbolize the set of natural numbers by tabulation.

A.3 Be able to tabulate standard sets and to find the cardinal numbers associated with each set.

A.4 Be able to define cardinality of a set.

A.5 Given an operation on a set, be able to perform it.

A.6 Be able to tell whether an operation is binary or not.

A.7 Be able to define binary operation.

A.8 Be able to tabulate the set of whole numbers.

A.9 Be able to decide if an object is an element of *W*.

A.10 Be able to decide if a set is a subset of *W*.

A.11 Be able to define addition on *W* using standard sets.

A.12 Be able to explain why disjoint sets must be used in the definition of addition.

A.13 Be able to define addition as a binary operation on *W*.

A.14 Be able to construct an addition table for *W*.

A.15 Be able to perform subtraction using standard sets.

A.16 Be able to define subtraction using standard sets.

A.17 Be able to define subtraction as a binary operation on *W*.

A.18 Be able to perform subtraction using the definition.

A.19 Be able to perform subtraction using an addition table.

A.20 Be able to define multiplication on *W* in terms of standard sets.

A.21 Be able to construct a multiplication table for *W*.

A.22 Be able to explain how one could define multiplication in terms of repeated addition.

A.23 Be able to define multiplication as a binary operation on *W*.

A.24 Be able to explain why addition and multiplication are called primary operations.

A.25 Be able to explain why subtraction and division are called secondary operations.

A.26 Be able to perform the operation of division using a multiplication table.

A.27 Be able to explain why division and subtraction are *not* closed on *W*.

A.28 Be able to show that division by zero is not possible.

A.29 Be able to define division as a binary operation on *W*.

A.30 Be able to define quotient.

A.31 Be able to explain why we cannot divide by zero.

DEVELOPMENT OF NATURAL NUMBERS

The *natural*, or *counting*, numbers were used by man before any symbolization or formal operations were developed. Counting is something that even the animals exhibit in a primitive way, although their concept of number is very limited. As you read in Chapter 2, the concept of zero did not develop until late in the history of civilized man. Thus, man began the long road to mathematical competence with the natural numbers as his original tool.

As we conjectured in Chapter 2, the early development of numeration probably started with the use of strokes ($/$), piles of stones, notched sticks, or knotted rope. No matter how or where the idea started, it always amounted to the same concept, that of a set with the following attributes:

■ **A.1** 1. There is a first element (in this case, 1).

2. Each element has a successor.

■ 3. The first element is not the successor of any element in the set.

Consider what such a set may look like:

$$N = \{/, //, ///, \ldots, (n + 1) \text{ strokes}\}$$

Now as we became more sophisticated and stopped using such simple notation as $////$ to mean 4, this set changed in notation only. We developed names and characters for the numbers:

$$\text{one} = 1 = /$$
$$\text{two} = 2 = //$$
$$\text{three} = 3 = ///$$

As a result of this notation change, we write the set of natural numbers as we did in Chapter 1:

■ **A.2** ■ $$N = \{1, 2, 3, \ldots\}$$

Particular subsets of *N*, called *standard sets*, are used to "measure" or "compare" sets of objects just as a ruler is a standard by which length is

■ **A.3** measured. When someone asks how many books are on the table, the answer "five" implies that the subset S_5, $S_5 = \{1, 2, 3, 4, 5\}$, can be placed in a 1–1 correspondence with the set of books on the table. We call the set S_5 a standard set because it is used as a standard for measuring sets of five elements. The last natural number in a standard set is called the *cardinal number* of a set with which it is being compared. Note that $n(S_5)$ is 5. Whenever we count, we are using the idea of a standard set and 1–1 correspondence. Listed below are standard sets which are subsets of the set of natural numbers. We have excluded the empty set, because $n(\varnothing)$, or 0, is not an element of the natural numbers.

$$S_1 = \{1\} \text{ and } n(S_1) = 1$$
$$S_2 = \{1, 2\} \text{ and } n(S_2) = 2$$
$$S_3 = \{1, 2, 3\} \text{ and } n(S_3) = 3$$
$$\vdots \qquad \vdots \qquad \qquad \vdots \quad \vdots$$
$$S_n = \{1, 2, 3, \ldots, n\} \text{ and } n(S_n) = n$$

In the illustration above, we see that the number property, or number of elements in a standard set, can be represented by the last element in the standard set.

This leads to the definition of cardinality or cardinal number of a given
■ set.

Definition

■ **A.4**   *The cardinal number, or cardinality, of a set is the number of elements in*
■ *the standard set corresponding to it.*

It should be obvious that the cardinality or cardinal number of a given set, A, is the number of elements in the set A, namely, $n(A)$. Thus, if $n(A) = n(S_7)$, the cardinal number or cardinality of A is 7.

EXERCISES

1. Write the standard set whose cardinality is 7.

2. Write the standard set whose cardinality is 8.

3. What relationship do the sets in Exercises 1 and 2 have?

4. If $a < b$, what can you say about S_a and S_b?

5. If $A \subseteq B$, how are the cardinalities of A and B related?

BINARY OPERATIONS

The concept of a binary operation was discussed in Chapter 1 with respect to operations on sets. Before we consider the attributes and definition of a binary operation in formal terms, we will consider several examples of operations. The operations discussed below will illustrate the idea of a binary operation more clearly. In order to understand exactly what a binary operation is, let us first look at some operations that are *not* binary.

Example 1

The Successor Operation. Given the set N and the operation (s), we define the operation on any number $a \in \dot{N}$ as producing the successor of a, or $a + 1$. Thus, $a(s) = a + 1$, or $a \to a + 1$.

Notice first that this operation is closed. In this case we have defined a *unitary* operation, since a single element in the set corresponds under the operation to a single element in the set. Notice from the definition that the element given corresponds uniquely to the second element, since every element has only one successor. Several examples are shown below:

■ A.5
1. $1(s) = 2$ or $1 \to 2$
2. $3(s) = 4$ or $3 \to 4$
■
3. $62(s) = 63$ or $62 \to 63$

Notice that two notations have been used. We read the arrow to mean "corresponds to."

Example 2

■ A.6   *The Digit Operation.* A second unitary operation, defined below, is one in which the correspondence is not unique. In other words, the given
■ number in the set corresponds to more than one number.

Given the set N and the operation $/\underline{}$, we define the operation on any number $a \in N$ as producing its digits.

Consider the examples below. Notice that in some cases the number corresponds to more than one number in N.

■ A.5
1. $/\underline{6} \longrightarrow 6$

2. $/\underline{16} \quad \begin{array}{l} \nearrow 1 \\ \searrow 6 \end{array}$

■
3. $/\underline{102} \begin{array}{l} \nearrow 1 \\ \to 0 \\ \searrow 2 \end{array}$

■ A.6   Now that we have seen two examples of operations that are not binary,
■ let us examine several binary operations.

Example 3

The First Number Operation. Given the set N and the operation $*$ for $a, b \in N$, we define the operation on the pair (a, b) to produce a. Thus, $a * b = a$, or $(a, b) \to a$.

Notice that this operation relates the pair (a, b) to a single unique element, which is determined by the first entry. Also notice that the notation below is shown in two ways. The use of $a * b = a$ is the more conventional or familiar notation. The correspondence notation can be read "(a, b) corresponds under the operation $*$ to a."

■ A.5
1. $(1, 2) \overset{*}{\to} 1$ or $1 * 2 = 1$
2. $(2, 1) \overset{*}{\to} 2$ or $2 * 1 = 2$
■
3. $(5, 7) \overset{*}{\to} 5$ or $5 * 7 = 5$

Example 4

The Greater Number Operation. Given the set N and the operation $\triangle$, we define the operation on pairs of elements from the set N in such a way that the result is the larger number of the pair. Thus, if $a < b$, then $(a, b) \xrightarrow{\triangle} b$ or $a \triangle b = b$. On the other hand, if $b < a$, then $(a, b) \xrightarrow{\triangle} a$ or $a \triangle b = a$.

■ A.6   Again, the operation is a binary operation. Notice that this operation is quite similar to the First Number Operation except for the name and definition. We still have an ordered pair which corresponds to a single ■ element by some rule.

■ A.5   1. $(1, 2) \xrightarrow{\triangle} 2$ or $1 \triangle 2 = 2$
           2. $(2, 1) \xrightarrow{\triangle} 2$ or $2 \triangle 1 = 2$
■   3. $(5, 7) \xrightarrow{\triangle} 7$ or $5 \triangle 7 = 7$

Example 5

The Between Operation. Given the set $E = \{2, 4, 6, 8, \ldots\}$ and the operation $\circ$ defined on pairs of elements of E in such a way that the pair (a, b) corresponds to the natural number between a and b, you will notice that the correspondence is unique; that is, each pair corresponds to only one natural number. However, the number that corresponds to a pair under the operation is not necessarily in the set E, because $(8, 10) \xrightarrow{\circ} 9$ and $9 \notin E$. As you know, this means that the operation is *not* closed. This does not mean that the operation is not a binary operation. Closure is simply a property that some operations have when defined on some sets.

◀ A.5   1. $(2, 4) \xrightarrow{\circ} 3$ or $2 \circ 4 = 3$
           2. $(4, 2) \xrightarrow{\circ} 3$ or $4 \circ 2 = 3$
■   3. $(6, 10) \xrightarrow{\circ} 8$ or $6 \circ 10 = 8$

Before we define a binary operation, let us consider the attributes of such an idea.

1. Pairs of elements are chosen from a given set.
2. The operation corresponds a unique element to the pairs.
3. The set of elements that corresponds to the pairs may or may not be in the given set.

Definition

◀ A.7   *A binary operation, $*$, defined on a set A is an operation that corresponds* ■ *to each ordered pair of elements (a, b), a unique element called $a * b$.*

EXERCISES

1. Which of the following operations are binary operations?
   a. squaring a natural number   b. adding natural numbers
   c. multiplying a natural number by 7

2. Define operations on N such that each pair corresponds to an element from the sets below.
   a. $\{0\}$ b. N c. $\{2, 4, 6, 8, \ldots\}$

3. Does the correspondence $a \to 2a$ define a unitary or a binary operation?

WHOLE NUMBERS

The set of whole numbers is much like the set of natural numbers in terms of its properties and its elements. The main difference is that the set of whole numbers, W, contains the element zero. Since $N = \{1, 2, 3, 4, \ldots\}$ and the set of whole numbers contains all the elements of N and in addition contains the number zero, we can tabulate W thus: $W = \{0, 1, 2, \ldots\}$.

■ A.8
■

In the beginning of this chapter we listed three characteristics of the natural numbers. These characteristics are based on a set of postulates known as *Peano's postulates*. In more advanced texts, Peano's five postulates are used as a basis for producing the integers, rational numbers, and finally the real numbers. Since such a development is quite advanced, we will simply state the postulates here for you to consider. Notice how these postulates are much the same as the characteristics which we have stated for the natural numbers and can be applied to the whole numbers.

P–1: There is a natural number 1.

P–2: Every natural number has a unique successor, n^+ or $n + 1$.

P–3: The natural number 1 is the successor to *no* natural number.

P–4: For natural numbers n and m, if $n + 1 = m + 1$, then $n = m$.

P–5: If N is a set of natural numbers with two conditions:
    1. $1 \in N$
    2. For each $n \in N$, $n + 1 \in N$; then N is all the natural numbers.

■ A.9
■

Notice that N is a proper subset of W. That is, all the elements of N are elements of W except for the element 0, which is in W but not in N.

To understand the relationship between the sets N and W and these elements, consider the following examples.

Examples

■ A.9
  1. $4 \in N$ and $4 \in W$

  2. $0 \notin N$ and $0 \in W$

  3. $N \notin W$ and $W \notin N$

▲ A.10
  4. $N \subseteq W$ and $W \nsubseteq N$

  5. $\{1, 2\} \notin N$ and $\{1, 2\} \notin W$

  6. $\{1, 2\} \subseteq N$ and $\{1, 2\} \subseteq W$

▲ ■
  7. $\{0, 1\} \nsubseteq N$ and $\{0, 1\} \subseteq W$

EXERCISES

1. Which of the following statements are true?

  a. $1 \in N$ b. $1 \in W$

  c. $N \subset W$ d. $W \subset N$

  e. $W \subseteq N$

2. Why do you think the set $\{1, 2, 3, \ldots\}$ is called the set of *natural numbers*?

3. Why is the name *whole numbers* reasonable for the set $\{0, 1, 2, \ldots\}$?

ADDITION OF NATURAL AND WHOLE NUMBERS

Since the primary difference between the two sets N and W in terms of their composition is that W contains 0 while N doesn't, we will discuss operations on the two sets at the same time. Later we will point out the importance of that single element, 0, in terms of properties which the set of whole numbers derives from having a zero.

We want now to make a formal definition of the operation of addition on the whole numbers. To do this, we must correspond pairs of whole numbers to some numbers. We must determine how this correspondence will be developed so that we can then easily define this correspondence to be an operation, specifically addition.

Not knowing exactly how to make our definition, we ask ourselves how can we add the numbers 3 and 5 using what we know about set union and its number property. Recall that

$$n(A) + n(B) = n(A \cup B) \text{ if and only if } A \cap B = \emptyset.$$

A.12 So it seems reasonable to use this approach. Since $n(S_3) = 3$ and $n(S_5) = 5$, take the union of S_3 and S_5 and call the sum of 3 and 5 the number property of this union. Nice, isn't it? Well, it won't work. Can you see why? The reason it doesn't work is that $S_3 \cup S_5 = S_5$ and $n(S_5) = 5$, so $n(S_3 \cup S_5) = n(S_5) = 5$. The problem here is that S_3 and S_5 are not disjoint. That is, $S_3 \cap S_5 = S_3$.

A.11 The way to get around this difficulty is to construct two disjoint sets, A and B, such that A is 1–1 with S_3 and B is 1–1 with S_5. Then take the union of A and B. Next find the standard set that is 1–1 with $A \cup B$ and call the number property of this standard set $3 + 5$. Now, let's do it.

Find the sum of 3 and 5. Let us construct two disjoint sets with cardinality 3 and 5, respectively.

$A = \{a, b, c\}, S_3 \sim A,$ and $n(S_3) = 3$
$B = \{q, r, s, t. u\}, S_5 \sim B,$ and $n(S_5) = 5$

Now $A \cup B = \{a, b, c, q, r, s, t, u\}$, and $A \cup B \sim S_8$. Because $n(A) + n(B) = n(A \cup B)$, then $n(S_3) + n(S_5) = n(S_8)$, or $3 + 5 = 8$. Thus, we would want to associate the number 8 with the pair $(3, 5)$; that is, $(3, 5) \overset{+}{\rightarrow} 8$.

In the example above, we used a rather long and unwieldy process to show how we would define the binary operation of addition in the set of whole numbers. If you consider this more carefully, you will realize that addition really is a shorter form of counting. What we do when we add two numbers is first produce two disjoint sets whose cardinalities are the given numbers. Then we count the objects in the union of these two sets. This process of counting sets to add is an immature process, but it is the basis for a later automatic or semiautomatic response to the number facts.

Consider how we would find the sum of 5 and 3 using this process. We count as follows: 1, 2, 3, 4, 5 then 6, 7, 8. The last number mentioned, 8, is the sum. If we wish to employ this idea along with the standard sets S_5 and S_3, we simply substitute an equivalent set for S_3 as follows:

$$S_5 \sim S_5 \quad \text{and} \quad S_3 \sim A = \{6, 7, 8\}$$

in this manner

$$n(S_5) = 5 \quad \text{and} \quad n(S_3) = n(A) = 3$$

now $S_5 \cup A = S_8 = \{1, 2, 3, 4, 5, 6, 7, 8\}$ and

$$n(S_5) + n(A) = n(S_5 \cup A) = n(S_8) = 8$$

thus

$$5 + 3 = 8$$

Consider some examples of this process

Example 1

Find the sum of 4 and 3 using set union.

SOLUTION $S_4 = \{1, 2, 3, 4\}$ and $S_3 \sim A = \{5, 6, 7\}$
Now $S_4 \cup A = S_7 = \{1, 2, 3, 4, 5, 6, 7\}$ and $n(S_4) + n(A)$
$= n(S_4 \cup A) = n(S_7) = 7$.
Thus, $4 + 3 = 7$.

Example 2

Find the sum of 6 and 2 using set union.

SOLUTION $S_6 = \{1, 2, 3, 4, 5, 6\}$ and $S_2 \sim A = \{7, 8\}$
Now $S_6 \cup A = \{1, 2, 3, 4, 5, 6, 7, 8\}$ and $n(S_6) + n(A)$
$= n(S_6 \cup A) = n(S_8) = 8$.
Thus, $6 + 2 = 8$

To find the number $a + b$, we find two disjoint sets A and B such that $n(A) = a$ and $n(B) = b$; then $a + b = n(A \cup B)$. Since we can always find these numbers in relation to standard sets, and since the union is unique, the set $A \cup B$ will always produce a unique number $n(A \cup B)$.

Definition

■ A.13 *Addition of whole numbers is a binary operation defined on the set of whole numbers which correspond to each pair (a, b) of whole numbers the unique*
■ *whole number $a + b$.*

In Figure 3.1 there are 100 basic facts that are the addition facts for the set of whole numbers. If we exclude the column and row headed by 0, we have 81 facts which also belong to the set of natural numbers. This table can also be used to produce all of the subtraction facts, as will be discussed in the next section.

◄ A.14

+	0	1	2	3	4	5	6	7	8	9
0	0	1	2	3	4	5	6	7	8	9
1	1	2	3	4	5	6	7	8	9	10
2	2	3	4	5	6	7	8	9	10	11
3	3	4	5	6	7	8	9	10	11	12
4	4	5	6	7	8	9	10	11	12	13
5	5	6	7	8	9	10	11	12	13	14
6	6	7	8	9	10	11	12	13	14	15
7	7	8	9	10	11	12	13	14	15	16
8	8	9	10	11	12	13	14	15	16	17
9	9	10	11	12	13	14	15	16	17	18

Figure 3.1

SUBTRACTION OF NATURAL AND WHOLE NUMBERS

The method we used to find the sum, $a + b$, of the numbers a and b was based upon set operations, namely, the relation $n(A) + n(B) = n(A \cup B)$ for A and B disjoint. In this section we will first describe subtraction in terms of operations on sets, and then we will formally define subtraction of whole numbers in terms of addition of whole numbers.

In the chapter on sets, we defined the set $A - B$. Recall that for the special case when $B \subseteq A$, $n(A) - n(B) = n(A - B)$. Consider the example used in the previous section, in which we found the union of sets A and B such that $n(A \cup B) = S_8$. Suppose we let $S_8 = A$ and $S_5 = B$; then $S_8 - S_5$ will be our $A - B$.

Example 1

◄ A.15   If $S_8 - S_5 = A - B = \{6, 7, 8\}$, then $(A - B) \sim S_3 = \{1, 2, 3\}$. It follows that $n(S_8 - S_5) = n(S_3) = 3$, or simply $8 - 5 = 3$.

Example 2

Looking at $8 - 3$, we see that $n(S_8 - S_3) = n(A)$, where $A = \{4, 5, 6, 7, 8\}$, and $A \sim S_5 = \{1, 2, 3, 4, 5\}$. Since $S_8 - S_3 = A$, we would say that ■ $8 - 3 = 5$.

A.16   The examples above point to the fact that subtraction can also be defined in terms of set operations. We really used a slight modification here. Instead of simply discussing the set $A - B$, we used the set $(A \cup B) - B$, which in the first example was $(A \cup S_5) - S_5$. This directly relates addition and subtraction.

A.24   We consider both of these operations, along with multiplication and A.25   division, as basic to arithmetic. We consider addition and multiplication *primary* operations, and subtraction and division *secondary* operations.

The reason for this distinction is that we can define the secondary opera-
▲ ■ tion in terms of the primary operation.

We now define subtraction as a binary operation in terms of addition.

Definition

■ A.17 *Subtraction of whole numbers is a binary operation which assigns to an ordered pair of numbers* (a, b) *a unique number* c, *or* $a - b$, *that satisfies*
■ *the equality* $b + c = a$.

■ A.18 Thus, the difference of two whole numbers, $a - b$, is that number which when *added* to b gives a. For example,

$$8 - 5 = 3 \quad \text{since } 5 + 3 = 8$$

and

$$8 - 3 = 5 \quad \text{since } 3 + 5 = 8$$

▲ A.27 However, $3 - 8$ is not possible in the set W, because there is no *whole*
▲ ■ number such that $3 = 8 + c$.

■ A.19 To use Figure 3.2 to find $a - b$, we simply choose the row beginning with b and move across that row until we reach the number a. Since the number heading this column is the number which when added to b produces a, that number is the difference $a - b$, or c. For example, if we move across the row headed by 5 on the left until we reach the number 8, we will find that 8 is in the column headed by 3. In other words, $8 - 5$
■ is 3, as shown in Figure 3.2. Try doing this by using the table in Figure 3.1.

	+	0	1	2	3	
						← difference here
	0	x	x	x	3	
	1	x	x	x	4	
	2	x	x	x	5	
	3	x	x	x	6	
	4	x	x	x	7	
enter here →	5	5	6	7	8	← move to here

Figure 3.2

EXERCISES

1. Explain why if $3 + 2 = 5$, then $5 - 3 = 2$.

2. Why does the definition of subtraction make the following statement true?
$3 + (7 - 3) = 7$

3. If a student understands addition, why should he be able to subtract?

MULTIPLICATION OF NATURAL AND WHOLE NUMBERS

We found in Chapter 1 that for the sets A and B we could produce the set $A \times B$ and that $n(A) \cdot n(B) = n(A \times B)$. We will use this fact as the basis for the operation of multiplication.

A.20   To multiply 2 and 3, we see that $n(S_2) = 2$; $n(S_3) = 3$; $S_2 \times S_3$ $= \{(1, 1), (1, 2), (1, 3), (2, 1), (2, 2), (2, 3)\}$; and $n(S_2 \times S_3) = 6$. So we define the product of 2 and 3, $2 \cdot 3$, to be $n(S_2 \times S_3)$. Thus, $2 \cdot 3 = 6$. In general, the product of the whole numbers a and b, $a \cdot b$, is $n(S_a \times S_b)$.

It might be interesting to note that the physical model of this cartesian set is a set of six points diagramed in Figures 3.3a and 3.3b.

$$S_2 \times S_3 = \{(1, 1), (1, 2), (1, 3), (2, 1), (2, 2), (2, 3)\}.$$

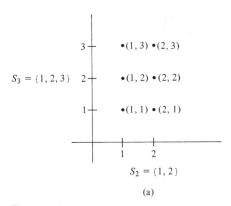

Figure 3.3a

Notice that

$$S_3 \times S_2 = \{(1, 1), (1, 2), (2, 1), (2, 2), (3, 1), (3, 2)\}$$

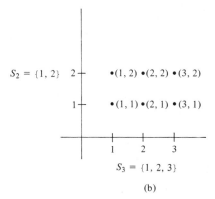

Figure 3.3b

We can see that this set of points, $S_2 \times S_3$, has the same number of points as the set $S_3 \times S_2$. We will discuss this further when we look at the properties of operations.

Figure 3.4 includes the multiplication facts for the whole and natural numbers. If we exclude the zero facts, which are represented by the row and column headed by 0, we have only those facts which belong to the natural numbers.

■ A.21

×	0	1	2	3	4	5	6	7	8	9
0	0	0	0	0	0	0	0	0	0	0
1	0	1	2	3	4	5	6	7	8	9
2	0	2	4	6	8	10	12	14	16	18
3	0	3	6	9	12	15	18	21	24	27
4	0	4	8	12	16	20	24	28	32	36
5	0	5	10	15	20	25	30	35	40	45
6	0	6	12	18	24	30	36	42	48	54
7	0	7	14	21	28	35	42	49	56	63
8	0	8	16	24	32	40	48	56	64	72
9	0	9	18	27	36	45	54	63	72	81

■

Figure 3.4

Whenever a table such as Figure 3.2 or 3.4 is constructed, it is interesting to consider emerging patterns within the table. In the last chapter, we discussed the symmetry around the main diagonal. In this case, the main diagonal contains the set 0, 1, 4, 9, 16, 25, 36, 49, 64, 81. This set in itself is of interest since it is a set of square numbers. We will discuss square numbers in some detail in Chapter 4.

You might notice that the columns headed by even numbers, 2, 4, 6, 8, contain only even numbers, whereas the columns headed by odd numbers contain both even and odd numbers alternately. Why?

If you consider the column headed by the numeral 9, you will notice that the right-hand digits produce the pattern 9, 8, 7, . . . , 1 if we exclude the zero. The ten's digit increases at the same time from 0 to 1, 2, 3, . . . , 8. If you sum the digits of any numeral in this column, you will notice that the sum is always 9, 1 + 8, 2 + 7, . . . , 8 + 1. Also notice the pairing of numerals such as 18 and 81, 27 and 72, etc.

If you look at each of the other columns, you will see many emerging patterns much like those discussed with respect to the column headed by 9. You will also find interesting patterns along any of the diagonals either from lower right to upper left, or from upper right to lower left.

■ A.22   Notice that the table was formed simply by adding the numeral which headed the column or row to zero, then to the result of that addition, and continuing the process. For example, if we look at the column headed by 5, we see that the first entry is 0. The next entry is $0 + 5$, or 5. The next entry is $5 + 5$, or 10, then $10 + 5$, $15 + 5, \ldots, 40 + 5$, or 45. This often leads one to define multiplication in terms of addition. We could say that ■ 7×5 means seven fives added together, or $5 + 5 + 5 + 5 + 5 + 5 + 5$. In fact, this is the way in which most young children begin to learn their multiplication facts, simply by addition. This type of definition is quite correct for the natural and whole numbers; however, it does cause some difficulty when one considers multiplication on other sets of numbers, such as the integers or rationals, which will be discussed in later chapters. We will simply define multiplication as we have defined the other operations, namely, as a binary operation on a set.

Definition

A.23   *Multiplication on the whole number is a binary operation that corresponds to every pair of whole numbers (a, b), a unique whole number, $a \cdot b$, where*
■ $a \cdot b = n(S_a \times S_b)$.

EXERCISES

1. Use the definition to find $4 \cdot 3$.

2. a. Write out the set $S_4 \times S_3$.
   b. How many elements are there in $S_4 \times S_3$?
   c. Can you give a convincing argument that multiplication of whole numbers is commutative?

DIVISION OF NATURAL AND WHOLE NUMBERS

A.24   We mentioned earlier that addition and multiplication are primary
A.25   operations and that subtraction and division are secondary operations. To emphasize this point, we will define division in terms of multiplication
▲ ■ analogously to our definition of subtraction in terms of addition.

With certain pairs (a, b) we will associate a single number c such that the product of c with b is equal to a. Note that c may not exist for every a and $b \in W$. The number c is called the *quotient* of a and b and is often symbolized by $a \div b$.

A.26   You may wish to look again at Figure 3.4 and try to apply what was shown in Figure 3.2 by substituting the number 15 for the number 8. Thus,
■ since $5 \times 3 = 15$, $15 \div 5$ will equal 3.

A.27   Using the table of multiplication, we can easily see that division is not defined for every pair of whole numbers. For example, there is no whole number that corresponds to the pair $(3, 2)$, since there is no whole number c such that $2 \cdot c = 3$. Thus, $3 \div 2$ is not defined if we restrict the set of quotients to the whole numbers. It should be obvious that if we want to define the operation of division for *all* pairs of whole numbers, we will
■ need a set that contains numbers other than whole numbers.

■ A.28   There is another property of division of whole numbers that is discernible from the table of multiplication facts. We can see from this table that for any whole number n, $n \cdot 0 = 0$. What does this imply about the quotient $2 \div 0$? That is, what number can we correspond to $(2, 0)$? In general, what number can we correspond to $(k, 0)$ for k any natural number? The answer to these questions is simply, No number. The reason is that there

■   is no whole number which when multiplied by 0 gives 2 or 3 or $k \neq 0$.

Definition

■ A.29   *Division is a binary operation which assigns to some ordered pairs of*

■   *whole numbers (a, b), a unique number c such that $a = b \cdot c$.*

So far we have been defining operations only. We are now ready for a slightly different type of definition. Let us define the object, c, in the above definition.

Definition

■ A.30   *For $b \neq 0$ and for a whole number a, $a \div b$ is the quotient of a and b if*

■   *and only if $a = b \cdot (a \div b)$.*

Notice the use of the phrase "if and only if." This phrase appears quite often in mathematical language. It is particularly found in the statement of definitions. In fact, all definitions either contain the phrase "if and only if" or an equivalent statement. The reason for this is that all definitions are reversible in the sense that a definition is really a correspondence between a word or symbol and its set of characteristics. That is, the word or symbol has the same meaning as the set of characteristics and the set of characteristics has the same meaning as the word or symbol.

For example, we might say that all "glomps" are green with red stripes and have two wheels and wings. The definition would be more precise if it stated the following: If the object is a glomp, then it has the color green with red stripes, two wheels, and wings; and if it has the color green with red stripes, two wheels, and wings, then it is a glomp. This implies that the word "glomp" means these characteristics, and anything with these characteristics must be a glomp.

A definition in mathematics does the same thing. In the above definition, given $a \div b$ means $a = b \cdot (a \div b)$; and given $a = b \cdot (a \div b)$, we have the existence of $a \div b$. Notice that the definition of division restricted b to being a non-zero number. In this case, b is always a natural number or a non-zero whole number. We should consider why this is necessary.

■ A.31   Suppose we wish to know the quotient when we divide 5 by 0. Then we must find a c such that $0 \times c = 5$. There doesn't seem to be such a c. It could be that our choice of 5 was unfortunate even though one counterexample is sufficient to make a proposition false. Let us consider the quotient produced by $0 \div 0$. You might think it produces a c such that $c = 1$ and that since $0 \times 1 = 0$, we have found a solution. *But* our definition requires that c is unique; in other words, there must be one and only one c to which $0 \div 0$ corresponds. Suppose we suggest that $0 \div 0$ is 54, in which case $0 \times 54 = 0$; of course, any c will suffice, which means that in this case c is not a unique solution. These two examples should raise serious doubts in your mind regarding division by zero. Try several examples to satisfy yourself that the quotient for a division example in which zero is the divisor does not exist.

EXERCISES

1. What are the primary operations?

2. What are the "inverse" operations which relate to the operations below?
a. moving forward   b. adding
c. doubling   d. dividing
e. squaring

3. Name the binary operation performed below.
a. $(5, 7) \to 35$ b. $(25, 5) \to 5$
c. $(13, 7) \to 6$ d. $(153, 3) \to 51$
e. $(187, 56) \to 243$ f. $(192, 6) \to 32$
g. $(45, 5) \to 225$

4. The statements below are either true or false. By using the definition of subtraction and division, check the statements for accuracy. Indicate why they are true or false.
a. $5 = 35 \div 7$ b. $18 = 30 - 14$
c. $39 \div 13 = 3$ d. $25 - 12 = 13$
e. $42 \div 6 = 7$ f. $20 - 14 = 6$
g. $625 \div 25 = 35$ h. $76 - 60 = 16$

5. The first number operation was defined earlier in the chapter. Perform the operation on the following pairs.
a. $(5, 7)$ b. $(62, 6)$ c. $(12, 31)$ d. $(32, 12)$

6. A binary operation on the set of whole numbers is described as doubling the first number and adding three times the second number to the result. Symbolically, $(a, b) = 2a + 3b$. If the pair is given as $(3, 2)$, we produce $2 \times 3 + 3 \times 2$, or $(3, 2) \to 12$. Find the following:
a. $(0, 2)$ b. $(2, 0)$ c. $(4, 1)$ d. $(1, 1)$ e. $(0, 0)$

SECTION B   PROPERTIES OF ADDITION AND MULTIPLICATION ON *N* AND *W*

BEHAVIORAL OBJECTIVES

B.1   Be able to show that *N* and *W* are closed under addition.

B.2   Be able to show that *N* and *W* are closed under multiplication.

B.3   Be able to state the closure property of *N* under addition.

B.4   Be able to state the closure property of *N* under multiplication.

B.5 Be able to state the closure property of W under addition.

B.6 Be able to state the closure property of W under multiplication.

B.7 Be able to give examples of sets that are or are not closed under specific operations.

B.8 Be able to give examples of sets that are commutative under specific operations.

B.9 Be able to give examples of sets that are not commutative under specific operations.

B.10 Be able to prove that N is commutative under addition.

B.11 Be able to prove that W is commutative under addition.

B.12 Be able to prove that N is commutative under multiplication.

B.13 Be able to prove that W is commutative under multiplication.

B.14 Be able to state the property of commutativity of addition on N.

B.15 Be able to state the property of commutativity of multiplication on N.

B.16 Be able to state the property of commutativity of addition on W.

B.17 Be able to state the property of commutativity of multiplication on W.

B.18 Be able to give an example of a set that is associative under a specific operation.

B.19 Be able to prove that addition on N is associative.

B.20 Be able to prove that addition on W is associative.

B.21 Be able to state the associative property of addition on N.

B.22 Be able to state the associative property of addition on W.

B.23 Be able to state the associative property of multiplication on N.

B.24 Be able to state the associative property of multiplication on W.

B.25 Be able to give an example of an operation that is not associative on a set.

B.26 Be able to determine the properties of a given operation on a set.

B.27 Be able to name a set that has an identity element under some specific operation.

B.28 Be able to find the identity of W for addition.

B.29 Be able to state the properties of zero under addition and multiplication.

B.30 Be able to find the identity of N and W under multiplication.

B.31 Be able to relate the distributive property of multiplication over addition to the set properties of intersection and union.

B.32 Be able to state the distributive property of multiplication over addition for N.

B.33 Be able to state the distributive property of multiplication over addition for *W*.

B.34 Be able to distinguish between the right and left distributive properties.

B.35 Be able to use the distributive property in problem situations.

B.36 Be able to summarize the properties of *N* and *W* under addition and multiplication.

B.37 Be able to distinguish *identity* from *right identity*.

B.38 Be able to find a right identity for subtraction on *W*.

B.39 Be able to find a right identity for division on *N* and *W*.

B.40 Be able to state the distributive property of multiplication over subtraction on *N* and *W*.

B.41 Be able to state the distributive property of division over addition and division over subtraction on *N* and *W*.

B.42 Be able to state the cancellation property of addition.

B.43 Be able to state the cancellation property of multiplication.

B.44 Be able to use the cancellation properties to solve equations.

PROPERTIES OF ADDITION AND MULTIPLICATION ON *N* AND *W*

CLOSURE

B.1 In Chapter 1 we stated that the operation of union always produces a set. Since union produces a set and addition was defined in terms of union of disjoint sets, we can state intuitively that addition in *N* is closed. Since $n(S)$ for any non-empty set *S* is a counting number, that is, $n(S) \in N$, $n(A \cup B)$ will always be a counting number. This means that the operation of addition on *N* is one which will always produce an element in *N*. Thus, *N* is closed under addition. If we include the empty set and zero, recalling that $n(\varnothing) = 0$, the union of two sets is again a set and $n(A \cup B) = n(A) + n(B)$, where *A* and *B* are disjoint. Thus, the set of whole numbers is closed under addition.

B.2 Multiplication has been defined in terms of $S_a \times S_b$. We know again that $n(S_a \times S_b)$ is a natural number if S_a and S_b are non-empty sets. Therefore, the natural numbers are closed under multiplication, since $a \times b = n(S_a \times S_b)$. If we allow S_a or S_b to be $\varnothing$, we produce a set $(S_a \times S_b)$ which is empty. This means that $n(S_a \times S_b) = 0$, and since 0 is an element of *W*, we infer that *W* is closed under multiplication.

From the previous discussion, we can state that the sets *N* and *W* are closed under the operations of multiplication and addition. It is important to note that it is meaningless to state that a set is closed or to state that an operation is closed. We must always indicate the *set* as well as the *operation* for closure and other properties. Since we have discussed closure in four instances, we will state each of these and abbreviate the word "closure" by *Cl*. We will abbreviate the operation by substituting + for addition and × for multiplication. Lastly, the letters *N* and *W* will be employed to represent the natural and whole numbers. Notice how we state these properties below.

■ **B.3**　1. Closure property for addition of natural numbers, or $Cl + N$: For all
■　　　$a, b \in N, a + b = c \in N$.

■ **B.4**　2. Closure property for multiplication of natural numbers, or $Cl \times N$: For
■　　　all $a, b \in N, a \times b = c \in N$.

■ **B.5**　3. Closure property for addition of whole numbers, or $Cl + W$: For all
■　　　$a, b \in W, a + b = c \in W$.

■ **B.6**　4. Closure property for multiplication of whole numbers, or $Cl \times W$: For
■　　　all $a, b \in W, a \times b = c \in W$.

The abbreviations which we are using above will be consistently employed throughout the text. In this manner we will be able to signify properties of operations on sets without causing confusion for the reader.

COMMUTATIVE PROPERTY

Before we consider the next property of addition and multiplication on the natural numbers, we will discuss several examples of sets and operations which may have some of the properties which follow. We will define operations on a set in two ways, by description and by table. Earlier we cited operations such as the first number operation or successor operation. These operations were defined in terms of their name and a description of the operation. We could define an operation on a set by simply giving the set and the table which shows the operation. Consider the set of elements $S = \{2, 4, 6\}$ in the table in Figure 3.5 under the operation ∘.

∘	2	4	6
2	2	3	4
4	3	4	5
6	4	5	6

Figure 3.5

The operation shown in the table is actually the between operation
■ **B.7**　defined earlier. Since the set was given as $\{2, 4, 6\}$, one can see that the operation on this set is not closed. This was pointed out when we first discussed the operation. Notice that we restricted our set to a finite
■　number of elements in this example. The use of finite sets to illustrate properties makes the investigation of those properties easier in some cases, since the possible combination of elements is restricted in number. If you look at the table closely, you will notice that there is symmetry

around the main diagonal. This is a hint that the operation on the set is commutative.

■ B.8   We can quickly check to see if the set is commutative under ∘.

$2 \circ 4 = 3$ and $4 \circ 2 = 3$;   therefore, $2 \circ 4 = 4 \circ 2$ Why?
$2 \circ 6 = 4$ and $6 \circ 2 = 4$;   therefore, $2 \circ 6 = 6 \circ 2$ Why?
$4 \circ 6 = 5$ and $6 \circ 4 = 5$;   therefore, $4 \circ 6 = 6 \circ 4$ Why?

Notice that we did not bother with $2 \circ 2$, $4 \circ 4$, and $6 \circ 6$ since these are trivially commutative. The main diagonal of any table will be composed of combinations of the form $a * a$ if the table is ordered in the same manner
■   on the rows and columns.

Consider a second operation on the set S, defined in the table in Figure 3.6.

*	2	4	6
2	2	2	2
4	4	4	4
6	6	6	6

Figure 3.6

This strange-looking operation is actually the first number operation. Notice that the way in which we use the table requires that our first entry from the ordered pair comes from the left-hand column. Thus, the pairs $(2, 2)$, $(2, 4)$, and $(2, 6)$ all correspond to 2. This is seen when you look at
◀ B.7   the row of 2's headed by the 2 in the left-hand column. This operation is closed for the set $\{2, 4, 6\}$ since the *only* entries in the table are elements
◀B.9 ■   of S. Notice that even though the set is closed, it is not commutative under this operation. We can see the lack of commutativity simply by comparing $(2, 4)$ and $(4, 2)$. Since $(2, 4)$ corresponds to 2 and $(4, 2)$ corresponds to 4, we know that $(2, 4) \neq (4, 2)$, which is sufficient evidence that the set is
▲   not commutative under the first number operation. One counter-example is enough to falsify any rule, law, or property. We often *prove* that something is false by simply finding a counter-example.

For A and B disjoint, $n(A) + n(B) = n(A \cup B)$ and $(A \cup B) = (B \cup A)$, it follows that $n(A \cup B) = n(B \cup A) = n(B) + n(A)$. This line of reasoning leads us to the conclusion that $n(A) + n(B) = n(B) + n(A)$. However, we have not given any reasons for these conclusions, nor have we directly shown the implications on this line of reasoning. We would like to show that the operations of addition and multiplication on the sets N and W are commutative.

Consider first the set N and the operation of addition. We would like to state that for all a and b in N, $a + b = b + a$. Consider the proof.

Example

B.10   Let $n(A) = a$ and $n(B) = b$. A and B are disjoint non-empty sets.

SOLUTION   To show that for all $a, b \in N$, $a + b = b + a$,

1. $n(A) = a$ and $n(B) = b$	1. given
2. $a + b = n(A \cup B)$	2. definition of $a + b$
3. $(A \cup B) = (B \cup A)$	3. commutative property of $\cup$
4. $n(A \cup B) = n(B \cup A)$	4. equal sets mean that they have the same elements and so must have the same number of elements
5. $n(B \cup A) = b + a$	5. definition of $b + a$
6. $a + b = b + a$	6. transitive property of equality through steps 2, 4, and 5

■ Thus, addition on the set of natural numbers is commutative.

■ B.11   To show that the addition is commutative on W, we need only show that zero commutes under addition with every element of N. Since the elements of N commute with each other under addition, we then have that all the elements of W commute under addition. We leave this to you

■ as an exercise, with the hint that $A \cup \emptyset = A$.

Consider next the commutativity of multiplication on N and W. The proof is based on the idea that $n(A \times B) = n(B \times A)$, but remember

■ B.12 that $A \times B \neq B \times A$. Consider the proof of the statement that for $a, b \in N$, $a \cdot b = b \cdot a$.

Example

Let $n(A) = a$ and $n(B) = b$. A and B are non-empty sets.

SOLUTION

1. $n(A) = a$ and $n(B) = b$	1. given
2. $a \cdot b = n(A \times B)$	2. definition of multiplication
3. $n(A \times B) = n(B \times A)$	3. property of the number of elements in the cartesian product of sets
4. $n(B \times A) = b \cdot a$	4. definition of multiplication
5. $a \cdot b = b \cdot a$	5. transitive property of equality using steps 2, 3, and 4

■

■ B.13   To complete this proof for the set W, we need only show that $n(A) \times n(\emptyset)$

■ for all A is 0. This will again be left as an exercise. We will state the properties below and abbreviate as follows: Co will stand for commutative property, and $Co + N$ will mean commutative property of addition on natural numbers.

■ B.14 ■   1. $Co + N$: For all a, b in N, $a + b = b + a$.

■ B.15 ■   2. $Co \times N$: For all a, b in N, $a \times b = b \times a$.

■ B.16 ■   3. $Co + W$: For all a, b in W, $a + b = b + a$.

■ B.17 ■   4. $Co \times W$: For all a, b in W, $a \times b = b \times a$.

As a result of the properties stated above, we can state several examples of commutativity in the set N and W under $+$ or $\times$.

Examples
$3 + 2 = 2 + 3$
$7 + 0 = 0 + 7$
$3 \times 2 = 2 \times 3$
$5 \times 0 = 0 \times 5$

EXERCISES

1. Which of the following sets are closed under the operation? Why?
   a. Even natural numbers under addition.
   b. Odd natural numbers under addition.
   c. Natural numbers of the form $3n$, where $n \in N = \{1, 2, 3, \ldots\}$ under addition.
   d. The operation $2a + b$, where a and $b \in N$.

2. Which of the tables below illustrate closure for the set $S = \{a, b, c\}$?

a.

$*$	a	b
a	a	b
b	a	a

b.

$\circ$	a	b	c
a	a	b	c
b	b	c	a
c	c	a	b

c.

$\oplus$	a	b	c
a	a	b	c
b	b	c	d
c	c	d	e

3. Name the property illustrated below.
   a. $2 + 3 = 3 + 2$
   b. $2 \times 3 = 3 \times 2$
   c. $2 + 3 = 5 \in S$
   d. $a \circ b = b \circ a$

4. Which of the tables below illustrate commutativity? Why?

a.

$\oplus$	a	b	c
a	a	b	c
b	b	c	a
c	c	a	b

b.

$*$	a	b	c
a	c	a	b
b	b	b	a
c	a	c	c

c.

$\circ$	a	b	c
a	a	c	a
b	c	b	b
c	b	b	c

5. Which of the following operations on N are commutative? Why?
   a. The operation a/b defined as $2a + 2b$ for a and $b \in N$.
   b. The operation $a\backslash b$ defined as $2a + 3b$ for a and $b \in N$.
   c. The operation $a \,\square\, b$ defined as $a \times a + b \times b$ for $a, b \in N$.

6. Show that $a \cdot b + a \cdot c = b \cdot a + c \cdot a$ for $a, b, c, \in N$.

7. If $a * b$ means to drive to work and walk home, and $b * a$ means to walk to work and drive home,
   a. is $a * b = b * a$ in terms of distance traveled? Why or why not?
   b. is $a * b = b * a$ in terms of time traveling? Why or why not?
   c. is $a * b = b * a$ in general? Why or why not?

8. Let $a \circ b$ mean "to put on a and then b." If a and b take on the meanings given below, compare $a \circ b$ and $b \circ a$.
   a. shoes and socks
   b. skirt and sweater
   c. coat and hat
   d. pants and belt
   e. shirt and tie

ASSOCIATIVE PROPERTY

Consider the operation defined by Figure 3.7 below on the set S = $\{2, 4, 6\}$. This operation is the greater number operation, $\triangle$, defined earlier.

$$
\begin{array}{c|ccc}
\triangle & 2 & 4 & 6 \\
\hline
2 & 2 & 4 & 6 \\
4 & 4 & 4 & 6 \\
6 & 6 & 6 & 6 \\
\end{array}
$$

Figure 3.7

One can see immediately that this operation is closed and commutative. We might next consider whether the associative property holds for this operation on S. We can do this by testing all possible examples of elements taken three at a time or by using the definition in some way. We might first test a few examples to see exactly how the operation works and if we feel that a counter-example exists.

$$(2 \triangle 4) \triangle 6 = 4 \triangle 6 = 6 \quad \text{and} \quad 2 \triangle (4 \triangle 6) = 2 \triangle 6 = 6$$

thus,

$$(2 \triangle 4) \triangle 6 = 2 \triangle (4 \triangle 6)$$

If we try several examples, we will continually find that this operation seems to be associative with respect to this set. The question is, How might we show this to be true? Consider some of the statements below.

1. The greatest element in the set is 6.
2. The element 6 operating on any element produces 6.
3. No matter how we group, if 6 is one of the elements operating, the final product will be 6.
4. The next greatest element is 4.
5. The element 4 operating on any element but 6 will produce 4.
6. No matter how we group, if 4 is one of the elements and 6 is not, then the final product will be 4.
7. The least element is 2.
8. The final product of any expression is 2 only if every element in the expression is 2.

■ **B.18** The conclusions which we should reach from the previous statements are as follows:

1. No matter what the grouping or elements are in an expression, if 6 is one of the elements, the product is 6.
2. Whenever an expression excludes 6 but includes 4, the product is always 4 no matter what the grouping.

3. We can have a product of 2 only if all elements are 2.

4. Therefore, grouping does not affect the operation on S.

So far we know that the operation $\triangle$ on S has the following properties:

▲ **B.26** 1. $Cl \triangle S$: For any a, b in S, $a \triangle b = c$ in S.

2. $Co \triangle S$: For any a, b in S, $a \triangle b = b \triangle a$.

■ 3. $A \triangle B$: For any $a, b,$ and c in S, $(a \triangle b) \triangle c = a \triangle (b \triangle c)$.

We might point out the following properties. There is a property of 6 which states that 6 operating on any element produces 6. There is a property of 2 which states that 2 operating on any element produces that element. The property of 2 is an identity property, or, more precisely, 2 is an identity element in this set. The property of 6 is much like a zero property of multiplication.

4. Identity property: For all a in S, $a \triangle 2 = a = 2 \triangle a$

▲ 5. 6 property: For all a in S, $a \triangle 6 = 6 = 6 \triangle a$

In the section which follows, we would like to show that natural numbers can be combined by addition or multiplication such that grouping in any manner will not affect the results. Consider first several examples of this idea.

1. $(2 + 3) + 6 = 5 + 6$
$\qquad\qquad\quad\ = 11$
$\ 2 + (3 + 6) = 2 + 9$
$\qquad\qquad\quad\ = 11$

thus,

$\quad (2 + 3) + 6 = 2 + (3 + 6)$

2. $(5 \times 2) \times 4 = 10 \times 4$
$\qquad\qquad\qquad = 40$
$\ 5 \times (2 \times 4) = 5 \times 8$
$\qquad\qquad\qquad = 40$

thus,

$\quad (5 \times 2) \times 4 = 5 \times (2 \times 4)$

The examples above are illustrative of the associative property of addition and multiplication on N and W. We may be more convincing if we show that these properties hold in general for all natural and whole numbers. We will again call on our development of set ideas from Chapter 1.

Example

B.19 Let $n(A) = a$, $n(B) = b$, and $n(C) = c$, and let A, B, and C be non-empty disjoint sets.

SOLUTION We wish to show that $(a + b) + c = a + (b + c)$ for all a, b, c in N.

1. $n(A) = a$, $n(B) = b$, and $n(C) = c$	1. given
2. $(a + b) + c = (n(A) + n(B)) + n(C)$	2. reflexive property of equality using different names
3. $(n(A) + n(B)) + n(C) = n(A \cup B) + n(C)$	3. property of number union of disjoint sets
4. $n(A \cup B) + n(C) = n((A \cup B) \cup C)$	4. number property for union of disjoint sets
5. $n((A \cup B) \cup C) = n(A \cup (B \cup C))$	5. associative property of union
6. $n(A \cup (B \cup C)) = n(A) + n(B \cup C)$	6. number property for union of disjoint sets
7. $n(A) + n(B \cup C) = n(A) + (n(B) + n(C))$	7. number property for union of disjoint sets
8. $n(A) + (n(B) + n(C)) = a + (b + c)$	8. replacing names from step 1
9. $(a + b) + c = a + (b + c)$	9. transitive property of equality, steps 2–8

■ **B.20**
■ We should again point out that if A, B, or C is an empty set, the proof will still hold. This proof would then apply to the whole numbers.

We will state the associative property of multiplication without proof. It can be shown that the sets $(A \times B) \times C$ and $A \times (B \times C)$ have the same number property and as a result $(n(A) \times n(B)) \times n(C) = n(A) \times (n(B) \times n(C))$. Since we have omitted discussion of the number property for such sets, we will simply assume that the operation of multiplication on the natural and whole numbers is an associative operation. We have abbreviated the associative properties by using the italic capital letter A. The properties discussed are stated in symbolic form below.

■ **B.21** ■ 1. $A + N$: For all a, b, c in N, $(a + b) + c = a + (b + c)$.
■ **B.22** ■ 2. $A + W$: For all a, b, c in W, $(a + b) + c = a + (b + c)$.
■ **B.23** ■ 3. $A \times N$: For all a, b, c in N, $(a \times b) \times c = a \times (b \times c)$.
■ **B.24** ■ 4. $A \times W$: For all a, b, c in W, $(a \times b) \times c = a \times (b \times c)$.

Consider an example of an operation on the set of whole numbers which is based on the operations of addition and multiplication. Using what we know about W under $+$ and $\times$, we can decide whether the properties discussed so far exist for this new operation on W. We define this new operation on W so that it corresponds the pair (a, b) under $\otimes$ to $2a + b$. For example, the pair $(3, 4)$ corresponds to $2 \times 3 + 4$, or 10.

■ **B.26**
■ We can show that the operation is closed as follows: For all a, b in W, show that $a \otimes b = c$ in W.

■ **B.6** 1. $a \otimes b = 2a + b$ 1. definition of $\otimes$
2. a in W and 2 in W 2. definition of W

3. $2a$ in W 3. $Cl \times W$

4. $2a$ in W and b in W 4. from step 3 and given

5. $2a + b$ in W 5. $Cl + W$

6. $a \otimes b \in W$ 6. replace $2a + b$ by $a \otimes b$ from step 1

■ Therefore, the operation is closed in W.

■ **B.9**   We can check to see if $\otimes$ is commutative in W. Since $(a, b) = 2a + b$ and $(b, a) = 2b + a$, we can see by counter-example that $\otimes$ is non-commutative in W. If we let $a = 3$ and $b = 4$, then we have

$$3 \otimes 4 = 2 \times 3 + 4, \text{ or } 10, \text{ and } 4 \otimes 3 = 2 \times 4 + 3, \text{ or } 11$$

Thus,

■ $3 \otimes 4 \neq 4 \otimes 3$

B.25   We can also check for associativity. Again we seek a counter-example first.

$$(a \otimes b) \otimes c = (2a + b) \otimes c = 2(2a + b) + c$$
$$a \otimes (b \otimes c) = a \otimes (2b + c) = 2a + 2b + c$$

If we substitute 2, 3, and 4 for a, b, and c, then we find

$$(a \otimes b) \otimes c = 2 \times (2 \times 2 + 3) + 4 = 14 + 4, \text{ or } 18$$
$$a \otimes (b \otimes c) = 2 \times 2 + 2 \times 3 + 4 = 14$$

We conclude that this operation is non-associative, since $(a \otimes b) \otimes c$
■ $\neq a \otimes (b \otimes c)$ for all choices of a, b, and c.

IDENTITY ELEMENTS

B.27   Figures 3.8 and 3.9 are illustrative of addition and multiplication in modulo 3. We will discuss modulo three in detail in Chapter 6. These tables are included to illustrate the identity elements under addition and multiplication on N and W. We will also point out the property which zero has under multiplication on W.

$\oplus$	0	1	2
0	0	1	2
1	1	2	0
2	2	0	1

Figure 3.8

Notice in Figure 3.8 that the row and column headed by 0 are identical to the row and column which represent the original entries. This indicates that the element 0 operating on any element a in the set will produce the element a. That is, $a \oplus 0 = a$ and $0 \oplus a = a$ for all a in the set on which the operation $\oplus$ is defined. Recall that any element with this property
■ is called an identity element on $\{0, 1, 2\}$ with respect to the operation $\oplus$.

■ B.27
▲ B.28   If you refer back to Figure 3.2, you will notice a similar situation. In that table, the element 0 acts as an identity element under addition. Since 0 is not an element of N but is included in W, we can state that the set W has an identity element, 0, under addition, while N has no identity element under addition. This is one of the major differences between the sets
▲ ■ N and W.

   In discussing identity elements for sets, we noted that the union operation was such that $\emptyset$ acted as an identity, since $A \cup \emptyset = A = \emptyset \cup A$.
■ B.37   We have stipulated that the identity element must commute. In an operation on a set which has the commutative property, we need not prove that the identity commutes, because the property assures us of this. On the other hand, if an identity does not commute, then we will call it a *right* or *left* identity depending on which side of an element it works as an identity. Many operations have a right or a left identity; some operations
■   may have both a right and a left identity that are not the same.

⊗	0	1	2
0	0	0	0
1	0	1	2
2	0	2	1

Figure 3.9

■ B.29   In Figure 3.9 you can see two patterns emerging. It is obvious that the row and column headed by 0 contain only the element 0. This is often stated as the *zero property of multiplication*. It is based on the fact that
■ $A \times \emptyset = \emptyset = \emptyset \times A$. Thus, $n(A) \times n(\emptyset) = 0$.
■ B.30   The second property you may observe is that the element 1 acts as an identity element. Notice that the shaded row and column contain the elements which were originally entered in the left column and top row. This implies that 1 operating on any element a will produce a. If you refer to Figure 3.9, you will find both the zero property and the identity property of 1. Since the number 1 is an element of N as well as W, the set N and the set W have an identity element under multiplication. Each of the properties is illustrated in the examples which follow the defini-
■   tions below.

■ B.27   *Identity Properties (I)*

▲ B.28   1. *I + W: There exists an element 0 in W such that for all a in W, a + 0*
▲   *= a = 0 + a.*
   2. *I × N: There exists an element 1 in N such that for all a in N, a × 1*
   = a = 1 × a.
   3. *I × W: There exists an element 1 in W such that for all a in W, a × 1*
■   *= a = 1 × a.*

B.29 *Zero Property (Z)*

■ *Z × W: For all a in W, a × 0 = 0 = 0 × a.*

Examples

$6 + 0 = 6$	and	$0 + 6 = 6$	$I + W$
$6 \times 1 = 6$	and	$1 \times 6 = 6$	$I \times W$
$6 \times 0 = 0$	and	$0 \times 6 = 0$	$Z \times W$

EXERCISES

1. Name the properties illustrated below for $S = \{\Box, \triangle, \bigcirc\}$ and the operation $*$.
   a. $\Box * \triangle = \triangle * \Box$
   b. $\Box * \bigcirc = \Box$ and $\bigcirc * \Box = \Box$
   c. $\Box * (\triangle * \bigcirc) = (\triangle * \bigcirc) * \Box$
   d. $\Box * (\triangle * \bigcirc) = \triangle * (\bigcirc * \Box)$
   e. $(\Box * \triangle) * (\bigcirc * \triangle) = (\bigcirc * \triangle) * (\Box * \triangle)$
   f. $(\bigcirc * \Box) * \triangle = (\Box * \bigcirc) * \triangle$

2. For each of the statements below, state the property or properties illustrated.
   a. $(a + b) + c = b + (a + c)$
   b. $(2 + 4) + (6 + 8) = (6 + 2) + (4 + 8)$
   c. $a \times (b \times c) = (a \times c) \times b$
   d. $(2 \times 4) \times 3 = (3 \times 2) \times 4$

3. Which of the properties discussed in this chapter does the operation $\circ$, defined below, have on the set $\{a, b\}$?

$\circ$	a	b
a	a	a
b	a	b

4. Which of the properties in this chapter hold for the operation shown in the table below?

$\times$	$\triangle$	$\bigcirc$	$\Box$
$\triangle$	$\bigcirc$	$\triangle$	$\triangle$
$\bigcirc$	$\triangle$	$\bigcirc$	$\Box$
$\Box$	$\Box$	$\Box$	$\bigcirc$

DISTRIBUTIVE PROPERTY

The properties discussed thus far were related to a single operation and were stated as properties of that operation on a set. We now want to develop a distributive property relating addition and multiplication. To do this we will first point out a corresponding property between cross-product and union, namely,

$$A \times (B \cup C) = (A \times B) \cup (A \times C)$$

That is, the cross-product distributes across union. For example,

$$A = \{1, 2\}, \qquad B = \{1, 3\},, \qquad C = \{2, 3\}$$
$$A \times (B \cup C) = \{1, 2\} \times \{1, 2, 3\} \quad \text{or}$$
$$\{(1, 1), (1, 2), (1, 3), (2, 1), (2, 2), (2, 3)\}$$
$$A \times B = \{1, 2\} \times \{1, 3\} = \{(1, 1), (1, 3), (2, 1), (2, 3)\}$$
$$A \times C = \{1, 2\} \times \{2, 3\} = \{(1, 2), (1, 3), (2, 2), (2, 3)\}$$

therefore,

$$(A \times B) \cup (A \times C) = \{(1, 1), (1, 2), (1, 3), (2, 1), (2, 2), (2, 3)\}$$

and

$$A \times (B \cup C) = (A \times B) \cup (A \times C)$$

■ **B.31**   This illustration serves only as an example to indicate how $\times$ distributes over $\cup$. We also point out that although we did not define $\times$ as an operation on two sets as we did with $\cup$ and $\cap$, this can be done without loss of meaning, because a new set is formed by the operation $\times$ when we produce $A \times B$.

Since we are stating that there is a distributive property of $\times$ over $\cup$, using the number properties of sets we can show that $n(A) \times [n(B) + n(C)] = [n(A) \times n(B)] + [n(A) \times n(C)]$, if $B \cap C = \varnothing$. If we let $n(A) = a$, $n(B) = b$, and $n(C) = c$, then it follows that

▲ **B.32** ▲ $Dx + N$: For all a, b, and c in N, $a \times (b + c) = a \times b + a \times c$.
● **B.33,** $Dx + W$: For all a, b, and c in W, $a \times (b + c) = a \times b + a \times c$.
● ■

The distributive property of multiplication over addition as shown
■ **B.34**   here is written as a left-hand distributive. We can easily write the right-hand distributive by using our commutative property for multiplication. A short proof follows.

Example
For all a, b, and c in W, $a \times (b + c) = (a \times b) + (a \times c)$. Show that $(b + c) \times a = (b \times a) + (c \times a)$.

PROOF

1. $a \times (b + c) = (a \times b) + (a \times c)$	1. $Dx + W$
2. $a \times (b + c) = (b + c) \times a$	2. $Co \times W$
3. $(b + c) \times a = (a \times b) + (a \times c)$	3. transitive property of equality
4. $(a \times b) + (a \times c) = (b \times a) + (c \times a)$	4. $Co \times W$
5. $(b + c) \times a = (b \times a) + (c \times a)$	5. transitive property of equality using steps 1–4

■

The distributive property enables us to compute and to do mental arithmetic more easily. Consider the following examples.

Examples

1. $8 \times 27 = 8 \times (20 + 7) = 8 \times 20 + 8 \times 7 = 160 + 56 = 216$
2. $15 \times 16 = 15 \times (10 + 6) = 15 \times 10 + 15 \times 6 = 150 + (10 + 5) \times 6$
   $= 150 + 60 + 30$, or 240

Before we conclude our discussion of the distributive property, consider a few physical examples of distributive property which may clarify this principle.

Example 1

■ B.35
A room with four walls of cinderblock is 40 blocks long, 25 blocks wide, and 18 blocks high. How many blocks are in the walls of the room if there are no windows?

SOLUTION We can find the number of blocks in one side and one front wall as follows:

Side wall $= 40 \times 18 = 720$, and since there are 2 walls, $2 \times 720 = 1440$
Front wall $= 25 \times 18 = 450$; again, there are 2 walls, so $2 \times 450 = 900$
All walls $= 1440 + 900 = 2340$

Since we can use the distributive property, the problem becomes

$$2 \times [18 \times (40 + 25)] \quad \text{or} \quad 2 \times (18 \times 65) = 2 \times 1170 = 2340$$

Notice that we had a choice of writing

$$2 \times (18 \times 40) + 2 \times (18 \times 25) \quad \text{or} \quad 2 \times [18 \times (40 + 25)]$$

which is $2 \times (18 \times 65)$.

Example 2

A movie house has three sections of seats. Each of the side sections is made up of 30 rows of 8 seats each. The center section contains 30 rows of 32 seats. How many seats are there in all?

SOLUTION We again have the choice of separately figuring the number of seats in each section as follows:

Side section $= 30 \times 8 = 240$
Side section $= 30 \times 8 = 240$
Middle section $= 30 \times 32 = 960$
Total number of seats $= 30 \times 8 + 30 \times 8 + 30 \times 32 = 1440$

Using the distributive property we combine this into a simpler computational problem as follows:

30 rows $\times$ (8 seats + 8 seats + 32 seats) per row
■ $30 \times (8 + 8 + 32) = 30 \times 48 = 1440$ seats

Another example where the distributive property is used even though it is not explicitly stated is in the multiplication algorithm. For example,

■ B.36

Property	NATURAL NUMBERS		WHOLE NUMBERS	
	Addition	Multiplication	Addition	Multiplication
Closure	$Cl + N$: for all $a, b \in N$, $a + b \in N$	$Cl \times N$: for all $a, b \in N$, $a \times b \in N$	$Cl + W$: for all $a, b \in W$, $a + b \in W$	$Cl \times W$: for all $a, b \in W$, $a \times b \in W$
Commutative	$Co + N$: for all $a, b \in N$, $a + b = b + a$	$Co \times N$: for all $a, b \in N$, $a \times b = b \times a$	$Co + W$: for all $a, b \in W$, $a + b = b + a$	$Co \times W$: for all $a, b \in W$, $a \times b = b \times a$
Associative	$A + N$: for all $a, b, c \in N$ $(a + b) + c = a + (b + c)$	$A \times N$: for all $a, b, c \in N$ $(a \times b) \times c = a \times (b \times c)$	$A + W$: for all $a, b, c \in W$ $(a + b) + c = a + (b + c)$	$A \times W$: for all $a, b, c \in W$, $(a \times b) \times c = a \times (b \times c)$
Identity	no identity	$I \times N$: for all $a \in N$ there exists a $1 \in N$ such that $a \times 1 = a$ $= 1 \times a$	$I + W$: for all $a \in W$ there exists a $0 \in W$ such that $a + 0 = a$ $= 0 + a$	$I \times W$: for all $a \in W$ there exists a $1 \in W$ such that $a \times 1 = a$ $= 1 \times a$
Zero	no zero	no zero	no zero	$Z \times W$: for all $a \in W$ there exists a $0 \in W$ such that $a \times 0 = 0 = 0 \times a$
Distributive	$Dx + N$: for all $a, b, c \in N$ $a \times (b + c) = (a \times b) + (a \times c)$		$Dx + W$: for all $a, b, c \in W$ $a \times (b + c) = (a \times b) + (a \times c)$	

Figure 3.10

$$153$$
$$\times\ 27$$

$$1071 \leftrightarrow\ \ 7 \times 153$$
$$3060 \leftrightarrow\ 20 \times 153$$
$$\overline{4131 \leftrightarrow (20 + 7) \times 153}$$

Figure 3.10 is a summary of the properties of addition and multiplication on the natural and whole numbers.

PROPERTIES OF SUBTRACTION AND DIVISION ON NATURAL AND WHOLE NUMBERS

CLOSURE FOR SUBTRACTION AND DIVISION

■ **B.7**   We stated earlier that a counter-example is sufficient for certain types of proof. We can show by counter-example that the operations of subtraction and division are not closed for *N* and *W*. This does not mean that the operations are faulty or incomplete. It simply means that the operations of subtraction and division when performed on *N* or *W* do
■ not necessarily produce an element of *N* or *W*.

We use the phrase "if and only if" in the statements below. Suppose *A* and *B* are two statements connected by "if and only if." That is, we have the sentence, "*A* if and only if *B*." This means "if *B*, then *A* and *A* only if *B*." Thus, if we have *B*, then we have *A* and we have *A* only when we have *B*. In other words, the statements *A* and *B* are equivalent.

Example 1

■ **B.7**   Notice that $2 - 6 = n$ if and only if there is a number in *N* or *W* such that $6 + n = 2$. Actually the number -4 is the solution, but $-4 \notin N$ or *W*. Therefore, the sets *N* and *W* are not closed under subtraction.

Example 2

Notice that $2 \div 6 = n$ if and only if there is a number in *N* and *W* such that $6 \times n = 2$. Since the solution to $6 \times n = 2$ is $1 \div 3$ and $1 \div 3$ is
■ not in *N* and *W*, we can state that *N* and *W* are not closed under division.

COMMUTATIVE AND ASSOCIATIVE PROPERTIES OF *N* AND *W* UNDER SUBTRACTION AND DIVISION

Again by counter-example, we can show that the commutative and associative properties do not hold for the operations of subtraction and division on *N* and *W*.

Example

◄ **B.9**   Let *a*, *b*, and *c* be 6, 4, and 2, respectively. Then $a - b = 6 - 4$ and $b - a = 4 - 6$. Since $6 - 4 \neq 4 - 6$, subtraction is not commutative

▲ **B.25** for N and W. Also $(a - b) - c = (6 - 4) - 2$, and $a - (b - c) = 6 - (4 - 2)$. Since $(6 - 4) - 2 \neq 6 - (4 - 2)$, subtraction is not associative for N and W.

▲ ■ It will be left to you to show that division is not commutative or associative for the sets N and W.

IDENTITIES FOR SUBTRACTION AND DIVISION ON N AND W

■ **B.37** The identity element for addition of whole numbers was shown to be 0. The way in which we have defined identity requires that an identity
▲ **B.38** element commute with every other element in the set. You will notice in the examples below that zero acts like a one-sided or a *right identity* and may be considered as a property of subtraction. However, it is not true using our definition that zero is an identity, since for all $a \in W, a - 0$
■ $\neq 0 - a$ unless $a = 0$.

Examples
1. $6 - 0 = 6$ but $0 - 6 \neq 6$.
2. $23 - 0 = 23$ but $0 - 23 \neq 23$.
3. $0 - 0 = 0$ and $0 - 0 = 0$. (This is a trivial case, since $a = a$ for
▲ any operation must equal $a = a$.)

■ **B.37** The identity for multiplication was 1. Again this element acts as an
▲ **B.39** identity on the right only for division on N and W. We can again state this as a property, but it is a special property of division, because for all
■ $a \in N$ and $W, a \div 1 \neq 1 \div a$ unless $a = 1$.

Examples
1. $6 \div 1 = 6$ but $1 \div 6 \neq 6$.
2. $23 \div 1 = 23$ but $1 \div 23 \neq 23$.
▲ 3. $1 \div 1 = 1$ and $1 \div 1 = 1$

We can summarize this section by stating the following special properties for subtraction and division on N and W.

$(RI - W)$ *Right identity for subtraction on W: There is an element 0 such that for all $a \in W$, $a - 0 = a$ but $0 - a \neq a$.*

$(RI \div W)$ *Right identity for division on N or W: There is an element 1*
and   *such that for all $a \in N$ or W, $a \div 1 = a$ but $1 \div a \neq a$.*
$(RI \div N)$

Notice that we have been very careful to point out that if a set has an identity element under some operation, then the identity has several properties:

1. It is a unique element (there is only one identity under the set and operation).

2. It operates on *all* elements in the set to produce the element on which it operates; that is,

$$[0 + a = a \quad \text{and} \quad 1 \times a = a \text{ for all } a \in N \text{ or } W]$$

3. The identity must commute with each element in the set; that is,

$$0 + a = a + 0 \quad \text{and} \quad 1 \times a = a \times 1$$

This must be true even if the operation on the set does not commute for all elements in the set.

As a result of the properties listed above, we do not consider the element 0 in W or 1 in N and W to be identities under subtraction and division, since they satisfy only two of the three properties. Which property is not satisfied under subtraction and division?

DISTRIBUTIVE PROPERTIES ON N AND W

We will state several distributive properties below. Each property will be stated as a right-hand distributive property. We will leave to you the task of deciding which of these properties can be written as left-hand properties as well. It may help if we note that multiplication is commutative whereas division is not.

B.40 $(D \times - N) \text{ and } (D \times - W)$ *Distributive property of multiplication over subtraction: For a, b, and c in N or W such that $a \geq b, (a - b) \times c = (a \times c) - (b \times c)$.*

Notice the restriction on a and b in the statement above. Can you see that we must require $a \geq b$; otherwise the difference $(a - b)$ is not an element of the set N or W? For example, $6 - 4 = 2 \in N$ and W, but $(4 - 6) \notin N$ or W.

Example
$(6 - 4) \times 3 = 2 \times 3$, or 6, and $(6 \times 3) - (4 \times 3) = 18 - 12$, or 6.

B.41 $(D \div + N) \text{ and } (D \div + W)$ *Distributive property of division over addition: For a, b, and c in N or W such that $a \div c$ and $b \div c$ exist, $(a + b) \div c = (a \div c) + (b \div c)$.*

Example
$(6 + 4) \div 2 = 10 \div 2$, or 5, and $(6 \div 2) + (4 \div 2) = 3 + 2$, or 5.

$(D \div - N) \text{ and } D \div - W)$ *Distributive property of division over subtraction: For a, b, and c in N and W such that $a \geq b$ and $a \div c$ and $b \div c$ exist, $(a - b) \div c = (a \div c) - (b \div c)$.*

Example
$(6 - 4) \div 2 = 2 \div 2$, or 1, and $(6 \div 2) - (4 \div 2) = 3 - 2$, or 1.

EXERCISES

1. a. If 0 is subtracted from a whole number, what is the result?
b. If a whole number is subtracted from 0, what is the result?
c. Is 0 an identity for subtraction?

2. a. For $a \in W, a \cdot 1 = ?$
b. For $a \in W, 1 \cdot a = ?$
c. Is 1 an identity element with respect to multiplication?

3. Find an example that shows that division is not commutative.

4. Find an example that shows that division is not associative.

5. Is there a left identity for division on N?

6. If we have an operation $*$ on W such that there is an element $z \in W$ and $z \times a = a$ for all $a \in W$, is the element z an identity? Why or why not?

7. If z is not an identity in Exercise 6 above, what condition would be necessary to allow z to be an identity?

8. What would we call an element $z \in W$ if
a. it is unique?
b. $z * a = a$ for all $a \in W$ under the operation $*$?

RELATIONS ON THE SET OF WHOLE AND NATURAL NUMBERS

EQUIVALENCE AND EQUALITY

The properties of an equivalence relation were discussed in Chapter 1. We again point out that equality is an equivalence relation, since it has the following properties:

1. Reflexive ($a\text{R}a$). For all $a, a = a$.
2. Symmetric (if $a\text{R}b$, then $b\text{R}a$). For all a and b, if $a = b$ then $b = a$.
3. Transitive (if $a\text{R}b$ and $b\text{R}c$, then $a\text{R}c$). For all a, b, and c, if $a = b$ and $b = c$, then $a = c$.

CANCELLATION LAWS

Equality has several other properties which will be used continually throughout the text. The first property is called the *cancellation property of addition* and states that we may add or subtract the same number to both sides of an equality.

■ **B.42**   *Cancellation Property of Addition: For all a, b, and c, whole numbers, a = b*
■   *if and only if a + c = b + c.*

The "if and only if" (often abbreviated "iff") statement above actually implies two ideas. The first idea is that we can add a number c to both members of an equality and we will still have an equality. The second idea is a little more difficult to see. It implies that if we have an equality $a + c$

$= b + c$, then we can eliminate c from both members and still have an equality, namely, $a = b$. That is, if $a + c = b + c$, then $a = b$.

Consider the examples below, which illustrate this property.

Examples
1. $5 = (3 + 2)$ if and only if $5 + 4 = (3 + 2) + 4$.
2. $7 = (2 + 5)$ iff $7 + 3 = (2 + 5) + 3$.

The cancellation property is normally used to solve or find solution sets to sentences or mathematical statements often called equations. Consider the examples below.

Example 1

B.44 Find a solution to $n + 5 = 17$.

SOLUTION $n + 5 = 12 + 5$ iff $n = 12$.

Example 2
Find a solution to $3 + n = 12$.

SOLUTION $3 + n = 3 + 9$ iff $n = 9$.

The next property of equality is much like the one we just discussed except that it deals with the operations of multiplication and division. We will state the property without further discussion and consider several examples as shown above.

B.43 *Cancellation Property of Multiplication: For all a, b, and $c \in W$, where $c \neq 0$, $a = b$ if and only if $a \times c = b \times c$.*

Consider the examples below.

Examples
B.44 1. If $n = 7$, then $5n = 5 \times 7$, or $5n = 35$.
2. If $5n = 35$, then $5n = 5 \times 7$ iff $n = 7$.
3. $3 = (1 + 2)$ iff $3 \times 2 = (1 + 2) \times 2$.

EXERCISES

1. Use the cancellation property to find the natural number that satisfies each of the following.

    a. $2 + n = 7$ b. $7n = 84$ c. $n + 7 = 7 + 2$

2. Supply reasons for each step in the proof of the cancellation property below.

    1. $a + c = b + c$ 1.
    2. $(a + c) - c = (b + c) - c$ 2.
    3. $a + (c - c) = b + (c - c)$ 3.
    4. $a + 0 = b + 0$ 4.
    5. $a = b$

SECTION C PROPERTIES OF ORDER ON *N* AND *W*

BEHAVIORAL OBJECTIVES

C.1 Given two numbers, be able to determine which is greater.

C.2 Be able to state the trichotomy property.

C.3 Be able to define $a < b$.

C.4 Be able to state which properties of an equivalence relation $<$ satisfies.

C.5 Be able to state the cancellation properties for $<$.

C.6 Given two natural numbers, be able to prove, using the definition, which is the greater.

C.7 Be able to find solution sets of sentences involving inequalities.

PROPERTIES OF ORDER
ON THE NATURAL AND WHOLE NUMBERS

We defined the set of natural and whole numbers essentially as a successor set. This implies that the set is ordered, since the successor of a given number is always one greater than the number which it succeeds. In the previous section, we dealt exclusively with the relation of equality between numbers.

If we choose two numbers, say a and b, from the set of natural or whole numbers, we know that these numbers are either equal, $a = b$, or not equal, $a \neq b$. It is this second concept that we wish to investigate in some detail. If we begin with 1 and take successors, that is, $1, 1 + 1, 2 + 1, 3 + 1, \ldots$, we will eventually reach first one of the numbers, say a, and then later the other number, b. In this case we say "a is less than b," or, in symbols, $a < b$. If it is the case that we reach b before we reach a in this listing of successors, we say that "a is greater than b" and write $a > b$. It is obvious that if $a < b$, then $b > a$. That is, if a is less than b, then b is greater than a.

The three possible relationships between two numbers is summarized in the statement of the trichotomy property.

■ **C.2** *Trichotomy Property: For a pair of numbers, a and b, one and only one of the following relationships holds:*

■ $a = b$ $a < b$ $a > b$

Let us consider the possible statements that we can write by using the trichotomy property.

$a = b$ is equivalent to $a \not< b$ and $b \not< a$.
$a < b$ is equivalent to $a \neq b$ and $b \not< a$, that is, $b \not\leq a$.
$a > b$ is equivalent to $a \neq b$ and $b \not< a$, that is, $b \not\leq a$.
$a \neq b$ is equivalent to $a < b$ or $b < a$.

$a \nless b$ is equivalent to $a = b$ or $b < a$, that is, $b \leqq a$.

$a \ngtr b$ is equivalent to $a = b$ or $a < b$, that is, $a \leqq b$.

$a \leqq b$ is equivalent to $a < b$ or $a = b$, that is, $a \ngtr b$.

$a \geqq b$ is equivalent to $b < a$ or $b = a$, that is, $a \nless b$.

Let us consider the single relation "less than." If a is less than b, we mean that b is at least the successor to a or the successor to the successor to ... the successor of b for a and $b \in N$ or W. This simply implies that there is a number which we can add to a such that we will produce the number b. We define "less than" in this way.

Definition

■ C.3
■   For a, b, and c in the set of natural or whole numbers, a is less than b, $a < b$, if there is a number c such that $a + c = b$ and $c \neq 0$.

From this definition it follows that if a is not less than b, then b is less than a or b equals a. We know that b is less than a if there is a natural or whole number c such that $b + c = a$ and $c \neq 0$.

■ C.1
▲ C.6   **Example 1**

Show in the pair (a, b) that a is less than b.

SOLUTION   1. Let $(a, b) = (2, 4)$. $2 < 4$, because $2 + 2 = 4$.

2. Let $(a, b) = (4, 7)$. $4 < 7$, because $4 + 3 = 7$.

Example 2

Show in the pair (a, b) that b is less than a.

SOLUTION   1. Let $(a, b) = (5, 3)$. $3 < 5$, because $3 + 2 = 5$.

▲ ■   2. Let $(a, b) = (9, 4)$. $4 < 9$, because $4 + 5 = 9$.

By the definition of "less than," we can now link inequalities to equalities. This link will allow us to prove that some of the properties of an equivalence relation hold for inequality. First, let us see which properties of an equivalence relation $<$ satisfies.

■ C.4   1. If "less than" reflexive? That is, is $a < a$? If a is less than a, then there is a natural number such that $a + c = a$. The only number which is a solution to this statement is the number 0; however, $0 \notin N$. Therefore, $<$ is not reflexive.

2. Is "less than" symmetric? That is, if $a < b$, is $b < a$? If $a < b$, then $b \nless a$ from the trichotomy property. Therefore, "less than" is non-symmetric.

3. Is "less than" transitive? That is, if $a < b$ and $b < c$, is $a < c$? Since $a < b$, there is a natural number d such that $a + d = b$. Since $b < c$, there is a natural number e such that $b + e = c$. Combining these statements, we see that

$$a + (d + e) = (a + d) + e$$
$$= b + e$$
$$= c$$

Since the natural numbers are closed under addition, $d + e$ is a natural number and $a + (d + e) = c$. By definition, $a < c$. Therefore, "less than" is a transitive relation.

Consider the cancellation laws given for addition and multiplication. We will state these properties for "less than" by replacing the equal sign by $<$.

■ **C.5** *Cancellation Property for Addition on "less than": For all a, b, and c in N or W, $a < b$ iff $a + c < b + c$.*

Cancellation Property of Multiplication on "less than": For all a, b, and c in N or W, $a < b$ iff $a \times c < b \times c$ and $c \neq 0$.

We will now prove the first of these properties and leave the second as an exercise.

PROOF OF CANCELLATION PROPERTY FOR "LESS THAN"

This proof requires that we show two things. The first is that if $a < b$, then $a + c < b + c$. The second is that if $a + c < b + c$, then $a < b$.

Part I

1. $a < b$	1. given
2. $a + d = b, d \in N$	2. definition of $<$
3. $(a + d) + c = b + c$	3. cancellation property of $=$
4. $(a + c) + d = b + c$	4. commutative and associative property of $+$
5. $a + c < b + c$	5. definition of $<$

Part II

1. $a + c < b + c$	1. given
2. $(a + c) + d = b + c$	2. definition of $<$
3. $(a + d) + c = b + c$	3. commutative and associative property of $+$
4. $a + d = b$	4. cancellation property of $+$ and $=$
5. $a < b$	5. definition of $<$

Consider how we can apply the cancellation laws for "less than" in the examples which follow.

Example 1

■ **C.7** Find the solution set for $n + 6 < 12$ and $n \in N$.

SOLUTION $n + 6 < 12$
$n + 6 - 6 < 12 - 6$ (using the cancellation property for $<$)
thus,
$n + 0 < 6$, or $n < 6$
In other words, the solution is $\{1, 2, 3, 4, 5\}$.

Example 2

Find the solution set for $x - 5 < 12$ and $x \in W$.

SOLUTION   Since $x - 5 < 12$, using the cancellation property we have
$$x - 5 + 5 < 12 + 5$$
or
$$x < 17 \quad \text{so that} \quad x = \{0, 1, 2, \ldots, 16\}$$

Example 3

Find the solution set for $n \div 3 < 15$ and $n \in N$.

SOLUTION $n \div 3 \times 3 < 15 \times 3$; then $n < 45$
or
$$\{n \in N \mid n = 1, 2, 3, \ldots, 44\}$$

Example 4

Find the solution set for $3n < 15$ and $n \in W$.

SOLUTION $3n \div 3 < 15 \div 3$; then $n < 5$
or
$$\{n \in W \mid n = 0, 1, 2, 3, 4\}$$

Since $a < b$ is equivalent to $b > a$, we can apply these cancellation laws to "greater than" as well as "less than." Before we indicate this by example, there are a few ideas in the examples above which may need clarification. Notice that in each problem above there is a variable and a constant which are joined by an operation. In each case, we have neutralized the effect of that constant by applying the inverse operation to both sides of the sentence. Essentially, this is the simplest approach to finding a solution set to an open sentence.

By "sentence" we mean a mathematical statement. An open sentence is a sentence which contains variables or unknown quantities. When a sentence contains such quantities, the truth or falsity of the sentence depends upon the number that replaces the variable. If we replace the variable with a known number, the statement may be true or false. For example, in the first example above, when n is 7, the sentence reads "$13 < 12$", which is false. If we replace n with 2, however, the statement reads $8 < 12$, which is true.

We are interested in all the values of n which render the sentence true. This set of values is called a *solution* or a *solution set*. We use the words "solution set" because there are often several solutions to a sentence. Consider the examples below in which the statement is written in the "greater than" form. Let $U = N$.

Example 1

C.7 $n - 5 > 12$; then $n - 5 + 5 > 12 + 5$, or $n > 17$.

Notice that this can' be done, because $12 < n - 5$ by definition and $12 + 5 < n + 5 + 5$, or $17 < n$. The solution sets produced, $\{n \mid n > 17$ and $n \in N\}$ and $\{n \mid 17 < n$ and $n \in N\}$, are equivalent, because $n > 17$ means $17 < n$.

Example 2

$3n > 12$; then $3n \div 3 > 12 \div 3$, or $n > 4$. Thus, the solution set is $\{n \,|\, n > 4 \text{ and } n \in N\}$.

If a sentence is more complex than this, we can still find the solution by applying these properties one at a time. Several examples are illustrated below. Let $U = W$.

Example 1

$2n - 6 > 16$

Since 6 is subtracted, we apply the inverse as follows:

$2n - 6 + 6 > 16 + 6$

which produces $2n > 22$. Since 2 is multiplied times n, we apply the inverse:

$2n \div 2 > 22 \div 2$

Then $n > 11$. The solution set is

$\{n \,|\, n > 11 \text{ and } n \text{ is a whole number}\}$,

or

$\{n \,|\, n = 12, 13, 14, \ldots\}$

Example 2

$n \div 5 + 3 > 8$

We undo the addition by applying subtraction:

$n \div 5 + 3 - 3 > 8 - 3$

Then $n \div 5 > 5$. Now we undo the division by multiplication:

$n \div 5 \times 5 > 5 \times 5$ or $n > 25$

or

$\{n \,|\, n = 26, 27, 28, \ldots\}$

EXERCISES

1. Name the properties shown below.
   a. For a, b, and $c \neq 0$, $a = b$ if and only if $ac = bc$.
   b. If $a < b$ and $b < c$, then $a < c$.
   c. For all a, b, and c, $a < b$ if and only if $a + c < b + c$.
   d. For all a and b, $a < b$, $a = b$, or $a \not> b$.

2. Write statements which are equivalent to the following.
   a. $a \not< b$ and $a \not> b$ b. $a \leq b$ c. $a \geq b$ d. $a \neq b$

3. Explain by using the definition of $<$ why the first element in each of the pairs is less than the second.
   a. $(2, 5)$ b. $(3, 7)$ c. $(4, 8)$

4. Find the solution sets for the following equalities.
   a. $n + 6 = 12$ b. $n - 6 = 14$ c. $2n + 6 = 12$
   d. $n \div 2 = 8$ e. $8n = 24$

5. Find the solution sets for the following order relations.
   a. $n + 5 < 6$ b. $2n + 5 < 15$ c. $3n - 6 < 18$
   d. $2n \geq 8$ e. $n \div 6 \geq 2$

6. Prove the cancellation property of multiplication on $<$.

CAPSULE II **DISCOVERY—ITS ROLE IN THE TEACHING OF ELEMENTARY SCHOOL MATHEMATICS**

The teaching of mathematics is a complex task which involves the teacher and student in a process which will ultimately lead to a knowledge of operations, applications, and structures, and, most importantly, to the knowledge and ability to attack and solve new or different problems. There are a variety of methods which can be used to accomplish these goals, and hopefully the student will not only accomplish the cognitive goals but will be able to enjoy the learning process. Children should be able to see how mathematics operates in their world.

One method of achieving the above goals is called the "discovery approach." Even though this label seems to imply the method to be used, there is a lack of agreement among experts regarding the definition and techniques that constitute the discovery approach in the mathematics classroom. We have all experienced the thrill of discovering the idea or concept which was new to us in some discipline. Children are constantly discovering new ideas in their home environment as well as in school. We believe that using the discovery approach along with other methods in the teaching of mathematics will add to the excitement of learning and increase the possibility of developing in the learner a willingness to explore problems independently.

Many writers distinguish between "pure discovery" and "guided discovery." However, since the students in any given class vary in many ways, such as ability, sex, interests, and socioeconomic factors, it is difficult for one to know whether a child has discovered on his own, discovered as a result of guidance, or, having prior knowledge of a concept, not discovered at all. We have not ruled out the idea of pure discovery in the classroom, but we do feel that most of the effective use of discovery as a method will fall into the realm of guided discovery in that the teacher will supply some guidance for the child or group *if it is required.* Let us consider the ingredients which are part of this discovery approach. All of these ingredients may not be present in every situation, but certainly most of them are necessary.

First, there must be a problem which is perceived by a child or group of children, and this problem should be clearly defined. The problem may arise from some need expressed by a child or group of children in the class, and the teacher must be willing to allow exploration and elicit response from the individual or group. This type of problem situation may be part of a project, planned trip, or discussion of subject matter which is not necessarily related to mathematics. For example, a discussion of conservation may lead to the idea that water is wasted daily in the home due to leaky faucets. Children may wish to measure this waste in their own homes or survey their neighborhood. Such a problem seems perfect for development of a long-range project which could be planned and executed by an interested group of children or by a single individual. Planning to solve this problem would draw on several mathematical topics, such as measurement of time, volume measure, averaging, graphing, and sampling. A sensitive teacher would be able to perceive the need for solution and allow the children to explore and plan an attack which might produce solutions to the problem. There are many situations such as the one mentioned which could involve long-range planning and skill development with little guidance from the teacher.

Not all problems develop as above, nor should all problems be of such a large scale. If we depended only upon situations such as the one mentioned, we would find that many important concepts within the structure of mathematics would not be discussed in the grades. If discovery is to take place, it is vital for the teacher to be sensitive to problem situations and to see himself in a guiding role rather than in an answer-giving role. Much of the discovery-type learning which can take place in the classroom must and can be motivated by the teacher. The learning which is initiated by the teacher must again bring to the children a problem situation which will allow some exploration and planning before solution is expected.

During the exploration and planning stages, the teacher should be relatively inactive, and guidance should only be offered if the children are absolutely blocked. In these stages the teacher must use intuition, knowledge of subject, and knowledge of children in a skillful manner. If children perceive the teacher primarily as the dispenser of knowledge, they may not continue to explore. The problem presented by the teacher, as well as the way in which the problem is posed, must be exciting and should draw on the natural curiosity of children. There should be a freedom in the classroom which allows children to work without fear of moral judgments. The teacher should be careful about giving facial or vocal clues which allow children to feel that their approach or solution is unworthy. It is important in this stage to have children checking solutions and making suggestions rather than to place this responsibility solely in the hands of the teacher. Asking good questions when needed, as well as suggesting alternative approaches to the problem at hand, is a key to structuring without giving answers.

It is important for the teacher to be aware of some of the prerequisite learning needed to attack problems, because this may determine whether the problem can be dealt with by the group. We do not imply that the teacher needs to know the answer but rather that he should be aware of possible blocks in the road to solution. If too many problems arise which are discouraging to the learners, the chance of keeping the children excited and interested in solving problems will diminish rapidly. After the exploration and development of plans of attack or hypotheses, then the group or groups may come together to evaluate their progress, or individuals may submit their solutions for evaluation or discussion by the class and/or the teacher. A means of dealing with multiple solutions could simply involve the class in checking solutions to see if they fit the given data. Thus, the steps described involve

1. a problem posed by children or teacher
2. a problem clearly defined
3. an exploratory period (with the teacher role inactive)
4. a planning stage in which possible solutions or hypotheses are posed by groups or individuals (with the teacher role inactive)
5. a set of possible solutions offered to the class for discussion.
6. a revision stage, if needed
7. an appraisal or evaluation of solutions by the class

One very important ingredient in this process is the teacher role. The teacher should play a guidance role and try to be as inactive as possible. Questions are often answered with questions or suggested strategies rather than with direct answers. Care must be taken not to expect early verbalization of principles, nor should one expect perfect solutions immediately. The balanced role which the teacher must play is of such importance that the wrong word, smile, or frown could cut off the process at a critical point. Therefore, the teacher must be open, supportive, interested, and excited, but careful not to be too judgmental or answer oriented. In many cases the process should be postponed at its height to keep the enthusiasm going. It is often better for children to ask for more than to sit back with the expression of "Not again" on their faces.

In the literature we read that children seem to lose their enthusiasm for learning somewhere between the intermediate and upper elementary grades. Children, particularly between the ages of 2 and 9, seem to have tremendous curiosity and excitement with regard to learning. Children can continue to be excited and to learn complex ideas or concepts at an intuitive, non-verbal, and informal level. The danger of early formalization is usually due to adult interference with the learning process. We suggest that the employment of discovery sequences as a part of the mathematics program will ultimately produce learners who are

1. excited about learning mathematics
2. willing to participate openly and with little anxiety
3. able to deal with complex concepts intuitively
4. able to retain the learning
5. able to attack new or different problems
6. able to transfer the learning more easily

To derive the benefits listed above, teachers must be open, flexible, and comfortable in their roles; students must be given the necessary security and respect due all thinking human beings. There must be support and cooperation among and from all groups, namely, faculty, administration, parents, and students. If all groups work toward this common goal there should be no loss of interest and enthusiasm on the part of students, and learning will be considered important.

Consider some of the criticisms of using the discovery approach as it has been described. It is argued that using discovery as a vehicle for learning is very time consuming. We agree that discovery takes time—often a great deal of time; however, in light of the benefits which may be derived, we suggest that the classroom teacher balance the time spent against the possibility of developing creativity and producing better learning.

Another major criticism is that all mathematics cannot be approached by using discovery sequences. This is probably true, although most important concepts which are considered part of the elementary school curriculum can be organized to allow for some discovery on the part of the learner. In fact, those who advocate extensive use of the discovery approach indicate that many of the ideas which would be deleted from the curriculum are not as interesting, important, and concept oriented as are the discovery sequences with which they might be replaced.

A third criticism is that many children feel threatened if they are not told exactly what to do and how to do it. This is certainly true, but we consider it a criticism if not an indictment of schools, parents, and society in general rather than a function of children. The sensitive teacher can balance the learning so that children who are overanxious regarding the learning environment can receive the support needed.

A final criticism offered is that school should not be a place to have fun, since learning is work and one must be seriously engaged in developing good work habits. The approach discussed implies that more work is actually done by the children, and further that work can be a positive and joyous experience. If it were possible for work to be exciting and enjoyable, life might be a more wonderful experience, because one often spends a great part of life engaged in work.

In the last few paragraphs we have suggested benefits and criticisms of the discovery approach. We have taken the point of view that the use of this approach is important because it will lead to better and more exciting learning. Research has not proven this to be the case, nor has research indicated that there is any loss in learning due to the approach. Our point of view is not that discovery is the answer of itself but that the discovery approach is another tool for the teacher. Each teacher must find a balance which is good for him, his class, and the individual students. We feel strongly that the teacher's role should diverge from that of *answer giver* and converge toward that of *agent of change*. This new role requires someone who is able to listen, learn, plan, and help rather than tell, know, dictate, and judge. Such a role requires a rather delicate balance of interest, enthusiasm, creativity, knowledge, and a feeling of security on the part of the teacher.

We have attempted to define an approach to teaching and indicate a role that this approach might take in the mathematics classroom. At the end of Chapter 5 we will suggest some discovery sequences to give the reader a more concrete idea of how this approach might be used. A list follows of books and articles which might be of help and interest to the reader.

SUGGESTED READINGS

Biggs, Edith E., and MacLean, James R. *Freedom to Learn: An Active Learning Approach to Mathematics*. Reading, Mass.: Addison-Wesley, 1969.

Bolding, J. "A Look at Discovery," *Mathematics Teacher*. Washington, D.C.: NCTM, February, 1964.

Corle, C. *Teaching Mathematics in the Elementary School*. New York: Ronald, 1964.

Davis, R. B. *Discovery in Mathematics: A Text for Teachers*. Reading, Mass.: Addison-Wesley, 1964.

Deans, E. *Elementary School Mathematics—New Directions*. Washington, D.C.: U.S. Dept. of Health, Education, and Welfare, 1963.

Hendrix, G. "Learning by Discovery," *Mathematics Teacher*. Washington, D.C.: NCTM, May, 1961.

Herman, G. "Learning by Discovery: A Critical Review of Studies," *The Journal of Experimental Education*, Fall, 1969.

Howard, C., and Dumas, E. *Teaching Contemporary Mathematics in the Elementary School*. New York: Harper & Row, 1966.

Johnson, H. C. "What Do We Mean by Discovery?" *Mathematics Teacher*. Washington, D.C.: NCTM, December, 1964.

Jones, P. S. "Discovery Teaching—From Socrates to Moderninity," *Mathematics Teacher*. Washington, D.C.: NCTM, October, 1970.

Kersh, B. Y. "Learning by Discovery: Instructional Strategies," *Arithmetic Teacher*. Washington, D.C.: NCTM, October, 1965.

Riedesel, C. A. *Guiding Discovery in Elementary School Mathematics*. New York: Appleton-Century-Crofts, 1967.

CHAPTER 4 **NUMBER THEORY**

SECTION A **FACTORIZATION**

BEHAVIORAL OBJECTIVES

A.1 Define a factor.

A.2 Define a prime number.

A.3 Define a composite number.

A.4 Given a number, be able to find its factorizations.

A.5 Define the prime factorization of a natural number.

A.6 Be able to use a "factor tree" in finding the prime factorization of a given number.

A.7 Be able to state and explain by example the fundamental theorem of arithmetic.

A.8 Be able to define an exponent.

A.9 Be able to use exponential notation in writing the factorization of a given number.

FACTORS AND FACTORIZATION

The natural numbers can be produced by performing operations on the number one. For example, two is the sum of one and one. In a similar fashion we could build the natural numbers by continued additions of one.

Some natural numbers can be produced using multiplication, for example, $2 \times 3 = 6$. The numbers two and three produce a product six. Two and three then are described as factors of six.

To define *factor*, we must be able to consider the attributes of the previous example and state them as a generalization which will be true under all circumstances. If we start with the numbers two and three and perform the operation of multiplication, we produce a third number, six. We say that two and three are factors of six.

We can state this clearly in a more general way.

Definition

■ **A.1**   *Given two natural numbers a and b such that their product is c, the num-*
■   *bers a and b are said to be factors of c. That is, if ab = c, then a is a factor of c and b is a factor of c.*

When we write a number as a product of factors, we say that we have *factored* the number, and the product of factors is the *factorization* of the number.

Consider some examples of this concept.

Example 1

A.4   What is the factorization of 6?

SOLUTION $6 = 2 \times 3$, or $6 = 6 \times 1$. Notice that both 2×3 and 6×1 are factorizations of the number 6.

Example 2

Find the factorization of 20.

SOLUTION $20 = 4 \times 5$, or $20 = 2 \times 10$, or $20 = 20 \times 1$, or $20 = 2 \times 2 \times 5$. In each case, 4×5, 2×10, 20×1, and $2 \times 2 \times 5$ are called factorizations of 20.

Through the ages man has been mystified by numbers and has attached great meaning to specific types of numbers. In the bible, reference is made to the perfection of the world in terms of the number six which is, as you will see in a later section, a perfect number. You might consider what is special about six. Try looking at the factors of six!

Mention is made of male (odd) numbers and female (even) numbers in some of the writings of numerologists. Often decisions were made by the ancients based on some numerical value corresponding to a name or event.

The Greeks were probably more interested in the lore, as well as the mathematics, of numbers than any other group of people in Western culture. They explored the properties of the natural numbers and considered questions which mathematicians find quite helpful and ingenious today. The subject of prime numbers was given considerable consideration by Greek mathematicians, particularly by Pythagoreas, who explored much of the mystery and beauty of mathematics.

EXERCISES

1. Find factors for the following numbers.
    a. 2   b. 6   c. 7   d. 27

2. Find *all* the factors of the following numbers.
    a. 6   b. 8   c. 16   d. 2

3. a. Add all the factors of 6 other than 6 itself.
    b. Add all the factors of 28 other than 28 itself.

PRIME AND COMPOSITE NUMBERS

The numbers 2, 3, 5, and 7 are examples of *prime numbers*. What do these numbers have as common attributes? They are all greater than one. They are not all odd numbers although most of them are. They are not consecutive numbers although they might follow a pattern. Actually, the idea of a prime number is a very simple but interesting concept. These numbers have factorizations which have common characteristics. Let us look at the factorizations of 2, 3, 5, and 7.

$$2 = 2 \times 1$$
$$3 = 3 \times 1$$
$$5 = 5 \times 1$$
$$7 = 7 \times 1$$

Any natural number has factors of itself and one. What makes these numbers so special? Consider the number two. Its *only* factors are two and one; there are no other factorizations of two. In fact, there are no other factorizations of three, five, or seven. This allows us to characterize the set of numbers two, three, five, and seven by the fact that their *only* factors are one and the number itself. The word "*only*" is the key to this definition. It is sufficiently restrictive so as to be used as a definition for prime numbers.

We must define prime numbers in a manner which will produce the essential attributes mentioned above. Let's list the attributes first.

1. A prime is a natural number greater than one.
2. A prime has factors of itself and one.
3. A prime has no other factors.

By combining these attributes we state the following definition.

Definition

■ A.2   *A prime number is a natural number greater than one whose only factors*
■   *are itself and one.*

Notice in the previous examples and definitions that the number one was excluded from the set of prime numbers. If we were to extract the number one and the set of prime numbers from the set of natural numbers, we would be left with a set of numbers which we characterize as *composite*. The simplest way to define this set of numbers is to consider the set of non-primes greater than one as the set of composite numbers. Thus, our definition for a composite number follows.

Definition

■ A.3   *A composite number is a natural number greater than one which is not a*
■   *prime number.*

From this definition we see that a composite number has the following attributes.

1. A composite number is a natural number greater than one.
2. A composite number has factors other than one and itself.

Using prime and composite numbers separates the set of natural numbers into three subsets: the set containing the number one, the set of primes, and the set of composites. Symbolically, we state that

$$N = \{1\} \cup P \cup C$$

where P is the set of primes and C is the set of composites.

We should note that the sets $\{1\}$, P, and C are disjoint. That is,

$$\{1\} \cap P = \phi$$
$$\{1\} \cap C = \phi$$
$$P \cap C = \phi$$

When a set is separated into disjoint subsets as above, we say that the set has been *partitioned*.

We have shown that we can multiply several factors together and produce a number. We have also shown that if the numbers which we multiply are natural numbers greater than one, then the number produced is a composite number.

EXERCISES

1. Identify the following numbers as composite or prime.
    a. 2   b. 7   c. 14
    d. 28   e. 29   f. 31

2. a. Is one a prime number?
    b. Find the first three smallest prime numbers.
    c. Find the eighth prime.

3. Is 27839254 prime? Why?

4. A number is called "abundant" if the sum of its factors, other than itself, is greater than itself. Find some abundant numbers. Can you find an odd one?

PRIME FACTORIZATION

We often want to find the factors of a composite number and are sometimes interested in those factors that are themselves prime.

Let us consider a few examples. The factors of six are two and three, and one and six. The only factors of six that are prime numbers are three and two. We can find no other prime factors of six.

What happens when we consider the factors of a number such as twelve? We can factor twelve in several ways, because

A.4
$$12 = 2 \times 6$$
$$12 = 3 \times 4$$
$$12 = 12 \times 1$$
■ $$12 = 2 \times 2 \times 3$$

If we restrict ourselves to only the factors of twelve that are prime numbers, we see that only one factorization of twelve consists of primes only, namely, $2 \times 2 \times 3$. We can find no other factors of twelve other than two and three that are themselves prime. In each of the other factorizations of twelve above, you will notice that one of the factors is a composite number. We know that a composite number can be written as the product of factors other than one. Let us consider what happens to each of the expressions listed above if we rewrite the factors that are composite numbers as the product of two or more factors. See Figure 4.1.

■ A.6

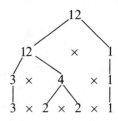

Figure 4.1

Note that we usually do not include 1 in the set of factors, because it is always a factor of a natural number.

In each of the examples above, we found the factorization to contain a set of factors which included two as a factor twice and three as a factor once. The only difference was in the order of the factors. If we check each of the factorizations of twelve, we find that the factors are prime numbers. We call such a factorization a *prime factorization* of the number twelve.

It is easy to make an error in the process of finding the prime factorization of a natural number. One often looks for a set of prime factors of the given number and not a factorization of the number. Notice that with the example above this approach would result in the factors two and three only; however, $12 \neq 2 \times 3$. It should be made clear that the product of the numbers included in the prime factorization must be the number being factored.

Consider the following examples of factorization:

Example 1

■ A.4

$36 = 2 \times 18$
$36 = 3 \times 12$
$36 = 4 \times 9$
$36 = 6 \times 6$
$36 = 36 \times 1$

Example 2

$11 = 11 \times 1$

Notice that when the natural number, n, is prime, the only factorization consists of a single pair of factors, 1 and n.

Example 3

■

$55 = 5 \times 11$

Notice when the number is composite, it is not necessary to list one as a factor.

The attributes of prime factorization are the same as those of factorization except that we have one additional attribute. That is, *all factors must be prime numbers.*

A.5   Definition

The prime factorization of a given natural number is a factorization of the number such that the factors are all prime numbers and that their product is the given number.

FUNDAMENTAL THEOREM OF ARITHMETIC

When one is teaching children to find the prime factorization of a given number, the use of a factor tree is often helpful. Consider Figure 4.2.

A.6

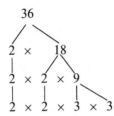

Thus, $36 = 2 \times 2 \times 3 \times 3$

(a)

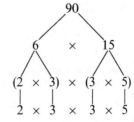

Thus, $90 = 2 \times 3 \times 3 \times 5$

(b)

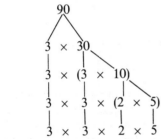

Thus, $90 = 3 \times 3 \times 2 \times 5$

(c)

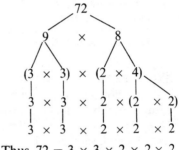

Thus, $72 = 3 \times 3 \times 2 \times 2 \times 2$

(d)

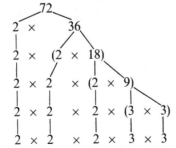

Thus, $72 = 2 \times 2 \times 2 \times 2 \times 3 \times 3$

(e)

Figure 4.2

■ **A.7**   Notice in the figure that the final set of prime factors of a given number is the same even though we started with a different pair of factors. This concept is named the *fundamental theorem of arithmetic*, which states:

■   *Except for the order of the factors, every natural number can be written as the product of prime factors in only* one *way.*

EXERCISES

1. Write each of the following numbers as a product of its prime factors.
  a. 232   b. 94   c. 180   d. 246

2. Let *A* be the set of natural numbers that leave a remainder of one when divided by three. That is, start with one and keep adding three.

$$A = \{1, 4, 7, 10, 13, \ldots, 52, 55, 58, \ldots\}$$

  a. Is this set closed under multiplication? Define prime numbers *in this system* as numbers greater than one which have no divisors in the system other than one and themselves.
  b. Show that ten is prime.
  c. Show that sixteen is not prime.
  d. Is twenty-five prime?
  e. Show that fifty-two is not prime.
  f. Find two different prime factorizations of one hundred.
  g. Does the fundamental law of arithmetic hold in this system?

EXPONENTIAL NOTATION

As we noticed in the examples above, the prime factorization of many natural numbers include primes repeated several times. For example, $72 = 2 \times 2 \times 2 \times 3 \times 3$. To make statements of this type, mathematicians have developed a language which is shorter than that found in the
■ **A.8**   example. This language is referred to as *exponential notation*. A number written to the right and above the given number is called an exponent and
  ■   indicates how many times the given number is used as a factor. In this way
■ **A.9**   72 can be written as $2^3 \times 3^2$, since 2 is used as a factor 3 times and 3 is used as a factor 2 times.

We can thus define (b^n) as *b* used as a factor *n* times, or

$$\underbrace{b^n = b \times b \times b \times \cdots \times b}_{n \text{ times}}$$

Examples
1. $3^2 = 3 \times 3$
2. $17^3 = 17 \times 17 \times 17$
3. $8^5 = 8 \times 8 \times 8 \times 8 \times 8$
4. $54 = 3 \times 3 \times 3 \times 2$, or $3^3 \times 2$
5. $72 = 3 \times 3 \times 2 \times 2 \times 2$, or $3^2 \times 2^3$
■ 6. $50 = 5 \times 5 \times 2$, or $5^2 \times 2$

EXERCISES

1. Write the following in exponential notation.
a. $500 = 5 \times 5 \times 5 \times 5 \times 2 \times 2$
b. $350 = 5 \times 5 \times 7 \times 2$
c. $144 = 2 \times 2 \times 2 \times 2 \times 3 \times 3$
d. $900 = 2 \times 2 \times 5 \times 5 \times 3 \times 3$

2. Write the following without the use of exponents.
a. $324 = 3^4 \times 2^2$
b. $1500 = 5^3 \times 2^2 \times 3$
c. $49,000 = 7^2 \times 5^3 \times 2^3$
d. $484 = 11^2 \times 2^2$

3. Define and give an example of the following terms.
a. factor
b. prime number
c. composite number
d. exponent

4. Find the prime factorization of the following using exponential notation where possible.
a. 132
b. 288
c. 294
d. 286

SECTION B DIVISIBILITY

BEHAVIORAL OBJECTIVES

B.1 Be able to describe the Sieve of Eratosthenes.

B.2 Be able to use the Sieve of Eratosthenes to find the primes in a given interval.

B.3 Be able to state the divisibility tests for 2 and powers of 2.

B.4 Be able to prove the divisibility tests for 2 and powers of 2.

B.5 Be able to state the divisibility tests for 3 and powers of 3.

B.6 Be able to prove the divisibility tests for 3 and powers of 3.

B.7 Be able to state the divisibility tests for 5 and powers of 5.

B.8 Be able to show that there are an infinite number of primes.

SIEVE OF ERATOSTHENES

Suppose we wanted to determine the prime numbers in a given interval. For example, between 90 and 100 there is one prime, namely 97. To find this prime we might have a lot of trouble unless we had some systematic method. In this section we will look at some of the methods which have been devised for this purpose and for the purpose of deciding whether a given number is prime.

One of the oldest methods for determining primes is the method of sieving. This method is often referred to as the Sieve of Eratosthenes and was developed by the man whose name the method bears. The method is **B.1** quite simple but tedious to apply. We consider first an interval of consecutive natural numbers. We then cross out all those numbers which are divisible by 2 except for 2 itself. Following this we begin to cross out multiples of 3, except for 3 itself. We continue this process using the next uncrossed number and thus eliminate the composites so that only primes are left. Figure 4.3 shows the primes between 1 and 100. Note that by definition 1 is not prime.

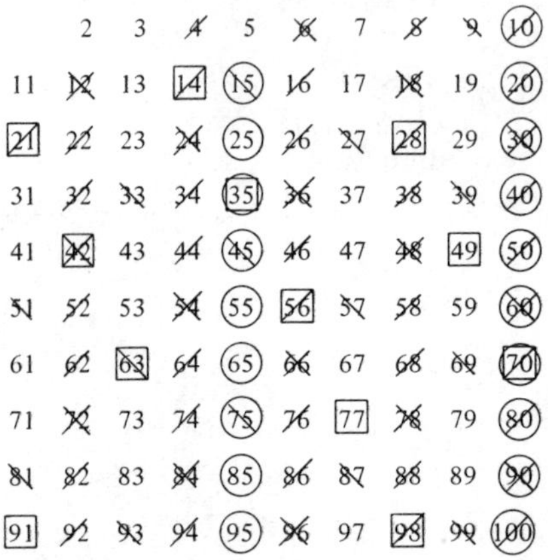

```
    2   3   4   5   6   7   8   9  (10)
11  12  13 [14](15) 16  17  18  19 (20)
[21] 22  23  24 (25) 26  27 [28] 29 (30)
31  32  33  34 (35) 36  37  38  39 (40)
41 [42] 43  44 (45) 46  47  48 [49](50)
51  52  53  54 (55)[56] 57  58  59 (60)
61  62 [63] 64 (65) 66  67  68  69 (70)
71  72  73  74 (75) 76 [77] 78  79 (80)
81  82  83  84 (85) 86  87  88  89 (90)
[91] 92  93  94 (95) 96  97 [98] 99 (100)
```

Figure 4.3 Sieve of Eratosthenes

Those numbers divisible by 2 are crossed out with a stroke to the right. Those numbers divisible by 3 are crossed out with a stroke to the left. Those numbers divisible by 5 are circled. Those numbers divisible by seven are contained within a square.

In Figure 4.3 all composites up to 100 were crossed out, and yet we did not need to use primes greater than 7. This leads us to the question, What is the largest prime that we need use in the sieve? To help answer the question let us look at some examples. To find the primes between 1 and 48 using the sieve technique we use only the primes 2, 3, and 5. Why don't we need to use 7? Using 7 would strike out the numbers 14, 21, 28, 35, 42, 49, etc. That is, we would strike out $7 \times 2, 7 \times 3, 7 \times 4, 7 \times 5, 7 \times 6, 7 \times 7, \ldots$ But 7×2 would have already been eliminated when we used 2, 7×3 would have been eliminated with 3, and 7×5 would have been eliminated with 5. Thus, those numbers that have factors of 7 and a prime less than 7 are eliminated using the prime, and 7 does not eliminate those numbers. For example, 7×4 is striken out before we use 7 because 4 is a composite, $4 = 2 \times 2$, having 2 as a factor when we use 2 in the sieve. With 7×6, 6 is a composite which has a prime less than 7 as a factor and would be eliminated when that prime would be used in the sieve. Thus, the first new number to be eliminated using 7 will have as factors 7 and the first prime greater than the last prime already used, namely, 7. So we see that the smallest new number that 7 will eliminate using the sieve technique is 7×7. Hence it is not necessary to use 7 when finding the primes up through 48.

Consider now what is necessary to find the primes between 1 and 28. Using 2, the first number we cross out is the product of 2 and the first prime, namely, 2×2, or 4. We then cross out all multiples of 2. With 3, the first new number to be crossed out is 3×3, or 9. With 5, the first new number to be crossed out is 5×5 or 25. From the previous example we know that the first new number that 7 eliminates is 7×7, or 49. Thus, to find all the primes less than 49 using the Sieve of Eratosthenes we need only use the primes 2, 3, and 5. This fact leads us to the following theorem.

Theorem
At least one of the prime factors of a composite number n is less than or equal to $\sqrt{n}$.

EXERCISES

1. Find the primes between 175 and 250 using the sieve and the theorem above.

2. Determine (1) which of the following numbers are prime; (2) which primes must be used to make this test.

   a. 51   b. 211   c. 331   d. 1001   e. 213

DIVISIBILITY BY POWERS OF TWO

We will now use the concept of factorization to introduce a new term. If a and b are factors of c, that is, $c = a \times b$, we say that a and b are *divisors of c* or that both a and b divide c. For example, since $10 = 2 \times 5$, 2 divides 10, 5 divides 10, 2 is a divisor of 10, and 5 is a divisor of 10. Thus, if 2 divides 6, then there is some number k such that $6 = 2 \times k$.

It is fairly common knowledge that 2 divides a number if the last digit of the number is divisible by 2. The reasoning behind this is not such common knowledge. Let us consider an example which should clarify the reasoning.

Example
$$3742 = 3 \times 10^3 + 7 \times 10^2 + 4 \times 10^1 + 2 \times 10^0,$$

We know that 2 divides 2×10^0. We can see that since 2 divides 10, it must divide 4×10 and since 2 divides 10 and 10^2, it divides 10^3 and must divide 3×10^3. Furthermore, since 2 divides each of the addends whose sum is 3742, it must be the case that 2 is a factor of each addend. Thus, by the distributive property, 2 is a factor of 3742. Therefore, 2 divides 3742.

B.4   We can generalize this statement by using it in the following way. Let $n = a_k 10^k + a_{k-1} 10^{k-1} + \cdots + a_1 10^1 + a_0$. Since 2 divides 10, 10^2, $10^3, \ldots, 10^k$, 2 divides each term of the form $a_i 10^i$ but not necessarily a_0. If a_0 is 0, 2, 4, 6, or 8, then 2 divides it. Therefore, 2 divides any natural number whose final digit is divisible by 2.

Let us next consider the set of natural numbers which are divisible by

4, {4, 8, 12, 16, 20, 24, 28, 32, ..., 100, 104, 108, ..., 200, 204, 208, ...}. You will notice that in each case the number formed by the last two digits of the number is divisible by 4. Observe that 4 does not divide 10 but that 4 divides 10^2, 10^3, 10^4, ... We now use the same method of reasoning as we did for divisibility by 2. Let $a_k 10^k + a_{k-1} 10^{k-1} + \cdots + a_2 10^2 + a_1 10^1 + a_0 10^0$ represent any number. Four divides any power of 10 that is greater than or equal to 2. Therefore, 4 divides the number $a_k 10^k + a_{k-1} 10^{k-1} + \cdots + a_2 10^2$ but does not necessarily divide $(a_1 10^1 + a_0 10^0)$. If 4 divides $(a_1 10^1 + a_0 10^0)$ (the number formed by the last two digits of the number), then it divides the number. It is left to you to show that a natural number is divisible by 8 if the number formed by the last 3 digits of the given number is divisible by 8.

To give you another way to understand the principles underlying the proofs above, let us consider the factors of 10 and the factors of positive powers of 10.

$$10 = 2 \times 5$$
$$10^2 = 10 \times 10 = 2 \cdot 5 \cdot 2 \cdot 5 = 2 \cdot 2 \cdot 5 \cdot 5 = 2^2 \times 5^2$$
$$10^3 = (2 \times 5)^3 = 2 \cdot 5 \cdot 2 \cdot 5 \cdot 2 \cdot 5 = 2 \cdot 2 \cdot 2 \cdot 5 \cdot 5 \cdot 5 = 2^3 \times 5^3$$

This should indicate that 10^k contains k powers of 2 as a factor and is therefore divisible by 2^k.

We have actually discussed three separate generalizations. Since these are quite similar in nature, they can be stated as a single theorem.

Theorem

B.3   *A natural number is divisible by*:
1.   *2 if the last digit of the number is divisible by 2.*
2.   *4 if the number formed by the last two digits of the number are divisible by 4.*
3.   *8 if the number formed by the last three digits of the number are divisible by 8.*

EXERCISES

1. Use the sieve method to find the primes between 200 and 300.

2. Write the set of primes which were needed to do Exercise 1.

3. Decide whether or not the following numbers are divisible by 2.
   a. 249   b. 77,784   c. 18,954   d. 22,425   e. 33,764

4. Repeat Exercise 3 by using 4 as the divisor.

5. Use the definition of "divides" to determine what numbers divide 0.

DIVISIBILITY BY POWERS OF THREE

We next consider the problem of whether or not a number is divisible by 3. Let us first consider the set of natural numbers divisible by 3,

$$\{3, 6, 9, 12, 15, 18, 21, 24, 27, 30, \ldots, 90, 93, 96, 99, 102, 105, \ldots, 168, 171, \ldots\}$$

Look at the sums produced when we add the digits of each numeral. This set of sums is

$$\{3, 6, 9, 3, 6, 9, 3, 6, 9, 3, \ldots, 9, 12, 15, 18, 3, 6, \ldots, 15, 9, \ldots\}$$

Notice that the sums produced are all multiples of 3. Can you find a number such that the sum of its digits is divisible by 3 and yet the number itself is not divisible by 3?

Now that we have looked at the patterns which relate to numbers divisible by 3, let us consider an example as we did with divisibility by 2. Suppose we have a number 4371.

$$4371 = 4 \times 10^3 + 3 \times 10^2 + 7 \times 10 + 1 \times 10^0$$

Since $10^3 = 1000 = 999 + 1$ and $10^2 = 100 = 99 + 1$ and $10 = 9 + 1$,

$$4371 = 4 \times (999 + 1) + 3 \times (99 + 1) + 7 \times (9 + 1) + 1$$

Now use the distributive property with each term.

$$4371 = (4 \times 999 + 4 \times 1) + (3 \times 99 + 3 \times 1) + (7 \times 9 + 7 \times 1) + 1$$

Reassociate the terms.

$$4371 = (4 \times 999 + 3 \times 99 + 7 \times 9) + (4 + 3 + 7 + 1)$$

We now have the first three terms of our expression divisible by 3 and the other four terms are the sum of the digits of our original number. That is,

$$4371 = \underbrace{(4 \times 999 + 3 \times 99 + 7 \times 9)}_{\text{divisible by 3}} + \underbrace{(4 + 3 + 7 + 1)}_{\text{sum of digits of 4371}}$$

Since $(4 + 3 + 7 + 1)$ is divisible by 3, 4371 is divisible by 3.

The thing to note about this example is that the divisibility of 4371 by 3 was guaranteed because the sum of its digits was divisible by 3.

We now state and prove a theorem for divisibility by 3 and for divisibility by 9.

Theorem

B.5   *A natural number is divisible by:*
■   1.   *3 if the sum of the digits of the number is divisible by 3.*
       2.   *9 if the sum of the digits of the number is divisible by 9.*

B.6   PROOF   Consider a natural number $n = a_k 10^k + a_{k-1} 10^{k-1} + \cdots + a_1 10^1 + a_0 10^0$. As shown in the example above, 10 to any power p can be written as a sum of two numbers. The digits of the first number are all 9 and the second number is 1. For example, 1,000 is $999 + 1$, 10,000 is $9,999 + 1$, and so forth. Therefore, n can be written in the following way.

$$n = a_k(\overbrace{99\ldots99}^{k} + 1) + a_{k-1}(\overbrace{99\ldots99}^{k-1} + 1) + \cdots + a_1(9 + 1) + a_0$$

Using the distributive property we have

$$n = (a_k \times \overbrace{99\ldots99}^{k} + a_k) + (a_{k-1} \times \overbrace{99\ldots99}^{k-1} + a_{k-1})$$
$$+ \cdots + (a_1 \times 9 + a_1) + a_0$$

and rearranging,

$$n = (a_k \times \overbrace{99\ldots99}^{k} + a_{k-1} \times \overbrace{99\ldots99}^{k-1} + \cdots + a_1 \times 9)$$
$$+ (a_k + a_{k-1} + \cdots + a_1 + a_0)$$

Now since the first part of the expression is divisible by 9, we can write $n = 3D + \{a_k + a_{k-1} + \cdots + a_1 + a_0\}$, where $3D$ is the term in the first parentheses above. Thus, if 3 divides $a_k + a_{k-1} + \cdots + a_1 + a_0$, then 3 must divide n and part 1 is proved.

It is left to you to show that the theorem is also true for the number 9.

EXERCISES

1. Decide if the following numbers are divisible by 3.
   a. 233   b. 1,770   c. 33,125   d. 243,432   e. 27,891

2. Repeat Exercise 1 by using 9 as a divisor.

3. Which of the following numbers are divisible by both 3 and 2?
   a. 232   b. 312   c. 33,771   d. 3,402   e. 56,310

4. Which of the following numbers are primes?
   a. 245   b. 211   c. 2,311   d. 321   e. 657

DIVISIBILITY BY POWERS OF FIVE

We next consider the pattern produced by the set of multiples of 5. The set includes the natural numbers $\{5, 10, 15, 20, \ldots, 155, 160, 165, \ldots, 1795, \ldots\}$. Notice that the last digit of each number shown in the set is either a 5 or a 0. Another way of discussing this would be to state that in each case if the last digit of a number is divisible by 5, then the number itself is divisible by 5. Can you show that 0 is divisible by 5? We will simply state the theorem regarding divisibility by 5 and 25, because the proof is relatively simple and the theorem itself is obvious.

Theorem

B.7   *A natural number is divisible by*
   1.   *5 if the last digit is divisible by 5.*
   2.   *25 if the last two digits are divisible by 25.*

For those students who wish to prove this theorem or wish to consider the proof, it might be instructive to look at the proof of the theorem for divisibility by 2 and its powers. The same proof can easily be applied here.

EXERCISES

1. Which of the following numbers are divisible by 5?
  a. 245   b. 350   c. 852   d. 2,300   e. 2,428

2. Which of the following numbers are divisible by both 8 and 9?
  a. 264   b. 2,772   c. 1,356   d. 3,024

3. Using divisibility tests, find the prime factorization of the following numbers.
  a. 243   b. 249   c. 324   d. 1,089

4. Under what conditions is a number divisible by 125?

INFINITUDE OF PRIMES

If we look at the list of primes in Figure 4.3, we note that as the primes get larger, their frequency in the set of natural numbers becomes less. That is, as the primes get larger, they are farther and farther apart. This leads to the question, Do the primes ultimately end, or is the set of primes infinite?

Let us consider several examples of a number formed by multiplying consecutive primes together and then adding one. If we multiply 2×3, the product is divisible by its factors 2 and 3. However, if we add one to the product, we produce the number $(2 \times 3) + 1$, or 7, which is a prime. We also know that 2 or 3 will not divide the number, because division by 2 or 3 will produce a remainder of 1. Let's consider the next number formed in the same manner, $(2 \times 3 \times 5) + 1$. We know that 2, 3, and 5 divide $2 \times 3 \times 5$. It follows that if we divide $(2 \times 3 \times 5) + 1$ by 2, 3, or 5, we will produce a remainder of one. *This does not imply that all numbers of this form are prime.* It simply demonstrates that all numbers formed in this way are not divisible by the consecutive primes used to form the number. Consider what the series would look like.

$n_1 = (2 \times 3) + 1$
$n_2 = (2 \times 3 \times 5) + 1$
$n_3 = (2 \times 3 \times 5 \times 7) + 1$
$n_4 = (2 \times 3 \times 5 \times 7 \times 11) + 1$
$n_5 = (2 \times 3 \times 5 \times 7 \times 11 \times 13) + 1$

.   .   .
.   .   .
.   .   .

$n_k \quad (2 \times 3 \times 5 \times 7 \times \ldots \times p_k \times p_{k+1}) + 1$

To answer the question of whether the set of primes is infinite, we see that if the set is *not* infinite, (i.e., it is finite) there must be a largest prime. So we formulate the following theorem.

Theorem

B.8   *There is no largest prime number.*

To prove this, we assume that there is a largest prime, and we represent it by p_{k+1}. By doing this, we can produce a number, n, in the form above.

That is, list all the primes up to p_{k+1} and form the number

$$n = (2 \times 3 \times 5 \times 7 \times \cdots \times p_k \times p_{k+1}) + 1.$$

We know that n is either a *prime* or a *composite* number. If n is a prime, then we are finished with the proof, since n is greater than p_{k+1}, which contradicts our assumption that there is a largest prime. On the other hand, n could be a composite number. If n is a composite, then there are at least two primes that divide it. We can say that $n = p \times q \times$ (a natural number). We have already shown that none of the primes in the list can divide n, since a remainder of 1 is always produced by such division. Therefore, p and q are primes other than those in the list. Thus, p and q must be larger than any number in the list, and $p \times q \leq n$. We have produced a contradiction of our original assumption in both cases. This means that our original assumption must be false. Thus, there is no largest prime.

EXERCISES

1. Make a sieve for the natural numbers from 1 to 40, and explain how the sieve operates.

2. Consider the set of numbers from 75 to 130.
a. How would you use the sieve to find the primes in this set?
b. Could you find the primes by using tests for divisibility?
c. What method or methods would be most efficient? Why?

3. State and give an example for the following.
a. divisibility by 2
b. divisibility by 3
c. divisibility by 5

4. In response to the numbers given below, indicate by marking them "a" if they are divisible by 2, "b" if they are divisible by 3, and/or "c" if they are divisible by 5.
a. 125   b. 48   c. 120   d. 141   e. 191   f. 270

5. Explain why a number of the form $(2 \times 3 \times 5 \times 7 \times 11 \times 13 \times 17) + 1$ is not divisible by the primes listed.

6. If we assume p to be the largest prime and then state that the number $n = (2 \times 3 \times 5 \times \ldots \times p) + 1$ is a prime, why have we contradicted our assumption?

7. If we state that the n in Exercise 6 is composite, how does that contradict our original assumption?

8. Show why the set of primes is an infinite set.

SECTION C **LEAST COMMON MULTIPLE AND GREATEST COMMON DIVISOR**

BEHAVIORAL OBJECTIVES

C.1 Define least common multiple, L.C.M.

C.2 Be able to find the least common multiple of two numbers.

C.3 Define the greatest common divisor, G.C.D.

C.4 Be able to find the greatest common divisor of two numbers.

C.5 Define relatively prime.

C.6 Be able to use Euclid's algorithm to find the G.C.D. of two numbers.

When we discuss operations on the rational numbers in Chapters 7 and 8, it will be necessary to know how to determine the Least Common Multiple and the Greatest Common Divisor given two natural numbers. We will refer to these concepts simply as L.C.M. and G.C.D. throughout the book.

There are several properties of both of these concepts which place them in this chapter. These concepts are properly included in the branch of mathematics which is known as number theory. However, we will not explore all of these properties in depth. Some will be stated, some proven, and others omitted.

LEAST COMMON MULTIPLE (L.C.M.)

Consider the concept of multiple again. We say that the natural number a is a multiple of the natural number b if there is a natural number c such that $a = bc$. For example, 30 is a multiple of 5 because there is a natural number 6 such that $30 = 5 \times 6$. We can also designate a set of multiples of a given number a in the following manner:

$$M_a = \{a, 2a, 3a, \ldots\} \quad \text{and} \quad a \in N$$

Suppose we consider two sets of multiples, for example, multiples of 3 and multiples of 4. The sets of multiples would look like this:

$$M_3 = \{3, 6, 9, 12, 15, 18, 21, 24, \ldots\}$$
$$M_4 = \{4, 8, 12, 16, 20, 24, 28, \ldots\}$$

C.2 The intersection of the two sets, $\{12, 24, \ldots\}$, is a set containing multiples of both 3 and 4. We call this set the set of common multiples of 3 and 4. $\{12, 24, 36, \ldots\}$ is in fact the set of multiples of 12. That is, $M_3 \cap M_4 = M_{12}$.

■ The least element in M_{12} is called the *least common multiple* of 3 and 4.

Thus, we can determine the least common multiple of two given numbers by first producing the set of multiples for the given numbers. Then performing the operation of intersection on the sets, we produce the set of multiples of both natural numbers. The least number in this set of multiples of both given numbers is the least common multiple of the given numbers.

This approach to finding the L.C.M. is quite simple for an introductory explanation; however, it would be a tedious operation if a and b were larger numbers. We therefore do not suggest using this as a method but simply as an introduction to the concept.

Let us consider one more example using the operations shown above. To find the L.C.M. for the numbers (12, 15), we first write M_{12} and M_{15}.

Example

■ C.2
$$M_{12} = \{12, 24, 36, 48, 60, 72, 84, \ldots\}$$
$$M_{15} = \{15, 30, 45, 60, 75, 90, \ldots\}$$

Then

$$M_{12} \cap M_{15} = \{60, 120, \ldots\}$$

The least element in $M_{12} \cap M_{15}$ is 60, which is called the least common
■ multiple (or L.C.M.) of 12 and 15.

This method of finding the L.C.M. of two given numbers is no more than an algorithm for finding the smallest number that is a multiple of both the given numbers.

We now give a formal definition of L.C.M.

Definition

■ C.1   *The L.C.M. of two natural numbers a and b is the smallest natural num-*
■ *ber that is a multiple of both a and b.*

Now we consider a different method and its rationale for finding the L.C.M. Let us use the numbers 12 and 15 of the previous example.

A multiple of 12 is any number whose factorization contains the factors of 12, and a multiple of 15 is any number whose factorization contains the factors of 15. We want a number that has in its factorization the factors of both 12 and 15 and is the smallest such number. Thus, its factorization will contain only the factors of 12 and 15.

Example

■ C.2 $12 = 2 \times 2 \times 3$ and $15 = 3 \times 5$

$\underline{2 \times 2 \times 3} \times 5$ contains the factors of 12
and

$2 \times 2 \times \underline{3 \times 5}$ contains the factors of 15

In the example above, if any factor were omitted, the number would not contain the necessary factors and would not be a multiple of both 12 and 15.

It is true that we could simple multiply the given numbers together and produce a common multiple. However, it would not necessarily be the L.C.M.

Let us consider two more examples.

Example

Find the L.C.M. for the pair (18, 20).

SOLUTION $18 = 3^2 \times 2$ and $20 = 2^2 \times 5$
$$\text{L.C.M.} = 3^2 \times 2^2 \times 5 = 180$$

Notice in the example above that we used powers of 2, 3, and 5. The powers used were the highest powers found in any given number.

Example

Find the L.C.M. for the pair (60, 225).

SOLUTION $60 = 2^2 \times 5 \times 3$ and $225 = 5^2 \times 3^2$
$$\text{L.C.M.} = 2^2 \times 5^2 \times 3^2 = 900$$

In the above example, 60×225 is a common multiple of 60 and 225, but it is considerably larger than 900.

EXERCISES

1. Find the set of multiples of
   a. 12 or M_{12} b. 16 or M_{16} c. 18 or M_{18}

2. Find the set of common multiples M_c by using intersection as follows:
   a. $M_{12} \cap M_{16}$ b. $M_{12} \cap M_{18}$ c. $M_{16} \cap M_{18}$

3. Find the L.C.M. for the following pairs, using the results of Exercise 2.
   a. (12, 16)   b. (12, 18)   c. (16, 18)

4. Find the L.C.M. for the following pairs, using the method described on pages 154 and 155.
   a. (14, 15)   b. (12, 24)   c. (32, 20)   d. (28, 36)

GREATEST COMMON DIVISOR (G.C.D.)

Recall that our definition of factor states that if $ab = c$, then a and b are factors of c, and that any number which is a factor of c is called a divisor of c.

Now we will consider how to find all divisors of a given number. We will use the notation D_a to indicate the set of divisors of the number a. What are the divisors of the number 12? We know that 12 divides itself, that is, $12 = 12 \times 1$. We know that 2 and 6 divide 12, since $12 = 2 \times 6$. We also know that 4 and 3 divide 12, since $12 = 4 \times 3$. Because this method of finding divisors is somewhat haphazard, we will try to find a more systematic method.

We consider the prime factorization of 12, which is $2^2 \times 3$. We can produce the divisors of 12 as follows:

1. Any number divides itself one time. This means that 12 and 1 are divisors.
2. Since 2^2 is a factor, we know that 2 and 2×2 are divisors.
3. Since 3 is a factor, we know that 3, 3×2, 3×4 are divisors.
4. By taking the union of the sets above, we produce all divisors of 12. Notice that the set D_{12} is not the factorization of 12, it is simply the set of all divisors of 12.

Let us consider an example using the method outlined above.

Example

Find the divisors of 54.

SOLUTION
1. We find the prime factorization $54 = 3^3 \times 2$.
2. Since a number divides itself once, 1 and 54 are divisors.
3. Since 3^3 is a factor, we know that 3×1, 3×3 and $3 \times 3 \times 3$ are divisors of 3, 9, and 27.
4. Since 2 is a factor, we know that 2×1, 2×3, 2×9, and 2×27 are divisors of 2, 6, 18, and 54.
5. The union of the sets above produces the set D_{54} $= \{1, 2, 3, 6, 9, 18, 27, 54\}$.

Consider now the set of divisors D_{12} and D_{54}. How might we find the G.C.D. of 12 and 54? This can be found by intersecting D_{12} and D_{54}. The intersection set will be noted as D_c.

■ C.4
$$D_{12} = \{1, 2, 3, 4, 6, 12\}$$
$$D_{54} = \{1, 2, 3, 6, 9, 18, 27, 54\}$$
$$D_c = D_{12} \cap D_{54} = \{1, 2, 3, 6\}$$

Since D_c is the set of common divisors, the largest element in that set is the greatest common divisor, or the G.C.D. of 12 and 54. Therefore,
■ the G.C.D. for the pair (12, 54) is 6.

When the L.C.M. was discussed in the last section, we used an approach similar to the one shown above. Again, it should be noted that this approach is accurate, understandable, and can be used to promote skills. It should be somewhat obvious that this method is very time consuming and after it is mastered should be replaced by a more direct approach.

Before going on to another method for finding the G.C.D., consider the following example.

Example
■ C.4
Find the G.C.D. for (24, 84).

SOLUTION
$$D_{24} = \{1, 2, 3, 4, 6, 8, 12, 16, 24\}$$
$$D_{84} = \{1, 2, 3, 4, 6, 7, 12, 14, 21, 28, 42, 84\}$$
$$D_c = D_{24} \cap D_{84} = \{1, 2, 3, 4, 6, 12\}$$
■
$$\text{G.C.D. of } (24, 84) = 12$$

We have not as yet defined the greatest common divisor. To find the G.C.D. of two natural numbers, we begin with two natural numbers. We find the divisors of each number. Then we choose the largest natural number which belongs to the set of common divisors.

Definition

■ **C.3** *The G.C.D. of two natural numbers, a and b, is the largest natural number*
■ *which divides both a and b.*

We should observe that *every common divisor of a and b itself divides the greatest common divisor.*

FINDING G.C.D. BY THE FACTOR METHOD

Consider next a method and rationale for finding the G.C.D., called the *prime factor method.* In the last example the pair $(24, 84)$ was used. The prime factorization of 24 is $2^3 \times 3$. Thus, the divisors of 24 are combinations of 1, 2, and 3. The prime factorization of 84 is $2^2 \times 3 \times 7$. Therefore, the divisors of 84 contain combinations of 1, 2, 3, and 7 in combination. The highest power of 2 which will divide both 24 and 84 is 2^2 because 2^3 does not divide 84. The highest power of 3 which will divide both 24 and 84 is 3^1 because it is a common factor. Since 7 is not a factor of 24, it will not divide both numbers. Taking those factors which divide both numbers, we produce the G.C.D. by taking their product. In the example below, we have written the factorization two ways to illustrate this idea.

Example

■ **C.4** Find the G.C.D. for $(24, 84)$.

SOLUTION $24 = 2^3 \times 3 = 2 \times \underline{2 \times 2 \times 3}$
$84 = 2^2 \times 3 \times 7 = \underline{2 \times 2 \times 3} \times 7$
G.C.D. of $(24, 84) = 12$

Notice that the product of common factors or shared factors produced the G.C.D. One more example is illustrated below.

Example

Find the G.C.D. for $(72, 108)$.

SOLUTION $72 = 2^3 \times 3^2 = 2 \times \underline{2 \times 2 \times 3 \times 3}$
$108 = 2^2 \times 3^3 = \underline{2 \times 2 \times 3 \times 3} \times 3$

■ Therefore, the G.C.D. of $(72, 108) = 36$.

You may have noticed that the factors which are common to both numbers are 2 and 3. It is also the case that the least power of the common factors found in each number as a factor is a factor of the G.C.D. Since 36 is $2^2 \times 3^2$, you should notice that the smallest power of 2 which can be found as a factor of the two numbers is 2^2, because it will divide 108. In the same way, 3^2 will divide 72.

When two natural numbers have a G.C.D. of 1, we know that they have no common prime factors. We call these numbers *relatively prime*.

Definition

■ C.5
■

The natural numbers a and b are said to be relatively prime if the G.C.D. of (a, b) = 1.

Consider the example below.

■ C.2   **Example**

▲ C.4   Find the G.C.D. and L.C.M. for (77, 45).

SOLUTION $77 = 7 \times 11$
$45 = 5 \times 3^2$

▲ ■   The G.C.D of (77, 45) is 1 and the L.C.M. is 3465 (77 × 45).

You will notice that the L.C.M. of (a, b) is the product of the two numbers, $a \times b$, when the G.C.D. of (a, b) is 1. If you consider that the L.C.M. is made of the factors of both numbers including the highest powers of common factors and the G.C.D. is made of the common factors including only the smallest powers, you will notice the relationship illustrated below.

■ C.2   **Example**

▲ C.4   Find the L.C.M. and G.C.D. for the pair (18, 20).

SOLUTION $18 = 3^2 \times 2$ and $20 = 2^2 \times 5$

■   L.C.M. $= 3^2 \times 2 \times 2 \times 5 = 180$

                  18   20

              $3^2 \times 2$ $2^2 \times 5$

▲

              G.C.D. $= 2$

Notice that the product of the L.C.M. and G.C.D. is equal to the product of the numbers. $180 \times 2 = 18 \times 20$.

EXERCISES

1. Find by any method the L.C.M. for the following pairs.
   a. (24, 36)   b. (45, 60)   c. (77, 44)

2. Find the G.C.D. for each pair in Exercise 1 by the factor method.

3. Using the results from Exercises 1 and 2 above, compare the product of the given numbers with the products of the G.C.D. and L.C.M.

4. Let a, b, and c be prime, and let $n = ab$ and $m = ac$.
   a. Find the G.C.D. of m and n.
   b. Find the L.C.M. of m and n.
   c. Is the product of the G.C.D. and L.C.M. of m and n equal to $m \times n$?

5. Show that every common divisor of 12 and 54 divides the greatest common divisor of 12 and 54.

6. Show that every common divisor of the natural numbers a and b divides the greatest common divisor of a and b.

7. Show that two different prime numbers are always relatively prime.

8. Are two composite numbers ever relatively prime?

EUCLID'S ALGORITHM

A method for finding the G.C.D. is called *Euclid's algorithm*. It is based on the division algorithm found in Chapter 3, which states that if a and b are two whole numbers, with $b \neq 0$, then unique whole numbers q and r exist such that $a = bq + r$, where $0 \leq r < b$. An example of this algorithm is shown below.

Example

■ **C.6**   Find the G.C.D. of $(84, 26)$.

SOLUTION 1. Divide one number by the other.
26 divides 84 with a remainder of 6.

2. Write the statement of the division algorithm for step 1.
$84 = 3 \times 26 + 6$

3. Divide the remainder into the last divisor.
6 divides 26

4. Write the statement of the division algorithm for the previous step.
$26 = 4 \times 6 + 2$

5. Divide the remainder into the last divisor.
2 divides 6

6. Write the statement of the division algorithm from the previous step.
$6 = 3 \times 2 + 0$

As long as this example seems, we have performed only three steps, which are simply repetitions of the first step. We summarize these as

1. $84 = (3 \times 26) + 6$
2. $26 = (4 \times 6) + 2$
3. $6 = (3 \times 2) + 0$

In performing this algorithm, the last remainder before the zero remain-
■ der is the G.C.D. In our example, 2 is the last remainder before 0. In the last step, we see that since 2 divides 3×2, then 2 divides 6. In the second

step, since 2 divides both 2 and 6, then it divides 26. In the first step, since 2 divides 6 and 26, it divides 84. This shows that 2 is a divisor of both 84 and 26 but does not ensure that 2 is the greatest common divisor.

To assure ourselves that 2 is the G.C.D., let us assume that there is a divisor d which is larger than 2. If there is such a divisor, then it divides both 84 and 26, which means it must divide 6, because $84 - (3 \times 26) = 6$. In the second step we can see that if d divides both 26 and 6, then it divides 2. Therefore, 2 is the G.C.D., because it fulfills the requirements of the definition. Notice that 2 divides 84 and 26; any other divisor of 84 and 26 also divides 2.

Let us do another example of finding the G.C.D., first using the factor method and then using Euclid's algorithm.

Example

■ C.4   1. Find the G.C.D. for the pair $(972, 45)$ using the factor method.

■
$$972 = 2^2 \times 3^5 \quad \text{or} \quad 2 \times 2 \times 3 \times 3 \times 3 \times 3 \times 3$$
$$45 = 5 \times 3^2 \quad \text{or} \quad 5 \times 3 \times 3$$

The G.C.D. of $(972, 45) = 9$.

■ C.6   2. Find the G.C.D. for the pair $(972, 45)$ using Euclid's algorithm.

$$972 = (21 \times 45) + 27$$
$$45 = (1 \times 27) + 18$$
$$27 = (1 \times 18) + 9$$
$$18 = (2 \times 9) + 0$$

■   The G.C.D. of $(972, 45) = 9$.

The factor method shown above simply involves finding common factors, whereas Euclid's algorithm is a process which uses successive division. Both methods use fundamental processes and can be taught easily to children in the intermediate grades. It would be instructive to teach and compare methods. Each method can be used to build skills.

EXERCISES

1. Using Euclid's algorithm, find the G.C.D. for each pair below.
   a. $(534, 162)$ b. $(66, 18)$
   c. $(84, 30)$ d. $(243, 522)$

2. From the example below, explain why the G.C.D. is a divisor of each remainder starting with the last one. Then explain why any divisor must divide the G.C.D.
$$612 = (3 \times 182) + 66$$
$$182 = (2 \times 66) + 50$$
$$66 = (1 \times 50) + 16$$
$$50 = (3 \times 16) + 2$$
$$16 = (8 \times 2) + 0$$

SECTION D **SPECIAL NUMBERS**

BEHAVIORAL OBJECTIVES

D.1 Be able to describe triangular numbers.

D.2 Be able to find triangular numbers.

D.3 Be able to describe square numbers.

D.4 Be able to find square numbers.

D.5 Be able to describe Pythagorean triples.

D.6 Be able to find Pythagorean triples.

D.7 Be able to describe prime twins.

D.8 Be able to describe perfect numbers.

D.9 Be able to describe abundant numbers.

D.10 Be able to describe deficient numbers.

This section on special numbers provides an investigation into the properties of some special types of numbers. Elementary school texts sometimes use these types of numbers to stimulate students to manipulate and investigate numbers.

POLYGONAL NUMBERS—TRIANGULAR NUMBERS

Figurate numbers are simply numbers represented by points or physical objects. The most interesting figurate numbers are called polygonal figurate numbers because they can be represented as geometric shapes that are polygons; that is, triangles, squares, and so on. The Pythagoreans studied the properties of these numbers in great detail. They will be discussed briefly here, and only the more well-known shapes will be dealt with.

The polygonal number on which many others are based is the triangular number. We will use the notation T_1, T_2, $T_3, \ldots, T_n$ to characterize specific triangular numbers and T to represent the set of all triangular numbers. Consider the examples shown in Figure 4.4. From the shapes formed, it will be obvious why the numbers are called triangular numbers. Thus, $T = \{1, 3, 6, 10, 15, 21, \ldots\}$. In order to find the next triangular number, let us look for a moment at the set of differences between successive triangular numbers. This set is $\{2, 3, 4, 5, 6, \ldots\}$. From this we might guess that we must add 7 to get the next triangular number, namely, 28. Thus, to get a new triangular number, we add to each triangular number a number one greater than the previously added number. For example, $T_5 = T_4 + 5 = 10 + 5$, and $T_7 = T_6 + 7 = 21 + 7$. In this manner one can produce the set T. However, if one wishes a particular element T_n, this method is not satisfactory because one must first know what the value of T_{n-1} is.

■ D.1

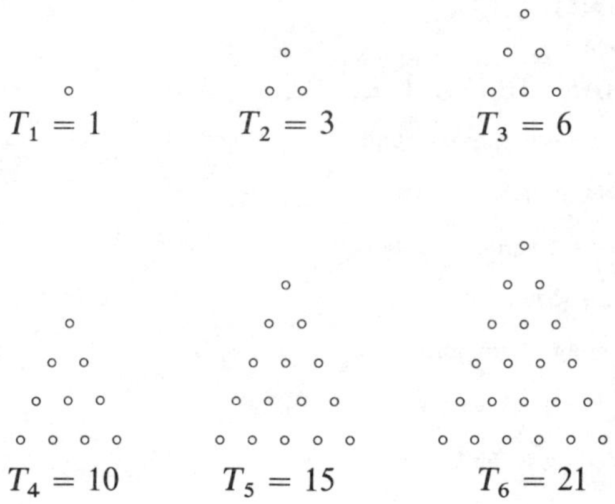

$T_1 = 1$ $T_2 = 3$ $T_3 = 6$

$T_4 = 10$ $T_5 = 15$ $T_6 = 21$

Figure 4.4

Consider the sequence whose sum produces a particular term. Notice that T_1 is 1, T_2 is $1 + 2$, T_3 is $1 + 2 + 3$, and T_4 is $1 + 2 + 3 + 4$. It seems intuitively reasonable to expect that T_n for any nth term can be found by some formula. Actually, the value of any term can be found by taking the sum of n consecutive natural numbers.

$T_5 = 1 + 2 + 3 + 4 + 5$, or 15
■ D.2
$T_6 = 1 + 2 + 3 + 4 + 5 + 6$, or 21

In general, T_n can be found as follows:

$$T_n = 1 + 2 + 3 + \cdots + n$$

We can use the formula above quite easily as long as n is small. If we wanted T_{40}, the job would then become tedious. There is a formula which can be used to find the sum of n consecutive numbers.

$$T_n = \frac{n(n + 1)}{2} = 1 + 2 + 3 + \cdots + n$$

Consider several examples below.

Example
Find T_n when n is 6, 9, and 12.

SOLUTION $T_6 = \dfrac{6(6 + 1)}{2} = \dfrac{42}{2} = 21$

$T_9 = \dfrac{9(9 + 1)}{2} = \dfrac{90}{2} = 45$

$T_{12} = \dfrac{12(12 + 1)}{2} = \dfrac{156}{2} = 78$

It is easy enough to fill the gaps between T_6, T_9, and T_{12} by adding the next consecutive numbers. This would demonstrate to you that you understand the concept.

■ D.3 SQUARE NUMBERS

The polygonal numbers of which students are most aware are called square numbers. To square a number we simply multiply the number by itself. We will call the square numbers S_1, S_2, S_3, ..., S_n and name the set S. The set $S = \{1^2, 2^2, 3^2, 4^2, \ldots, n^2\}$, or $\{1, 4, 9, 16, \ldots, n^2\}$. Consider Figure 4.5, which illustrates the reasons why these numbers are characterized as squares.

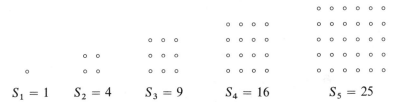

$$S_1 = 1 \qquad S_2 = 4 \qquad S_3 = 9 \qquad S_4 = 16 \qquad S_5 = 25$$

Figure 4.5

From Figure 4.5 it should be obvious that

■ D.4

$$S_1 = 1 \times 1 = 1$$
$$S_2 = 2 \times 2 = 4$$
$$S_3 = 3 \times 3 = 9$$
$$S_4 = 4 \times 4 = 16$$

.

.

.

$$S_n = n \times n = n^2$$

Another way of looking at square numbers is to characterize them as the sum of consecutive odd numbers. This can be shown by the use of the statements below.

$$S_1 = 1$$
$$S_2 = 1 + 3$$
$$S_3 = 1 + 3 + 5$$

There are many other types of polygonal numbers which can be considered. However, we leave this to you and your instructor. Each type of number mentioned in this section is related to the triangular number. In this way, they are all interrelated. All relationships have not been explored; look for other relationships on your own. It might be interesting to consider numbers formed by adding square and triangular numbers or triangular and triangular numbers. If the numbers are chosen so that they fit geometrically, the results can be interesting.

EXERCISES

1. Using the formula for triangular numbers, find
   a. T_7 b. T_9 c. T_{10}

2. Which square numbers are found by adding the following?
   a. $T_3 + T_4$ b. $T_6 + T_7$

3. What triangular number   will result from the following operation?
   a. $S_6 - T_5$ b. $S_4 - T_4$

4. If $T_n = 1 + 2 + 3 + \ldots + n$, obviously $T_n = n + \ldots + 3 + 2 + 1$.
   a. Show that $2T_n = n(n + 1)$.

   b. Show from a. that $T_n = \dfrac{n(n + 1)}{2}$.

5. a. Show that $S_3 = T_2 + T_3$.
   b. Can you show that $S_n = T_{n-1} + T_n$?

6. a. How many lines can you draw to connect two points?
   b. How many lines can you draw to connect three points?
   c. How many lines connect four points? Five points?

7. What set of numbers is produced by connecting points as described in Exercise 6?

PYTHAGOREAN TRIPLES

One of the most well-known accomplishments of the Pythagoreans or maybe of Pythagoras himself was the theorem which bears his name, *the Pythagorean theorem.* The theorem states that the sum of the squares of the measures of the legs in a right triangle is equal to the square of the measure of the hypotenuse (longest side) of the triangle. The numbers which relate in this way are called Pythagorean triples and are written in the form (a, b, c). The relationship can be described as follows:

$$a^2 = b^2 + c^2$$

Thus, the triple (a, b, c) is called a Pythagorean triple if $a^2 = b^2 + c^2$. Consider the triple $(5, 4, 3)$.

$$5^2 = 4^2 + 3^2$$

or

$$25 = 16 + 9$$

A general formula for finding Pythagorean triples was given by Euclid and is stated below:

$$(u^2 + v^2)^2 = u^4 + 2u^2v^2 + v^4$$

or

$$(2uv)^2 + (u^2 - v^2)^2$$

where u and v are relatively prime.

The derivation of the formula will be omitted even though it requires very little mathematics. It is left as an exercise. Consider examples below of Pythagorean triples by using Euclid's formula.

Examples

■ **D.6** 1. Let $u = 2$ and $v = 1$. Then

$$(2uv)^2 = 16, (u^2 - v^2)^2 = 9, (u^2 + v^2)^2 = 25$$

which produces the triple (5, 4, 3).

2. Let $u = 5$ and $v = 2$. Then

$$(2uv)^2 = 20^2, (u^2 - v^2)^2 = 21^2, \text{ and } (u^2 + v^2)^2 = 29^2$$

■ which produces the triple (29, 21, 20). Note that $29^2 = 841$, $20^2 = 400$, and $21^2 = 441$.

PRIME TWINS

■ **D.7** Prime twins are primes of the form p and $p + 2$. We could characterize them as two consecutive odd primes. We know that there is only one pair of consecutive primes, the pair (2, 3), because the only even prime is 2. There are many prime twins, and it is conjectured that the set is infinite. If you look back at Figure 4.3 you will note that between 1 and 100 there ■ are eight pairs, for example, (3, 5) and (5, 7).

As mentioned earlier, the number of primes decreases in frequency between each successive hundred numbers. The number of prime twins also decreases in frequency. The question, Are there always twin primes? has not been answered.

PERFECT, ABUNDANT, AND DEFICIENT NUMBERS

■ **D.8** A perfect number is a natural number such that the sum of its factors ■ (except for the number itself) equals the number. The number six is a perfect number and was first mentioned in relation to the Creation by Aurelius Augustinus (A.D. 534–430), who said that God created the world in 6 days (rather than 9 or 13) because the Creation and six are both perfect. The earth was to have been created in 6 days. The number twenty-eight is also a perfect number and has been associated with the length of the lunar cycle.

A formula for finding perfect numbers was given by Euclid in his *Elements*. The formula states that $2^{n-1}(2^n - 1)$ is a perfect number as long as $(2^n - 1)$ is a prime. Consider the numbers produced when n is 2 or 3.

$$2^{n-1}(2^n - 1) = 2 \times 3 = 6 \quad \text{(for } n = 2\text{)}$$
$$2^{n-1}(2^n - 1) = 4 \times 7 = 28 \quad \text{(for } n = 3\text{)}$$

When n is 4, $(2^n - 1)$ is not a prime. When n is 5, $(2^n - 1)$ is the prime 31. It will be left as an exercise to find the next perfect number.

If a number has factors whose sum (except for the number) is greater

■ **D.9**   than the number, then the number is called *abundant*. The number twelve is an example of an abundant number. The factors of 12, excluding 12, are 1, 2, 3, 4, 6. Notice that the sum of the factors is 16, which is greater
■   than 12.

D.10   A number which is not a perfect number or an abundant number is called a *deficient number*. The sum of its factors (except for the number)
■   is less than the number and is thus deficient. The number eight has factors 1, 2, and 4, if we exclude 8 from the set. The Second Creation from Noah is not a perfect creation, according to Dark Age philosophers, since there were eight people aboard the ark and eight is a deficient number. The sum of the factors is 7, which makes the number eight a deficient number.

EXERCISES

1. Which of the triples below are Pythagorean triples?
   a. (6, 8, 10)   b. (3, 5, 6)   c. (13, 12, 5)   d. (9, 4, 8)

2. Find a Pythagorean triple.

3. Find a triple that is not Pythagorean.

4. Using the formula given in this section, find Pythagorean triples when u and v are given as follows:
   a. u is 3 and v is 2
   b. u is 3 and v is 1
   c. u is 5 and v is 2

5. Find a perfect number using the formula given in this section when $n = 5$.

6. Which of the following are deficient numbers?
   a. 24   b. 31   c. 32

7. a. The first four primes are 2, 3, 5, 7. Show that numbers of the form $2 \cdot 3 \cdot 5 \cdot 7 + n$ for $n = 1, 2, 3, 4, 5, 6, 7$ are not prime.
   b. How many natural numbers are there in the above set?
   c. Can you find 30 consecutive integers that are not prime?

CHAPTER 5 **INTEGERS**

SECTION A **NUMBER LINES**

BEHAVIORAL OBJECTIVES

A.1 Be able to construct a number line.

A.2 Relate the sets *N*, *W*, and *J*.

A.3 Describe the set *J* on the number line.

A.4 Describe the J^+ on the number line.

A.5 Describe the set J^-.

A.6 Describe the set J^o.

A.7 Describe a partition on the set *J*.

A.8 Define a partition in general.

A.9 Give an example of a partition.

THE NUMBER LINE

The number line is used to graphically represent numbers. We could have used a number line earlier in the textbook as a representation of the natural and whole numbers. The number line, or number ray, is used in the elementary grades to represent the natural or whole numbers, for counting, and to diagram the solution of simple sentences.

The number line consists of a straight line with a point or mark noted as 0. To the right of the 0 mark is a mark for 1. This convention is used to define the unit length on the line. Since the distance from the 0 mark to the 1 mark is an arbitrary distance, the unit varies in length from one number line to another. In a sense, this distance defines the number one. Once we choose a distance, we call it the *unit distance*. We then determine all other distances based upon it. In all measure systems, we have a unit distance convention except that the units of measure have been standardized so that a unit of length in the English system does not vary from one ruler to another.

Figure 5.1 shows the steps one goes through in making a number line.

■ A.1

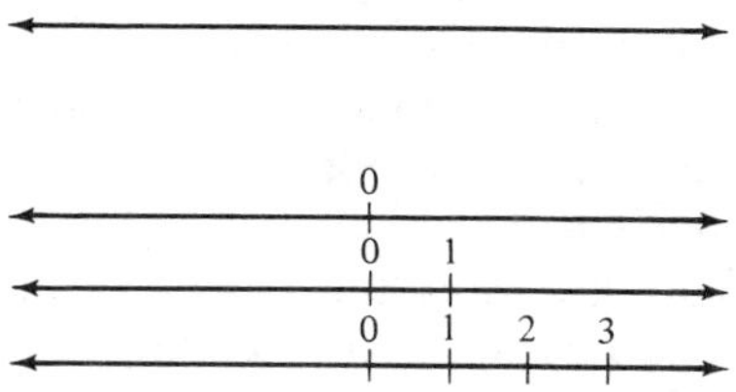

draw a line placing arrows on both ends to indicate that the line continues

choose a zero point

choose a unit point

place other numbers one unit away from the preceeding number

■ Figure 5.1

You will notice that the steps followed in Figure 5.1 are as follows:

1. Draw a line placing arrows on each end.
2. Choose a zero point.
3. Choose a unit.
4. Place succeeding whole numbers one unit from the preceding number.

We place arrows at both ends to show that the line is infinite in both directions. It is a good idea to use a complete line rather than a ray, which would have an endpoint at zero and not include negative numbers. Children should be made aware of the fact that there are numbers corresponding to points on the left of the 0 as well as on the right. A number ray would look like Figure 5.2.

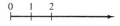

Figure 5.2

Once we decide on a unit, each number is measured one unit to the right of its predecessor; we could state that the successor of each number is found by measuring to the right one unit. This convention fits our definition of natural or whole number.

If we follow this convention on the left and name the first number to the left of the zero point -1, then -2, -3, -4, etc., we will eventually have represented the set of integers on a number line and used the standard notation which you probably learned in school. In this way, the set of integers, J, and the number line would look like Figure 5.3.

■ **A.3** $J = \{\ldots, -3, -2, -1, 0, 1, 2, 3, \ldots\}$

■

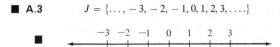

Figure 5.3

Notice that the set does not have a first or last element, it extends infinitely far in both directions. We show this when tabulating J by using ellipses on both the right and the left. When we make a number line, the arrowhead replaces the three dots.

In this chapter we will discuss the set of integers in a manner somewhat different from the standard notation. This will be done for two reasons; first, the use of the notation above causes some difficulty relating to the use of sign numbers in operations, and second, proofs will be clearer with the notation we will introduce. Too often people confuse "negative three," written -3, with "minus 3," (subtract 3) written -3. Many times the rules for multiplying integers interfere with understanding. We feel that our approach may minimize this difficulty.

Before we define the set of integers, we would like to point out that the integers can be considered as the union of three disjoint sets. The sets can be represented as J^+, J^-, and J^0, that is, positive integers, negative integers, and the zero integer. The set J^+ is equal to N and the set $J^+ \cup J^0$ is equal to W. Both N and W are subsets of J. Consider Figure 5.4.

A.2, 4, 5, 6

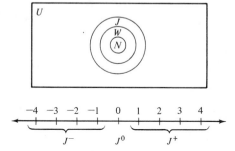

Figure 5.4

It should be clear from these diagrams that

$$J - W = J^- \quad \text{and} \quad J - N = J^- \cup J^0$$

You will notice from the number line that for every number to the right of 0 there corresponds one to the left of 0 and that these numbers are the same distance from 0. These number pairs are called *inverses* and have a specific property which we will discuss more fully later. In standard notation, the inverse of any number a is written $-a$. This means that the inverse of 3 is -3 and the inverse of -3 is $-(-3)$, or 3. We will show later why $-(-a)$ is a.

PARTITIONS

You will notice that J^+, J^-, and J^0 are related as follows:

A.7   1. All three sets are disjoint from each other.
2. The union of all three sets produces J.
3. Each set is non-empty.

Definition

A.8   *When subsets of a given set are designated in such a way that each subset is non-empty, is disjoint from each other subset, and the union of all the sets produces the original set or the universe, then we call this a partition on the set.*

Consider several examples of partitions.

Example 1

A.9   Let $N = U$. Then the set of even natural numbers, E, and the set of odd natural numbers, O, form a partition on N, because

$$E \cap O = \varnothing \quad \text{and} \quad E \cup O = \text{U or } N$$

Example 2

In a classroom, the set of children with blond hair, with red hair, with black hair, and with brown hair form a partition on the set, because

blond $\cap$ brown $\cap$ red $\cap$ black $= \varnothing$

and

blond $\cup$ brown $\cup$ red $\cup$ black $= U =$ the set of students in the classroom

EXERCISES

1. Construct a number line.

2. Which of the statements below are true? Why? Why are the others not true?
   a. $N \subseteq J$ b. $N \subseteq J^-$ c. $N \subseteq J^+$
   d. $W \subseteq J^+$ e. $W \subseteq J$ f. $N \cup W = J$

3. Tabulate the set J^+, J^-, and J.

4. Which of the descriptions of sets below are partitions on the natural numbers?
   a. $E = \{2, 4, 6, 8, \ldots\}$; $M_3 = \{3, 6, 9, 12, \ldots\}$
   b. $P =$ the set of primes; $C =$ set of composites
   c. $S =$ the set of square numbers; $T =$ the set of triangular numbers
   d. $A = \{1, 4, 7, 10, \ldots, 1 + 3n, \ldots\}$; $B = \{2, 5, 8, 11, \ldots, 2 + 3n, \ldots\}$; and
   $C = \{3, 6, 9, \ldots, 3 + 3n, \ldots\}$

SECTION B EQUIVALENCE CLASSES

BEHAVIORAL OBJECTIVES

B.1 Describe the characteristics of an equivalence class $I\,(a, b)$.

B.2 Define integer in terms of equivalence class.

B.3 Define the set I^+.

B.4 Define the set I^-.

B.5 Define the set I^0.

B.6 Translate an integer $I\,(a, b)$ to an integer in J.

B.7 Translate an integer in J to an integer in $I\,(a, b)$.

B.8 From a pair (a, b) produce the integer $I\,(a, b)$.

B.9 Given two equivalent pairs, find two equal sums.

B.10 Given two equal sums, find two equivalence pairs.

B.11 Find the basic pair in an equivalence class given a pair.

B.12 Define equivalent pair.

B.13 Show that $\sim$ for pairs is reflexive.

B.14 Show that $\sim$ for pairs is symmetric.

B.15 Show that $\sim$ for pairs is transitive.

B.16 Define equality for I.

INTEGERS AS SETS

In this section we define integers as sets of ordered pairs. Remember, by ordered pair we mean that the order of the elements is significant; that is, $(1, 2)$ is different from $(2, 1)$. Each integer will be represented as a set, and the set will be described in terms of a basic pair or simplest form. Before we define integers formally, consider the three sets below. See if you can list or state some characteristics which each set has in common.

$$I(0, 1) = \{(0, 1), (1, 2), (2, 3), (3, 4), \ldots, (a, a + 1), \ldots\}$$
$$I(0, 2) = \{(0, 2), (1, 3), (2, 4), (3, 5), \ldots, (a, a + 2), \ldots\}$$
$$I(0, 3) = \{(0, 3), (1, 4), (2, 5), (3, 6), \ldots, (a, a + 3), \ldots\}$$

The sets listed above represent the integers 1, 2, and 3. You should notice the following characteristics.

B.1 1. Each set is disjoint with the other sets.
2. The pairs contain entries which are whole numbers.
3. The first number in each pair consists of the whole numbers in order, namely $0, 1, 2, \ldots$.
4. In each pair the second number relates to the first in a specific manner throughout; namely, the difference between them is constant.
5. The first pair is always $(0, n)$, this entry is called the *basic pair*.

Consider next some examples of negative integers written in this form.

$$I(1, 0) = \{(1, 0), (2, 1), (3, 2), (4, 3), \ldots, (a + 1, a), \ldots\}$$
$$I(2, 0) = \{(2, 0), (3, 1), (4, 2), (5, 3), \ldots, (a + 2, a), \ldots\}$$
$$I(3, 0) = \{(3, 0), (4, 1), (5, 2), (6, 3), \ldots, (a + 3, a), \ldots\}$$

It is probably apparent that the sets listed above represent the integers -1, -2, and -3. Notice how similar to the first listing of sets the second listing is. Rather than list the characteristics of the sets $I(1, 0)$, $I(2, 0)$, and $I(3, 0)$, we will point out the essential difference between the first and second listing of the sets;

The pairs in each set of the second listing are the reverse of the pairs in corresponding sets of the first listing.

Consider how we might summarize the common characteristics of both these listings of sets.

■ **B.1** 1. Each set of pairs is disjoint with the other sets.

2. The pairs contain entries from the set of whole numbers.

3. The first entry in each set is $(0, n)$ or $(n, 0)$, this is called the basic pair.

4. Positive integers begin with a basic pair $(0, n)$, where $n \in N$, and negative integers begin with a basic pair $(n, 0)$, where $n \in N$.

■ 5. In any one set, the difference between the ordered pairs is a constant.

You will notice that characteristic 4 requires that n be a natural number. Let us consider what set would be produced if our first entry included $n = 0$.

$$I(0, 0) = \{(0, 0), (1, 1), (2, 2), (3, 3), \ldots, (a, a), \ldots\}$$

The set shown above represents the zero integer. This is the only set in the set of integers which does not have a leading entry $(0, n)$ such that n is a natural number. Notice that this set follows every listed characteristic except no. 4.

We are now ready to define the set of integers. We will use the letter I to represent the set and to differentiate from the set J discussed in the last section. Remember that we described the set J by using the number line.

Definitions

■ **B.2**   *The set of integers I is a set such that each element in the set can be repre-*
■ *sented by $I(a, b) = \{(a, b) | a \text{ and } b \in W\}$.*

■ **B.3**   *The set of positive integers I^+ is a subset of I such that each element of*
■ *I^+ can be represented by $I(a, a + n)$ where $I(a, a + n) = \{(a, a + n) | a \in W$*
■ *and $n \in N\}$.*

■ **B.4**   *The set of negative integers I^- is a subset of I such that each element in*
■ *I^- can be represented by $I(a + n, a)$ where $I(a + n, a) = \{(a + n, a) | a \in W$*
■ *and $n \in N\}$.*

■ **B.5**   *The zero integer I^0 is a subset of I such that each element of I^0 can be*
■ *represented by $I(a, a) = \{(a, a) | a \in W\}$.*

Thus, $I = \{\ldots, I(3, 0), I(2, 0), I(1, 0), I(0, 0), I(0, 1), I(0, 2), I(0, 3), \ldots\}$

The definitions above require that the following statements be considered true. Since all the elements in the sets I^+, I^-, and I^0 are sets and since these sets are disjoint, it should follow from the definition that there is a partition on I formed by the sets I^+, I^-, and I^0. Therefore, we can state that

$$I^+ \cup I^- \cup I^0 = I \quad \text{and} \quad I^+ \cap I^- \cap I^0 = \phi$$

Note that any integer can be named by any ordered pair belonging to the set. In this way, $I(1, 5)$ can represent the set $\{(0, 4), (1, 5), (2, 6), \ldots\}$ or the integer $4 \in J$. However, we will general represent an integer by the the basic pair. Thus, we usually represent the integer $\{(0, 4), (1, 5), (2, 6), \ldots\}$ by $I(0, 4)$ instead of $I(1, 5)$ or $I(19, 23)$.

Notice that the basic pairs were chosen in such a way that we can easily translate a basic pair in I to an integer in J or directly indicate its position on the number line. We designated integers as $I(0, n)$ in I^+, because we translate them directly into the integer $n \in J^+$, which will lie to the right of 0 on the number line. In the same way, $I(n, 0)$ is in I^-, translates directly to $-n \in J^-$, and lies n units to the left of 0 on the number line. If we continue in this manner, we should see that $I(0, 5)$ can be translated to mean $5 \in J$ and $I(5, 0)$ would read $-5 \in J$. Since $I(0, 0)$ is the zero integer, we would translate this to mean $0 \in J$. Consider a few examples of this translation below.

Example 1

B.6   Translate $I(0, 3)$ into J.

SOLUTION   Since 3 is to the right of 0, $I(0, 3)$ translates to $3 \in J$.

Example 2

Translate $I(4, 0)$ into J.

■   SOLUTION   Since 4 is to the left of 0, $I(4, 0)$ translates to $-4 \in J$.

From the examples above, we should see that given an integer in J, we can translate it to the notation in I by considering its position on the number line and placing it in a relative position to 0 in the basic pair. Consider $-6 \in J$, which would be to the left of 0 on the number line. Since -6 is six units to the left of 0, we write $I(6, 0)$, placing the numeral 6 to the left of the numeral 0 in the pair. Now consider the examples below.

Example 1

B.7   Translate $4 \in J$ into the notation used in the set I.

SOLUTION   Since 4 is to the right of 0 on the number line, we write $I(0, 4)$
            to represent the integer.

Example 2

Translate $-3 \in J$ into the notation used in the set I.

SOLUTION   Since -3 is to the left of 0 on the number line, we write $I(3, 0)$
■   to represent the integer.

To summarize our notation, let us make the following statements. *An integer is a set of ordered pairs*, specifically, a set of ordered pairs where the numbers comprising the pairs are whole numbers. We sometimes represent an integer by writing any one of the pairs from the set that is the integer, whereas we generally represent the integer by a specific pair called the basic pair.

So we have defined a set of objects and have called them integers. We defined this set by using whole numbers. Before we can actually claim that this set is the same as J, we must show that our objects act in the way we really expect the integers to act. So it will be necessary to define operations on our set of integers. The remainder of this section is devoted to the groundwork necessary to define addition and multiplication.

EXERCISES

1. Represent the integers from the set J as sets of ordered pairs in the set I.
   a. 2   b. -3 c. 0

2. From the given pair produce the set which represents the integer in the set I.
   a. $(5, 9)$ b. $(7, 1)$ c. $(4, 4)$

3. Decide which of the following are in I^+, I^-, or I^0.
   a. $I(5, 7)$ b. $I(7, 7)$ c. $I(7, 9)$ d. $I(8, 2)$

4. Which of the following pairs belong to the same set in I?
   a. $(1, 2)$ b. $(3, 5)$ c. $(4, 5)$ d. $(2, 2)$ e. $(4, 4)$
   f. $(7, 3)$ g. $(8, 2)$ h. $(5, 7)$ i. $(8, 4)$ j. $(9, 3)$

INTEGERS AS SETS OF EQUIVALENT PAIRS OF WHOLE NUMBERS

In the exercise set you were asked to find pairs which represent the same set. We will consider several examples below before we define equality for integers. Consider the set which is represented by the basic pair $(0, 5)$.

$$I(0, 5) = \{(0, 5), (1, 6), (2, 7), (3, 8), \ldots\}$$

We state that $(0, 5) \sim (1, 6) \sim (2, 7) \sim (3, 8), \ldots$. In other words the pairs are equivalent. Notice we are *not* calling them equal pairs. What characteristic do you see that makes them alike? Note that in a pair (a, b) either $a - b$ or $b - a$ is a whole number. In this case $b - a$ is a whole number 5. This is one characteristic that we require for equivalence.

We choose, however, to employ a different characteristic of the set which will allow us more generality later. The pairs $(1, 6)$ and $(3, 8)$ can be related by the equation $1 + 8 = 6 + 3$. Notice that this is also true for $(0, 5)$ and $(2, 7)$, because $0 + 7 = 5 + 2$. We are really adding like this:

Try relating the other pairs in this manner. Consider the examples below:

Examples
■ **B.9** 1. If $(2, 5) \sim (4, 7)$, then $2 + 7 = 5 + 4$. Notice that $(5, 2) \sim (7, 4)$ and $(7, 5) \sim (4, 2)$. Also, $(2, 4) \sim (5, 7)$.

2. If $(3, 1) \sim (8, 6)$, then $3 + 6 = 1 + 8$. If $1 + 8 = 3 + 6$, then $(1, 3) \sim (6, 8)$. If $8 + 1 = 3 + 6$, then $(8, 3) \sim (6, 1)$, and if $3 + 6 = 8 + 1$, then $(3, 8) \sim (1, 6)$.

B.10 ■

3. If $4 + 2 = 5 + 1$, then $(4, 5) \sim (1, 2)$. Notice that $2 + 4 = 1 + 5$, $2 + 4 = 5 + 1$, and $4 + 2 = 5 + 1$.

4. If $3 + 2 = 1 + 4$, then $(3, 1) \sim (4, 2)$. Also, $2 + 3 = 4 + 1$ implies that $(2, 4) \sim (1, 3)$, $2 + 3 = 1 + 4$ implies that $(2, 1) \sim (4, 3)$, $3 + 2 = 4 + 1$ implies that $(3, 4) \sim (1, 2)$.

▲

We can check each of the examples above by finding the basic pair for the set and checking to see if both pairs are in the set. To do this we need to know how to find the basic pair from a given pair. For the pair

B.11 $(2, 5)$ we know that the basic pair is in the form $(0, a)$. If the relationship above holds, then we can find a as follows:

$$\text{if } (2, 5) \sim (0, a), \text{ then } 2 + a = 5 + 0, \text{ or } a = 3$$

■

This means that the set contains $(0, 3)$ as a leading element. It follows that the set which contains $(2, 5)$ is $\{(0, 3), (1, 4), (2, 5), (3, 6), (4, 7), \ldots\}$. This demonstrates that $(0, 3), (2, 5)$, and $(4, 7)$ are in the same set and can be used to represent that set.

B.11 ◄

You might try finding basic pairs for a given pair in this manner. Consider the operations in finding the pairs in an integer.

B.8 ▲

1. You are given a pair, for example, $(7, 3)$.

2. From the given pair you must decide if the basic pair is in the form $(a, 0)$, $(0, a)$, or (a, a). In this case, since the first entry is larger than the second, the pair will be $(a, 0)$.

3. Apply the definition we have developed: If $(7, 3) \sim (a, 0)$, then $(7 + 0 = 3 + a$, or $a = 4$.

Again we can check our results by writing the set to which these pairs belong. Since the leading pair is $(4, 0)$, the set contains $(4, 0), (5, 1), (6, 2)$,

▲ $(7, 3), \ldots$ and you can see that both pairs are in the set $I(4, 0) = \{(4, 0)$,

■ $(5, 1), \ldots\}$.

We are ready to define equivalent pairs of whole numbers at this point.

Definition

B.12 *For pairs (a, b) and (c, d), $(a, b) \sim (c, d)$ if and only if $a + d = b + c$.*

■

Notice that our definition implies two specific requirements:

1. If $(a, b) \sim (c, d)$ then $a + d = b + c$
2. If $a + d \sim b + c$ then $(a, b) = (c, d)$

From the second requirement, we can find equivalent pairs from sums which are given as equal. Consider the examples below.

Examples

■ **B.9** 1. If $2 + 5 = 3 + 4$, then $(2, 3) \sim (4, 5)$. Notice that the first pair comes from the leading number in each sum in the order written, but the second pair comes from the second number in each sum written in reverse order.

■ 2. If $7 + 1 = 8 + 0$, then $(7, 8) \sim (0, 1)$.

We must point out in the above examples that if $2 + 5 = 3 + 4$, then $5 + 2 = 3 + 4$ by the commutative property for whole numbers. Thus, $(5, 3) \sim (4, 2)$. Similarly, using the commutative property again we have that since $5 + 2 = 4 + 3$, $(5, 4) \sim (3, 2)$. Thus, we are able to manipulate the form of an addition fact to suit our convenience.

Let us now apply equivalence to integers. If $(a, b) \sim (c, d)$, then (a, b) and (c, d) are in the same equivalence class, that is, either of them could be used to represent the integer. Thus, the definition of integer permits us to say that

■ **B.16** ■ $I(a, b) = I(c, d)$ if and only if $(a, b) \sim (c, d)$.

EXERCISES

1. Construct a set of equivalent pairs using the definition for the given pairs.
   a. $(2, 5)$ b. $(7, 1)$ c. $(3, 2)$ d. $(4, 4)$

2. Find the basic pair for the given pairs using the definition of equivalent pairs.
   a. $(5, 12)$ b. $(8, 3)$ c. $(4, 8)$

3. From the given sums find equivalent pairs by using the definition.
   a. $5 + 7 = 12 + 0$ b. $8 + 5 = 4 + 9$
   c. $3 + 2 = a + 4$ d. $a + 5 = x + 6$

4. Under what conditions will the following pairs be equivalent?
   a. (a, b) and $(2, 4)$
   b. $(a + 2, a)$ and $(6, 4)$
   c. (a, b) and $(3, d)$

5. In $I(0, 4)$
   a. why are $(7, 11)$ and $(11, 15)$ not equal pairs?
   b. why are $(7, 11)$ and $(11, 15)$ equivalent pairs?

6. From the equivalent pairs stated, find three other equivalent pairs by using the entries in the stated pairs.
   a. $(2, 5) \sim (1, 4)$
   c. $(a, 6) \sim (b, 9)$
   b. $(5, 6) \sim (7, 8)$
   d. $(a, b) \sim (c, d)$

7. Match the sums on the left with the pairs on the right by using the definitions of equivalence.
   a. $5 + 6 = 9 + 2$ 1. $(2, 5) \sim (6, 9)$
   b. $3 + 6 = 4 + 5$ 2. $(3, 4) \sim (5, 6)$
   c. $2 + 9 = 5 + 6$ 3. $(5, 9) \sim (2, 6)$
   d. $4 + 5 = 3 + 6$ 4. $(4, 3) \sim (6, 5)$

EQUIVALENCE PROPERTIES

We have defined equivalence for ordered pairs of whole numbers. The question which should normally follow this definition is, Is equivalence for pairs of whole numbers an equivalence relation? We will answer this question in the affirmative.

If equivalence of pairs of whole numbers is an equivalence relation, then it is *reflexive, symmetric,* and *transitive.* We will show that it is both reflexive and transitive and leave it to you to show that the relation is symmetric.

PROOF OF REFLEXIVE PROPERTY: for all (a, b), $(a, b) \sim (a, b)$

B.13
1. a and $b \in W$	1. definition of ordered pairs of whole numbers
2. $a + b = b + a$	2. commutative property of addition for whole numbers
■ 3. $(a, b) \sim (a, b)$	3. definition of $\sim$ on I

We will state below a proof for the transitive property of equivalent pairs of whole numbers. We consider that we are given the following: $(a, b) \sim (c, d)$ and $(c, d) \sim (e, f)$. If the relation is transitive, then it should follow that $(a, b) \sim (e, f)$. Our job is to *prove* from the given statements that $(a, b) \sim (e, f)$.

PROOF OF TRANSITIVE PROPERTY

B.15
1. $(a, b) \sim (c, d)$ and $(c, d) \sim (e, f)$	1. given
2. $a + d = b + c$ and $c + f = d + e$	2. definition of $\sim$ on I.
3. $(a + d) + (c + f) = (b + c) + (d + e)$	3. reflexive property of equality, and substitute from step 2
4. $(a + f) + (d + c) = (b + e) + (d + c)$	4. Co and $A + W$
5. $d + c = d + c$	5. reflexive of $=$ for W
6. $a + f = b + e$	6. cancellation law for $+$ W
■ 7. $(a, b) \sim (e, f)$	7. definition of $\sim$ on I

EXERCISES

B.14 **1.** Show by example that if $(a, b) \sim (c, d)$, then $(c, d) \sim (a, b)$. Use definition of $\sim$ on I and the pairs $(2, 3)$ and $(4, 5)$.

■ **2.** Prove that the symmetric property holds for $\sim$ on I.

3. Using the properties of equivalence, make statements about the pairs below.
a. $(2, 4)$ and $(5, 7)$
b. $(6, 1)$, $(8, 3)$, and $(9, 4)$

4. If (a, b) and $(c, d) \in I(a, b)$, show that $(a, b) \sim (c, d)$.

SECTION C   ADDITION

BEHAVIORAL OBJECTIVES

C.1   Given two integers in J, find the sum.

C.2   Given two integers in I, find the sum.

C.3   Compare sums in I to sums in J.

C.4   Define $+$ for I.

C.5   Show that I is closed for $+$.

C.6   Show that I is commutative for $+$.

C.7   Show that I is associative for $+$.

C.8   Show that an identity exists in I.

C.9   Define additive inverse.

C.10 Show that inverses exist in I.

C.11 Use the cancellation laws for integers under addition to solve equations.

C.12 Prove cancellation for I.

ADDITION OF INTEGERS

We have constructed the set I so that it will be in 1–1 correspondence with the set J. In this manner, we would expect that an operation performed on elements in I and on the corresponding elements in J would produce corresponding results. Before we consider such operations, let us look at several examples of operations on elements in both sets simultaneously.

■ **C.1**   **Example 1**
▲ **C.3**   Find the sum of 3 and 2. In the set J we know that $3 + 2 = 5$. In the set I we expect that $I(0, 3) + I(0, 2) = I(0, 5)$. That is,

$$3 + 2 = 5 \in J$$

Example 2
■   Find the sum of $-4 + (-3)$. In the set J we know that $-4 + (-3)$
▲ $= -7$. In the set I we expect that $I(4, 0) + I(3, 0) = I(7, 0)$.

But suppose we do not use the basic pair to represent the operation. Let $I(4, 0)$ be represented as $I(8, 4)$, and let $I(3, 0)$ be represented by $I(9, 6)$. You can check to see that these pairs are equivalent. Then we expect that $I(8, 4) + I(9, 6) = I(7, 0)$. Notice that if we add the first elements in each pair and the second elements in each pair, we produce the pair $(17, 10)$ which is equivalent to $(7, 0)$.

Consider one last example. In this case we will choose a positive and a negative number from J. We will give the sum in J without explanation. If you remember your algebra from high school, you will recognize the result.

Example 3

C.1 ■   Find the sum of 5 and -8. In the set J, $5 + (-8) = -3$. In the set I we
C.3 ■ expect that $I(0, 5) + I(8, 0) = I(3, 0)$.

    The operation of addition of pairs would give $I(0, 5) + I(8, 0) = I(8, 5)$, which is equivalent to $I(3, 0)$, so we should be satisfied with the example.

    The examples shown above have implied a definition for addition. We are ready to define addition formally at this point.

Definition

■ C.4   *For integers represented by $I(a, b)$ and $I(c, d)$, the addition operation cor-*
■   *responds under $(+)$, $I(a, b) + I(c, d)$, with $I(a + c, b + d)$.*

    Before we discuss the properties of this operation, we would like to consider a few more examples in which we apply the definition. Notice that the definition actually employs the operation of addition on the set of *whole numbers* because a, b, c, and d are all whole numbers. In this way we do not actually operate on ordered pairs but only on the entries in the pairs which are numbers from another set.

    We must note that in the definition above we are using the symbol $+$ in two different contexts. The first is as addition of integers, as in $I(a, b) + I(c, d)$, the second is as addition of whole numbers, as in $I(a + c, b + d)$. We have defined a new operation, "$+$ of integers," in terms of an old operation, "$+$ of whole numbers." ·

Examples

◄ C.2 1. Find the sum of $I(2, 5)$ and $I(3, 8)$.

$$I(2, 5) + I(3, 8) = I(5, 13), \text{ or } I(0, 8)$$

Notice that we write the result in basic form. We could have translated the original pairs to basic pairs first and then added as follows:

$$I(2, 5) = I(0, 3), \text{ and } I(3, 8) = I(0, 5), \text{ then, } I(0, 3) + I(0, 5) = I(0, 8).$$

2. Find the sum of $I(3, 5)$ and $I(5, 3)$.

$$I(3, 5) + I(5, 3) = I(8, 8), \text{ or } I(0, 0).$$

Notice the relationship between these pairs. Their sum is zero. We will discuss the significance of this later.

3. Find the sum of $I(3, 4)$, $I(4, 2)$, and $I(5, 3)$.

    a. $I[(3, 4) + (4, 2)] + I(5, 3) = I(7, 6) + I(5, 3) = I(12, 9), \text{ or } I(3, 0)$
■   b. $I(3, 4) + I[(4, 2) + (5, 3)] = I(3, 4) + I(9, 5) = I(12, 9), \text{ or } I(3, 0)$

The final result was the same even though we grouped differently. We will look at this when we consider the properties of addition on I.

    The previous examples have used and demonstrated an important quality of our definition of integers: The result of adding integers does not depend upon the pair chosen to represent the integer. For example, $I(0, 5)$ can be represented by $I(1, 6)$ or by $I(7, 12)$; $I(0, 9)$ can be represented by $I(7, 16)$ or by $I(9, 18)$.

We know that $I(0, 5) + I(0, 9)$ should be $I(0, 14)$. Something would be drastically wrong with our definition if $I(1, 6) + I(7, 16)$ did not represent the same integers as $I(7, 12) + I(9, 18)$.

By the definition of addition

$$I(1, 6) + I(7, 16) = I(8, 22)$$

and

$$I(7, 12) + I(9, 18) = I(16, 30)$$

Now all our work on equivalent pairs comes in handy, because $I(8, 22) = I(0, 14)$, and $I(16, 30) = I(0, 14)$, so $I(8, 22) = I(16, 30)$ since both these pairs represent the integer $I(0, 14)$.

Thus, the sum of two integers is independent of the choice of pairs to represent the integers. The proof of the general case of this statement is left as an exercise.

EXERCISES

1. Find the sums indicated below, and write your answers in simplest form.
  a. $I(3, 4) + I(5, 7)$ b. $I(5, 7) + I(3, 4)$
  c. $I(6, 8) + I(8, 6)$ d. $I(9, 7) + I(5, 1)$
  e. $I[(7, 2) + (1, 6)] + I(2, 4)$ f. $I(7, 2) + I[(1, 6) + 2, 4)]$

2. Can you state a property which Exercise 1a and b may represent?

3. Find the following sums. What property do the sums imply?
  a. $I(2, a) + I(b, 3)$ b. $I(b, 3) + I(2, a)$
  c. $I(a, b) + I(c, d)$ d. $I(c, d) + I(a, b)$

4. Find the following sums. What property might these sums represent?
  a. $I(2, 3) + I(0, 0)$ b. $I(0, 0) + I(5, 4)$
  c. $I(4, 2) + I(3, 3)$ d. $I(3, 3) + I(4, 4)$
  e. $I(a, b) + I(c, c)$

5. Find the sums below and describe what the problems have in common.
  a. $I(3, 5) + I(5, 3)$
  b. $I(6, 4) + I(4, 6)$
  c. $I(a, b) + I(b, a)$

6. Find the following sums as indicated. Describe the property which may be illustrated in these problems.
  a. $I[(3, 5) + (6, 4)] + I(9, 7)$ b. $I(3, 5) + I[(6, 4) + (9, 7)]$
  c $I[(7, 8) + (3, 4)] + I(8, 6)$ d. $I(7, 8) + I[(3, 4) + (8, 6)]$
  e. $I[(a, b) + (c, d)] + I(e, f)$ f. $I(a, b) + I[(c, d) + (e, f)]$

7. Find an argument which shows that the sum of two integers in I will always produce an integer in I.

8. Suppose that (a, b) and (c, d) are two different pairs in $I(0, n)$ and that (e, f) and (g, h) are two different pairs in $I(0, k)$.

a. $(a, b) \sim ?$ $(e, f) \sim ?$
b. $a + d = ?$ $e + h = ?$
c. Find two different pairs that represent $I(0, n) + I(0, k)$.
d. Show that the two pairs in c are equivalent.

CLOSURE PROPERTY OF ADDITION ON *I*

The first property we will discuss is closure. Remember, to show that a set is closed, we must indicate that the operation on the set always produces elements in the set. We indicated in Chapter 3 that the whole numbers were closed under addition. Thus, the sum of two whole numbers is always another whole number.

In the set I, each integer is a set of equivalent ordered pairs. We will use the notation $I(a, b)$ for the integer that is the set of ordered pairs equivalent to (a, b). Thus $I(a, b) = \{(x, y)|(x, y) \sim (a, b)\}$.

More commonly, we use basic pairs to represent integers so that we will have integers written as $I(a, 0)$, $I(0, a)$, $I(a, a)$, etc.

Examples

1. The integer $-5 \in J$ corresponds to $I(5, 0)$, where $I(5, 0) = \{(x, y)|(x, y) \sim (5, 0)\}$.

2. The integer $3 \in J$ corresponds to $I(0, 3)$, where $I(2, 5) = I(0, 3)$ $= I(17, 20)$, or $(2, 5) \sim (0, 3) \sim (17, 20)$.

In Example 2 we cannot write $(2, 5) = (0, 3)$, because the *pairs* are obviously *not equal*. Notice that in a given pair the entries are always whole numbers. If we consider the sets represented by $I(2, 8)$ and $I(6, 5)$, by definition the sum $I(2, 8) + I(6, 5)$ is found by adding together the first entries and then adding together the second entries. In this way we produce the integer $I(8, 13)$. Notice that because the original pairs were made of whole numbers, the sums produced are also whole numbers (because whole numbers are closed under addition). Before using this new notation, let us repeat the definition of addition using the new notation.

Definition

C.4
■   For integers $I(a, b)$ and $I(c, d)$, the addition operation corresponds $I(a, b) + I(c, d)$ with the integer $I(a + c, b + d)$.

We would like to show formally that the operation of addition on I is closed. To do this, we must show that the sum of any two integers, represented by $I(a, b)$ and $I(c, d)$, is an integer; that is, a set of ordered pairs.

PROOF   Given $I(a, b)$ and $I(c, d)$ such that a, b, c, and d are in W, show that $I(a, b) + I(c, d)$ is an integer.

C.5   1. $I(a, b) + I(c, d) = I(a + c, b + d)$ 1. definition of $+$ in I
2. a, b, c, d in W 2. given

3. $a + c$ in W and $b + d$ in W 3. $Cl + W$

■ 4. $I(a + c, b + d)$ is an integer   4. definition of I

Notice in the proof above that we used the closure property for addition of whole numbers. This is often confusing for students, because it seems that we are using that which we are trying to prove. This is not really the case; the integers as we defined them are not whole numbers but rather sets of ordered pairs of whole numbers. The reason for defining integers in terms of whole numbers is simply that it enables us to use all the properties of whole numbers in proving the properties of the integers. Make note that we are using properties of W to prove properties in I. Notice that W is a subset of J and that in some sense I and J might be considered identical except for the way they are represented. We call sets like I and J *isomorphs*, and we say that I is isomorphic to J. We will discuss isomorphs in Chapter 6.

COMMUTATIVE PROPERTY OF ADDITION ON *I*

In the previous section we pointed out that we use the properties of whole numbers to prove properties for the set I. This practice will aid us in our proof of commutativity. We already know that whole numbers are commutative under the operation of addition. Since adding within the set I simply involves the use of addition in W, we can call on this property in W. Consider the examples below as well as those in the exercise set just preceding this section, particularly Exercise 3.

Examples
1. Find the sum of $I(4, 5)$ and $I(7, 2)$ as well as the sum of $I(7, 2)$ and $I(4, 5)$. Are the sums the same? Why?

$$I(4, 5) + I(7, 2) = I(11, 7) = I(4, 0)$$
$$I(7, 2) + I(4, 5) = I(11, 7) = I(4, 0)$$

2. Find the sums $I(2, a)$ and $I(b, 6)$ as well as the sum of $I(b, 6)$ and $I(2, a)$. Are the sums the same? Why?

$$I(2, a) + I(b, 6) = I(2 + b, a + 6)$$
$$I(b, 6) + I(2, a) = I(b + 2, 6 + a)$$

Notice in Example 1 that both sums were $I(11, 7)$, because $4 + 7 = 7 + 4$ and $5 + 2 = 2 + 5$. This is due to the commutativity of W under addition. In Example 2 you can see that $2 + b$ and $b + 2$ are the same whole number, as are $6 + a$ and $a + 6$. Again our reasons are that the numbers 2, a, b, and 6 are all in W, and W is commutative under addition.

To prove conclusively that the commutative property holds for I under addition, we must be able to completely generalize the proof. We do this by representing any two integers as $I(a, b)$ and $I(c, d)$. This ensures that our results will apply to any two integers in I. We will use the fact that the entries in the pairs are in W and that W is commutative under addition. With these facts in mind, consider the proof which follows below.

■ **C.6** TO PROVE For any $I(a, b)$ and $I(c, d)$ in I, $I(a, b) + I(c, d) = I(c, d) + I(a, b)$.

1. $I(a, b) + I(c, d) = I(a + c, b + d)$	1. definition of $+$ on I
2. a, b, c, d in W	2. definition of integer representation
3. $a + c = c + a$ and $b + d = d + b$	3. $Co + W$
4. $(a + c, b + d) \sim (c + a, d + b)$	4. definition of $\sim$ and reflexive property of equality from step 3
5. $I(a + c, b + d) = I(c + a, d + b)$	5. definition of $=$ on I
6. $I(c + a, d + b) = I(c, d) + I(a, b)$	6. definition of $+$ on I
7. $I(a, b) + I(c, d) = I(c, d) + I(a, b)$	7. transitive property of steps 1, 5, and 6

■

ASSOCIATIVE PROPERTY OF ADDITION ON I

Before we prove that addition is associative in I, we will produce several examples. You may also refer to the exercise set at the end of the preceding section, particularly Exercise 6. Remember, we have indicated in Chapter 3 that the whole numbers are associative under addition; this fact will be used to prove associativity for I under addition. Consider the examples which follow.

Example 1

Does $[I(3, 5) + I(2, 6)] + I(8, 5) = I(3, 5) + [I(2, 6) + I(8, 5)]$?

$$[I(3, 5) + I(2, 6)] + I(8, 5) = \underline{I(5, 11) + I(8, 5)}$$
$$= I(13, 16) = I(0, 3)$$
$$I(3, 5) + [I(2, 6) + I(8, 5)] = \underline{I(3, 5) + I(10, 11)}$$
$$= I(13, 16) = I(0, 3)$$

Notice that underlined steps in each example are different and yet the final result is the same. Can you explain why?

Example 2

Does $[I(a, 5) + I(2, b)] + I(c, d) = I(a, 5) + [I(2, b) + I(c, d)]$? Adding on the left;

$$[I(a, 5) + I(2, b)] = I(a + 2, 5 + b) \quad \text{and} \quad \underline{I(a + 2, 5 + b) + I(c, d)}$$
$$= I((a + 2) + c, (5 + b) + d)$$

Adding on the right,

$$I(a, 5) + [I(2, b) + I(c, d)] = \underline{I(a, 5) + I(2 + c, b + d)}$$
$$= I(a + (2 + c), 5 + (b + d))$$

In Example 2 we show as first entries $(a + 2) + c$ and $a + (2 + c)$, and as second entries $(5 + b) + d$ and $5 + (b + d)$. In this case the use of parentheses indicates grouping. The difference in grouping should be more apparent in Example 2 than in Example 1, in which we performed rather than simply indicated the operation. As you will remember, the numbers a, 5, 2, b, c, and d belong to the set W by definition. Thus, the associative property for whole numbers can be used to show that $(a + 2) + c = a + (2 + c)$ and that $(5 + b) + d = 5 + (b + d)$. This use of the associative property for whole numbers to prove associativity for I is an example of what we meant. This means that $(a + 2) + c$ is equal to $a + (2 + c)$ and that $(5 + b) + d$ is equal to $5 + (b + d)$. A formal proof of associativity in I for addition follows. We have left it for you to supply reasons for some steps.

■ **C.7** TO PROVE For $I(a, b)$, $I(c, d)$, $I(e, f)$ in I, $[I(a, b) + I(c, d)] + I(e, f) = I(a, b) + [I(c, d) + I(e, f)]$.

1. $[I(a, b) + I(c, d)] + I(e, f) = I(a + c, b + d) + I(e, f)$ 1. definition of $+$ in I

2. $I(a + c, b + d) + I(e, f) = I[(a + c) + e, (b + d) + f]$ 2. Why?

3. a, b, c, d, e, f in W 3. Why?

4. $[(a + c) + e, (b + d) + f] \sim [a + (c + e), b + (d + f)]$ 4. reflexive property of $\sim$ and associative property on W

5. $I[(a + c) + e, (b + d) + f] = [a + (c + e), b + (d + f)]$ 5. definition of $=$ on I

6. $I[a + (c + e), b + (d + f)] = I(a, b) + I(c + e, d + f)$ 6. definition of $+$ in I

7. $I(a, b) + I(c + e, d + f) = [I(a, b) + I(c, d)] + I(e, f)$ 7. Why?

8. $[I(a, b) + I(c, d)] + I(e, f) = I(a, b) + [I(c, d) + I(e, f)]$ 8. transitive property and steps 1, 2, 5, 6, and 7

■

EXERCISES

1. For the problems below, consider the set S, which contains ordered pairs (a, b) such that a and b are in N. An operation on S is called $*$, such that $(a, b) * (c, d) = (a \cdot c, b \cdot d)$, where $\cdot$ means multiply. Two pairs are equal by definition of equivalence for I.
   a. Show that the set is closed under $*$.
   b. Is the operation commutative in S?
   c. Is the operation associative in S?

2. Below are several mathematical statements using symbols from J. Rewrite the statements using the notation of this chapter; that is, use symbols from I and then give the name of the property that the statements demonstrate.
   a. $2 + 3 = 3 + 2$
   b. $(-2 + 3) + 5 = (-2) + (3 + 5)$
   c. $(-3) + (-2) = (-2) + (-3)$

IDENTITY ELEMENT IN *I*

We defined an identity element with respect to an operation in a set as an element which operates on all other elements in the set without producing a change in the other elements. For example, we have indicated that 0 is the identity for *W* under addition. You may have noticed that we have always stated properties for a set and an operation. It makes no sense to say that a set *S* is commutative unless we state the operation. It is also meaningless to state that an operation commutes without specifying a set. Therefore, properties should always be stated in terms of set and operation. Thus, 0 is the identity for *W* under $+$, since $0 + a = a$ and $a + 0 = a$ for all *a* in *W*. In plain English, 0 plus any number in the set *W* produces that number. We also stipulated that this must be true whether the identity acts on the right or on the left.

In the set *I*, we have an identity under addition. You may have guessed that the identity is $I(0, 0)$. We will indicate some examples below and then attempt to show two things: first, that an identity exists, and second, that it is *unique*. The proof of existence is simple, because we need only show that $I(0, 0)$ acts as an identity to prove that an identity exists. Also, we wish to know if there are many identities or if this is the only one in the set under $+$.

Examples

1. Does $I(3, 4) + I(0, 0) = I(3, 4)$, and does $I(0, 0) + I(3, 4) = I(3, 4)$?
   Since we know that *I* is commutative under $+$, we can answer the question by showing that $I(3, 4) + I(0, 0) = I(3, 4)$.

   $$I(3, 4) + I(0, 0) = I(3 + 0, 4 + 0) = I(3, 4)$$

2. Does $I(7, 8) + I(5, 5) = I(7, 8)$? $I(7, 8) + I(5, 5) = I(12, 13)$, which is equal to $I(7, 8)$. Remember that $(7, 8) \sim (12, 13)$ if $7 + 13 = 8 + 12$.
3. Does $I(5, 9) + I(a, a) = I(5, 9)$?

   $$I(5, 9) + I(a, a) = I(5 + a, 9 + a)$$

   and

   $$(5, 9) \sim (5 + a, 9 + a) \quad \text{if} \quad [5 + (9 + a)] = [9 + (5 + a)]$$
   which is true by *A* and *Co* $+$ *W*.

From the examples above, you should be fairly certain that (a, a), or its equivalent $(0, 0)$, represents the integer which acts as an identity under $+$ on *I*. To really convince you, we will formally show that this is the case.

C.8 TO PROVE $I(0, 0)$ is an identity in *I*, that is, for any $I(a, b) \in I$, $I(a, b) + I(0, 0) = I(a, b)$ and $I(0, 0) + I(a, b) = I(a, b)$.

1. $I(a, b) + I(0, 0) = I(a + 0, b + 0)$	1. Why?
2. $I(a + 0, b + 0) = I(a, b)$	2. 0 is an identity in *W*

3. $I(a, b) + I(0, 0) = I(a, b)$ 3. Why?

4. $I(0, 0) + I(a, b) = I(a, b)$ 4. commutative property in I on step 3

We now want to show that this identity is unique. That is, any element that acts as an identity is really $I(0, 0)$.

TO PROVE If $I(x, y)$ is an identity, then $I(x, y) = I(0, 0)$.

1. $I(x, y) + I(0, 0) = I(0, 0)$ 1. hypothesis

2. $I(x, y) + I(0, 0) = I(x, y)$ 2. $I(0, 0)$ is an identity on I

3. $I(x, y) + I(0, 0) = I(x, y) + I(0, 0)$ 3. Why?

4. $I(0, 0) = I(x, y)$ 4. transitive property of equality, steps 1, 2, and 3

ADDITIVE INVERSES IN I

We discussed the concept of inverse or opposite operations in Chapter 3. We might consider elements in a set as having opposite elements in a very special sense. It is only possible to discuss inverses in a set which has an identity element under a given operation. In other words, *the identity property is a necessary condition for having inverses.* We define inverse by asking the question, Are there pairs of elements from the set whose sum (in this case) is the identity element? Consider the examples below to help clarify this idea.

Example 1

Is there an element in I which when added to $I(0, 2)$ will produce $I(0, 0)$? Since we see that $I(0, 2) + I(2, 0) = I(2, 2) \sim I(0, 0)$, the answer is Yes.

Example 2

What element added to $I(2, 0)$ will produce $I(0, 0)$? Notice that $I(2, 0) + I(0, 2) = I(2, 2)$, or $I(0, 0)$.

If an inverse operates only on the right or only on the left, then we call it a *right inverse* or a *left inverse*. However, we will reserve the term *inverse* for those elements which produce the identity in combination with another element whether they act on the right or on the left. Before we formally define inverse, consider a few more examples.

Example 3

Find the inverse of $I(6, 4)$. We want an integer $I(x, y)$ such that $I(6, 4) + I(x, y) = I(0, 0)$. If this is true, then $(6 + x, 4 + y) \sim (0, 0)$. Thus, $6 + x = 4 + y$ by definition of equivalence and by $Co + W$ we have $x + 6 = y + 4$. Now, by definition of equivalence we see that $I(x, y) = I(4, 6)$. So the desired inverse is $I(4, 6)$. Notice that $I(6, 4) + I(4, 6) = I(10, 10) = I(0, 0)$, since $(10, 10) \sim (0, 0)$.

Example 4

Find the inverse of $I(2, 8)$. We want an integer $I(x, y)$ such that $I(2, 8)$ $+ I(x, y) = I(0, 0)$. Therefore, $2 + x = 8 + y$ and $x + 2 = y + 8$, which means that $I(x, y) = I(8, 2)$.

Notice in the examples above that we found the inverse of $I(6, 4)$ to be $I(4, 6)$ and the inverse of $I(2, 8)$ to be $I(8, 2)$. If this can be generalized, we might indicate that the inverse of $I(a, b)$ is an integer with the entries in reverse order, that is, $I(b, a)$. Let us check this generalization.

Example 1

If the inverse of $I(3, 8)$ is $I(8, 3)$, then the sum produced is $I(0, 0)$. $I(3, 8)$ $+ I(8, 3) = I(11, 11) = I(0, 0)$, also $I(8, 3) + I(3, 8) = I(11, 11) = I(0, 0)$. So we see that $I(8, 3)$ is the inverse of $I(3, 8)$.

Example 2

If the inverse of $I(a, b)$ is $I(b, a)$, then $I(a, b) + I(b, a) = I(0, 0) = I(b, a)$ $+ I(a, b)$. Notice that the sums produce $I(a + b, b + a)$ and $I(b + a, a + b)$ and that both of these are equal to $I(0, 0)$.

Definition

■ C.9 $I(x, y)$ is called the additive inverse of $I(a, b)$ if and only if $I(x, y) + I(a, b)$
■ $= I(0, 0)$ and $I(a, b) + I(x, y) = I(0, 0)$.

The above definition in terms of equivalent pairs that make up the integers reads

$I(x, y)$ is the additive inverse of $I(a, b)$

iff

$I(x, y) + I(a, b) = I(0, 0)$ and $I(a, b) + I(x, y) = I(0, 0)$

In Examples 3 and 4 we indicated a method for producing the additive inverse of a given integer. We will again show the steps involved in a more general manner. You should be able to furnish reasons for the proof.

PROOF   Given $I(a, b)$, find an $I(x, y)$ such that $I(a, b) + I(x, y) = I(0, 0)$.

■ C.10

1. $I(a, b) + I(x, y) = I(0, 0)$ Why?
2. $I(a + x, b + y) = I(0, 0)$ Why?
3. $(a + x, b + y) \sim (0, 0)$ Why?
4. $a + x + 0 = b + y + 0$ Why?
5. $a + x = b + y$ Why?
6. $x + a = y + b$ Why?
7. $(x, y) \sim (b, a)$ Why?

■   8. $I(x, y) = I(b, a)$ Why?

We summarize the properties of addition on I below. The first four properties listed are called *group properties*. If a set under an operation has these properties, then we say that the set and operation form a group. If the set is commutative under the operation, we call the structure a *commutative group* or an *abelian group*. (The latter is named after the mathematician Abel.) Therefore, we can state that the integers under addition form a commutative group. The properties of addition on I are

1. closure property
2. associative property
3. identity element
4. additive inverse
5. commutative property

EXERCISES

1. Find the inverse of the integers given below.
   a. $I(7, 9)$ b. $I(6, 6)$ c. $I(x, 8)$ d. $I(3, y)$ e. $I(x, y)$

2. Consider an operation on ordered pairs (a, b) and (c, d), where $a, b, c,$ and d are in W. The operation $*$ is defined as follows: $(a, b) * (c, d) = (a + b, c + d)$.
  a. Is the set closed under $*$?
  b. Is there an identity in the set? Explain.
  c. Are there inverses in the set? Explain.

3. Consider the same set defined in Exercise 2. The operation $\diamondsuit$ is defined as $(a, b) \diamondsuit (c, d)$ $= (a + d, b + c)$.
  a. Is the set closed under $\diamondsuit$?
  b. Is the set commutative?
  c. Is there an identity in the set?

CANCELLATION LAWS FOR I UNDER ADDITION
We showed in Chapter 3 that adding the same number to both sides of an equality does not affect the equality. We also showed that subtracting from both sides of an equality does not affect the equality. By combining these statements, we developed *cancellation laws* for the sets N and W. We will now consider the cancellation laws for I. We will consider both statements together; essentially, then, adding to or subtracting from both sides of an equality with the same number will produce an equivalent statement.

Before we prove this to be true, consider the examples below.

Examples
1. $I(2, 0) + I(3, 0) = I(5, 0)$
  $[I(2, 0) + I(3, 0)] + I(4, 0) = I(5, 0) + I(4, 0)$ (adding $I(4, 0)$ to both sides)

$$I(5, 0) + I(4, 0) = I(9, 0)$$
$$I(9, 0) = I(9, 0)$$

2. $I(2,0) + I(3,0) = I(5,0)$
$[I(2,0) + I(3,0)] + I(0,3) = I(5,0) \; + \; I(0,3)$ (adding $I(0,3)$ to both sides)

$$I(2,0) + [I(3,0) + I(0,3)] = I(2,0)$$
$$I(2,0) + I(3,3) = I(2,0)$$
$$I(5,3) = I(2,0)$$

Notice in the second example that the effect of adding $I(0,3)$, the additive inverse of $I(3,0)$, to both sides of the equality was much like that of subtracting $I(3,0)$ from both sides. Since we have not yet discussed subtraction, it can be considered here only on an intuitive level. We will discuss subtraction in the next section from this point of view.

Consider now a proof of the property discussed above.

Theorem

For $I(a,b)$, $I(c,d)$, and $I(x,y)$ in I, $I(a,b) = I(c,d)$ if and only if $I(a,b) + I(x,y) = I(c,d) + I(x,y)$.

Notice that this implies two things:

1. The "only if" part: If $I(a,b) = I(c,d)$, then $I(a,b) + I(x,y) = I(c,d) + I(x,y)$
2. The "if" part: If $I(a,b) + I(x,y) = I(c,d) + I(x,y)$, then $(a,b) = (c,d)$

We prove the "only if" part as follows:

■ **C.12**
1. $I(a,b) = I(c,d)$ 1. hypothesis
2. $I(a,b) + I(x,y) = I(a,b) + I(x,y)$ 2. reflexive property of equality
3. $I(a,b) + I(x,y) = I(c,d) + I(x,y)$ 3. replace $I(a,b)$ on the right side by its other name from step 2

The proof of the "if" part is nothing like the above proof.

1. $I(a,b) + I(x,y) = I(c,d) + I(x,y)$ 1. hypothesis
2. $I(a,b) + I(x,y) + I(y,x) = I(c,d) + I(x,y) + I(y,x)$ 2. part 1 of theorem
3. $I(a,b) + I(x+y, y+x) = I(c,d) + I(x+y, y+x)$ 3. definition of $+$ on I
4. $I(x+y, y+x)$ is the additive identity   4. previous work on $I(0,0)$
5. $I(a,b) = I(c,d)$ 5. from step 3 using identity property
■

The above theorem is used in solving equations, as the examples below illustrate.

Example 1

■ **C.11**   If $I(3,0) + I(x, y) = I(5,0)$, find $I(x, y)$.

SOLUTION $I(0, 3) + I(3, 0) + I(x, y) = I(0, 3) + I(5, 0)$
$$I(x, y) = I(5, 3)$$
$$I(x, y) = I(2, 0)$$

since $(5, 0) \sim (2, 0)$.

Notice that we added the inverse of $I(3,0)$ to both sides.

Example 2

$I(x, y) + I(6, a) = I(5, a)$. Find $I(x, y)$.

SOLUTION $I(x, y) + I(6, a) + I(a, 6) = I(5, a) + I(a, 6)$
$$I(x, y) = I(5, a) + I(a, 6)$$
$$(5 + a, a + 6) \sim (5, 6)$$
$$\sim (0, 1)$$

so

■
$$I(x, y) = I(0, 1).$$

Notice in the examples above that we associated without showing the changes in grouping. We also added inverses and assumed that $I(0,0)$ was produced by the addition. Since $I(0, 0)$ is the identity, we simply skipped several steps and wrote our results. In this way, the expression $I(2, 3) + I(3, 2) + I(5, 7)$ automatically becomes $I(5, 7)$. We do this now because we know that $I(2, 3) + I(3, 2)$ is $I(5, 5)$, or $I(0, 0)$. Since we know that $I(0, 0)$ is the identity, we do not bother to show that $I(0, 0) + I(5, 7)$ is $I(5, 7)$. This saves us considerable work.

EXERCISES

1. Find the inverse of the following integers.
   a. $I(3, 7)$ b. $I(8, 2)$ c. $I(9, 0)$ d. $I(e, 5)$ e. $I(c, d)$

2. Find $I(x, y)$ in the following sentences.
   a. $I(x, y) + I(4, 6) = I(2, 4)$ b. $I(3, 0) + I(x, y) = I(7, 5)$
   c. $I(a, b) = I(x, y) + I(a, b)$ d. $I(a, 2) = I(x, y) + I(a, b)$

3. Simplify the following expressions.
   a. $[I(a, b) + I(2, 3)] + I(b, a)$ b. $[I(3, 4) + I(9, 0)] + I(4, 3)$
   c. $[I(3, 4) + I(5, 4)] + I(8, 9)$ d. $[I(c, d) + I(d, c)] + I(a, b)$

4. The following equations are written in terms of J.

   $2 + n = 5$
   $n + 3 = 7$
   $n + 7 = 5$
   $5 + n = 1$

   a. Rewrite the equations using the set notation of I.
   b. Solve each equation.
   c. Translate the solutions back to the notation of J.

5. The following equations are written in terms of *J*.

$$4 + n = 9$$
$$4 + n = 3$$
$$n + 7 = 9$$
$$n + 7 = 5$$

Solve the equations *without* using the set notation of *I*.

SECTION D **SUBTRACTION**

BEHAVIORAL OBJECTIVES

D.1 Explain subtraction in *J*.

D.2 Explain subtraction in *I*.

D.3 Define subtraction in *I*.

D.4 Show that subtraction is closed in *I*.

D.5 Define zero property for subtraction on *I*.

SUBTRACTION ON *I*

Before we define subtraction, we must consider our definition of subtraction for *W*. In *W* subtraction was not closed. We found that there were no solutions to questions such as $3 - 5$ or $2 - 3$ in the whole numbers. Intuitively, one may realize that if we take 5 from 3 we are moving from right to left on the number line, as shown in Figure 5.5.

Figure 5.5

This type of explanation seems sufficient for such operations as $3 - 5 = -2$. However, we start to have difficulty when we subtract -3 from 5 and write $5 - (-3)$. Several writers have solved this problem by first differentiating between $-$, which means minus or subtract and is an operation, and $^-$, which means negative and refers to the size of a number, such as $^-3$. We could then state that subtraction is the inverse of addition and $a - b = c$ if and only if $c + b = a$. Let us consider an example using this point of view.

■ **D.1**

1. Find the difference $5 - (^-3)$. Let $c = 5 - (^-3)$. The solution can be found by solving the sentence $5 = {}^-3 + c$, or $c = 8$

2. Find the result when $^-4$ is subtracted from $^-5$. If $^-5 - (^-4) = c$, then $^-5 = {}^-4 + c$, or $c = {}^-1$

■

The development shown above does not differ greatly from the development which we plan to use; however, we find that it causes some difficulty in terms of the concept of — meaning subtract in one case and negative in another. The distinction is an important one and can more easily be made using ordered pair notation.

Consider again the question raised earlier of finding the difference $3 - 5$. Suppose we translate to ordered pair notation in the set I. Then the question reads, Find the difference $I(0, 3) - I(0, 5)$. To answer this we need only answer the question, "What must we add to $I(0, 5)$ to produce $I(0, 3)$?" Thus, we will define subtraction in terms of addition. The solution to this question follows: Find $I(0, 3) - I(0, 5)$. Make $I(0, 3) - I(0, 5)$ equal to $I(a, b)$. Then $I(a, b)$ is the integer that must be added to $I(0, 5)$ to produce $I(0, 3)$. That is, $I(0, 3) = I(0, 5) + I(a, b)$. Solving this by adding the inverse of $I(0, 5)$ to both sides of the sentence,

$$I(0, 3) + I(5, 0) = I(a, b) + [I(0, 5) + I(5, 0)]$$

Since

$$I(0, 5) + I(5, 0) = I(0, 0)$$

we produce

$$I(0, 3) + I(5, 0) = I(a, b)$$

The key to the above problem is adding the inverse of $I(0, 5)$. That is, if $I(0, 3) - I(0, 5) = I(a, b)$, then $I(a, b) = I(0, 3) + I(5, 0)$. This idea is the basis for our definition of subtraction. Before making a formal definition, let us see some examples.

Example 1

Find the difference $5 - (^-3)$. In the notation of the set I, this problem reads: Find $I(0, 5) - I(3, 0)$. Let $I(0, 5) - I(3, 0) = I(a, b)$; then adding the inverse as above, we have

$$I(0, 5) + I(0, 3) = I(a, b)$$
$$I(0, 8) = I(a, b)$$

Thus, $I(0, 5) - I(3, 0) = I(0, 8)$, or

$$5 - (^-3) = 8$$

Example 2

Subtract -4 from -5. That is, find $I(5, 0) - I(4, 0)$.

$$I(5, 0) - I(4, 0) = I(5, 0) + I(0, 4)$$
$$= I(5, 4)$$
$$= I(1, 0)$$

The answer is (-1).

We can summarize the results of our investigation as follows:

■ **D.2** 1. Subtraction can be defined as the inverse of addition. That is, $I(a, b) - I(c, d) = I(x, y)$ if and only if $I(a, b) = I(c, d) + I(x, y)$.

2. We can simplify this operation by using additive inverse.

3. Any subtraction on the integers can be translated to addition.

■ 4. The statement $I(a, b) - I(c, d)$ is equivalent to $I(a, b) + I(d, c)$, where $I(d, c)$ is the additive inverse of $I(c, d)$.

Let us use property 4 instead of property 1 to define subtraction on I. This will save one step in performing subtraction.

Definition
■ **D.3** *Subtraction on the set I is a binary operation that assigns to a pair of*
■ *integers $I(a, b) - I(c, d)$ the integer $I(a, b) + I(d, c)$.*

Thus, we have from the definition of subtraction the statement

$$I(a, b) - I(c, d) = I(a, b) + I(d, c)$$

CLOSURE OF SUBTRACTION ON *I*
In the previous section we showed that any subtraction problem can be translated to mean "add the inverse of the number being subtracted." In this way we can state that subtraction must be closed, because any subtraction can be performed by adding, and addition has been shown to be closed for I.

A proof of the above statement follows:

PROOF OF CLOSURE FOR SUBTRACTION ON I

■ **D.4** Given $I(a, b)$ and $I(c, d)$ in I, show that $I(a, b) - I(c, d)$ is in I.

1. $I(a, b)$ and $I(c, d)$ are in I	1. given
2. $I(d, c) \in I$	2. every element in I has an additive inverse in I
3. $I(a, b) + I(d, c) \in I$	3. addition is closed
4. $I(a, b) - I(c, d) = I(a, b) + I(d, c)$	4. definition of subtraction
■ 5. $I(a, b) - I(c, d) \in I$	5. placing step 4 in step 3

PROPERTIES OF SUBTRACTION ON *I*
We will state without proof the properties I has under subtraction. It will be left as an exercise for you to show that the properties stated are correct.

We have shown that subtraction is not commutative or associative for N or W by indicating counter-examples. Consider how you can show that this is also the case for the set I. We will simply state that I is *not* commutative or associative under subtraction.

It is often the case that writers call $I(0, 0)$ the identity element for subtraction. As we mentioned earlier, we require that an identity element commute under an operation. In this way, if $I(0, 0)$ were an identity for subtraction, it would be the case that $I(a, b) - I(0, 0) = I(a, b)$ and $I(0, 0) - I(a, b) = I(a, b)$. Check these sentences. You will notice that $I(a, b) - I(0, 0)$ is $I(a, b)$ but that $I(0, 0) - I(a, b) \neq I(a, b)$. Why? Thus, subtraction does not have an identity.

The property illustrated above is a zero property of subtraction. It can be simply stated as follows: Zero subtracted from any element in the set I will produce that element. Formally we state the property as follows.

■ **D.5** *Zero Property for Subtraction on I: For all $I(a, b)$ in I, $I(a, b) - I(0, 0)$*
■   *$= I(a, b)$.*

Examples

1. $I(2, 4) - I(0, 0) = I(2, 4) + I(0, 0) = I(2, 4)$
2. $I(5, 3) - I(0, 0) = I(5, 3) + I(0, 0) = I(5, 3)$
3. $I(a, b) - I(0, 0) = I(a, b) + I(0, 0) = I(a, b)$

Can you prove the zero property stated above?

Because we have stated that subtraction does not have an identity element, it follows that inverses do not exist for subtraction. Why?

EXERCISES

1. a. Find an example of two whole numbers that do not commute under subtraction.
   b. Translate the above example into the notation of this chapter.
   c. Find an example of two integers that do not commute under subtraction.

2. Find an example that shows that the associative property does not hold when the operation is subtraction on integers.

3. Prove the zero property for subtraction on I.

4. Explain why we do not have inverses for subtraction on integers.

SECTION E MULTIPLICATION

BEHAVIORAL OBJECTIVES

E.1 Compare muliplication in W to multiplication in I.

E.2 Be able to multiply in I.

E.3 Define multiplication in I.

E.4 Be able to multiply in J.

E.5 Prove that I is closed under multiplication.

E.6 Prove that I is commutative under multiplication.

E.7 Prove that *I* is associative under multiplication.

E.8 Show that an identity exists in *I* under multiplication.

E.9 Given an illustration of a property, identify the property.

E.10 Explain why integers do not have inverses under multiplication

E.11 Define the zero property for multiplication on *I*.

E.12 Define a distributive property of multiplication over addition on *I*.

E.13 Prove a distributive property of multiplication over addition on *I*.

DEFINITION OF MULTIPLICATION ON *I*

Before we define multiplication on *I*, it might be helpful to consider whole number multiplication. Our definition for multiplication on *I* must be devised in such a way that it is consistent with the operation already defined for *N* and *W*, since we can show a correspondence between *N* and a subset of *I* by corresponding any element *a* in *N* to the element *I*(0, *a*). With this in mind, consider the examples of multiplication shown below.

■ E.1 Multiplication on *N* and *W* gives

$$4 \times 5 = 20$$
$$3 \times 0 = 0$$
$$2 \times 4 = 8$$

We would expect the corresponding multiplication on *I* to give

$$I(0,4) \times I(0,5) = I(0,20)$$
$$I(0,3) \times I(0,0) = I(0,0)$$
■ $$I(0,2) \times I(0,4) = I(0,8)$$

The examples above suggest that the product of $I(0, a)$ with $I(0, b)$ is $I(0, ab)$. Essentially this is true; but suppose the pairs which represent the integers in question are not in simplest form? We must consider a definition which will not require that we simplify our pairs before we multiply. Consider the same examples again where we have used other pairs to represent the set.

$$4 \times 5 = 20$$
$$3 \times 0 = 0$$
$$2 \times 4 = 8$$

We would expect the corresponding multiplication on *I* to give

$$I(0,4) \times I(2,7) = I(0,20)$$
$$I(5,8) \times I(3,3) = I(0,0)$$
$$I(5,7) \times I(1,5) = I(0,8)$$

Notice that there seems to be no apparent connection between the pairs given above and the pairs given as the product. We simply have assumed that this is the only possible result, since multiplication is an operation which produces unique products for given factors. We realize that the product $I(0, 20)$ does not immediately relate to the factors $I(0, 4)$ and $I(2, 7)$, but let us consider the set of pairs comprising $I(0, 20)$ and see if any particular pair seems to relate to the factors.

$$I(0, 20) = \{(0, 20), (1, 21), (2, 22), (3, 23),$$
$$(4, 24), (5, 25), (6, 26), (7, 27), \underline{(8, 28)}, \ldots\}$$

We have underlined the pair $(8, 28)$ even though it may not be obvious that it relates directly to the factors $I(0, 4)$ and $I(2, 7)$. Notice that 4×7 is 28 and 2×4 is 8. This may give you some hint about the relationship that exists in given pairs and their product. Notice also that 0×7 is 0 and 0×2 is 0. In this way we have found all products of these four whole numbers.

Next let us consider the second example in the same manner. If we write the two pairs and the possible products of the entries in each pair with each of the entries in the other pair, the result looks like the products below.

$(5, 8)$ and $(3, 3)$ give the products $\begin{array}{l} 5 \times 3 = 15 \\ 8 \times 3 = 24 \end{array} \Big\rangle\ 15 + 24$

$\begin{array}{l} 5 \times 3 = 15 \\ 8 \times 3 = 24 \end{array} \Big\rangle\ 15 + 24$

We expect to produce the set represented by $I(0, 0)$ and note that the pair $(15 + 24, 15 + 24)$ is equivalent to $(0, 0)$. How we define multiplication of integers may not yet be clear until we consider the third example.

We found that $I(5, 7) \times I(1, 5)$ should equal $I(0, 8)$. Listing the products below may help us to put the pieces together.

$I(5, 7)$ and $I(1, 5)$ give the products $\begin{array}{l} 5 \times 5 = 25 \\ 7 \times 1 = 7 \end{array} \Big\rangle\ 32$

$\begin{array}{l} 5 \times 1 = 5 \\ 7 \times 5 = 35 \end{array} \Big\rangle\ 40$

Notice the sums produced, and relate them to the ordered pair $(32, 40)$, which is equivalent to $(0, 8)$.

Let us summarize our findings above and consider the application of these findings in the examples which follow this summary. We will state the relationship between the given pairs and their product in a step-by-step procedure.

1. To produce the first element of the product:
   a. Find the product of the first element in the leading pair with the second element in the other pair; that is, $(\underline{7}, 5)$ and $(5, \underline{1}) \to 7 \times 1 = 7$.
   b. Find the product of the last element in the leading pair with the first element in the other pair; that is, $(7, \underline{5})$ and $(\underline{5}, 1) \to 5 \times 5 = 25$.
   c. Find the sum of these products, namely, $(7 \times 1) + (5 \times 5) = 7 + 25$, or 32.

2. To produce the second element of the product:

   a. Find the product of the first element in each pair; that is, (7, 5) and (5, 1) → 7 × 5 = 35.

   b. Find the product of the second element in each pair; that is, (7, 5) and (5, 1) → 5 × 1 = 5.

   c. Find the sum of these products, namely, $(7 \times 5) + (5 \times 1) = 35 + 5$, or 40.

3. The ordered pair produced is $(32, 40) \sim (0, 8)$.

We might expect that the translation into J would produce the same result, namely $I(7, 5) \rightarrow -2 \in J$ and $I(5, 1) \rightarrow -4 \in J$. If $I(7, 5) \times I(5, 1) = I(0, 8)$, then $(-4) \times (-2) = 8 \in J$.

Before we consider more examples, look at the diagram in final form. The steps are summarized below by using regular type to indicate the products whose sum is the first entry and boldface for those products whose sum is the second entry.

$$I(7, 5) \times I(5, 1) = I([(7 \times 1) + (5 \times 5)], [(7 \times 5) + (5 \times 1)])$$
$$= I(32, \mathbf{40})$$

Consider the examples below in which we consider this method, and compare the results to multiplication in W.

■ E.3
. E.1 ▲ **Example 1**

   $5 \times 2 = 10$ in the set W, or $I(2, 7) \times I(2, 4)$ should be $I(0, 10)$.

$(2 \times 4) + (7 \times 2) = 22$

$(2 \times 2) + (7 \times 4) = 32$

The pair $(22, 32)$ is equivalent to $(0, 10)$.

Example 2

E.1 ▲ $6 \times 4 = 24$ in the set W, or $I(2, 8) \times I(3, 7)$ should be $I(0, 24)$.

$(7 \times 2) + (8 \times 3) = 38$

$(2 \times 3) + (8 \times 7) = 62$

■   The pair $(38, 62)$ is equivalent to $(0, 24)$.

The question which seems to arise is whether this process will hold for both positive and negative integers. If you remember your high school mathematics, you probably recall learning that the product of a positive and negative number produced a negative number, and the product of two negative numbers produced a positive number. We have tried to avoid the use of such rules, because they often cause misunderstanding in terms of the operations involved. We may consider these in light of our development as a check on our results. Consider the examples below to see if the results of our method are consistent with the rules you remember. These rules would apply to the operation on *J*.

■ E.3 **Example 1**

▲ E.4 ▲ $-7 \times 3 = -21$ in the set *J*, or $I(8, 1) \times (3, 6)$ should be $I(21, 0)$.

$\underline{(1 \times 3) + (8 \times 6) = 51}$

$(1 \times 6) + (8 \times 3) = 30$

The pair $(51, 30)$ is equivalent to $(21, 0)$.

Example 2

▲ E.4 ▲ $-5 \times -6 = 30$ in the set *J*, or $I(7, 2) \times I(9, 3)$ should be $I(0, 30)$.

$\underline{(2 \times 9) + (7 \times 3) = 39}$

$(2 \times 3) + (7 \times 9) = 69$

■ The pair $(39, 69)$ is equivalent to $(0, 30)$.

We have used several examples to give you an understanding of the definition that we plan to use for multiplication in *I*. The examples do not in any way prove that the definition is sound, they serve only to help you understand the definition. Definitions cannot be proven. We can show that a definition is faulty if we find a counter-example which shows the definition to be inconsistent. We are now ready for a formal definition of multiplication on *I*.

Definition

■ E.3   *Multiplication on the set I is a binary operation which corresponds to a*
■   *pair of integers I(a, b) and I(c, d) the integer I(ad + bc, bd + ac).*

Several examples are shown below. Try doing them on your own and check your results. Notice the comparison of results in *J*.

■ E.2 **Examples**

▲ E.4 1. $I(5, 2) \times I(2, 3) = I(2 \times 2 + 5 \times 3,$ 1. $(^-3) \times 1 = (^-3)$
                              $2 \times 3 + 5 \times 2)$
                      $= I(19, 16)$
                      $= I(3, 0)$

2. $I(0, 3) \times I(6, 9) = I(0, 9)$ 2. $3 \times 3 = 9$

3. $I(0, 5) \times I(4, 0) = I(20, 0)$ 3. $5 \times (^-4) = (^-20)$

4. $I(8, 6) \times I(5, 4) = I(0, 2)$ 4. $(^-2) \times (^-1) = 2$

5. $I(0, 0) \times I(9, 8) = I(0, 0)$ 5. $0 \times (^-1) = 0$

▲ ■ 6. $I(0, 1) \times I(2, 0) = I(2, 0)$ 6. $1 \times (^-2) = {}^-2$

Notice that Example 5 illustrates the fact that the element $I(0, 0)$ multiplied by another integer will produce $I(0, 0)$. This is the zero property and is consistent with out findings in the set of whole numbers. Notice in Example 6 that there is an element which acts like an identity for multiplication. The product of $I(0, 1)$ with any other element in the set will produce that other element. This too is consistent with our findings regarding identity elements in N and W under multiplication. We will show later that $I(0, 1)$ is the identity for multiplication on I.

EXERCISES

1. Find the following products:
   a. $I(2, 0) \times I(3, 0)$ b. $I(0, 2) \times I(3, 0)$
   c. $I(2, 0) \times I(0, 3)$ d. $I(0, 2) \times I(0, 3)$

2. Translate the products in Exercise 1 from the set I to the set J.

3. State rules which are implied by Exercise 2 about multiplication on J.

4. Find the following products:
   a. $I(2, 0) \times I(2, 3)$ b. $I(2, 3) \times I(2, 0)$
   c. $I(0, 3) \times I(2, 3)$ d. $I(2, 3) \times I(0, 3)$

5. Do you think that I is commutative under multiplication? Why or why not?

6. Find the following products:

   a. $I(2, 3) \times I(3, 5)$ b. $I(3, 2) \times I(5, 3)$
   c. $I(a, b) \times I(c, d)$ d. $I(b, a) \times I(d, c)$

7. Does the order in given pairs affect the product of the given pairs in Exercise 6? Why or why not?

8. Find the following products:
   a. $[I(2, 3) \times I(3, 2)] \times I(5, 3)$ b. $I(2, 3) \times [I(3, 2) \times I(5, 3)]$
   c. $[I(3, 0) \times I(5, 1)] \times (2, 0)$ d. $I(3, 0) \times [I(5, 1) \times I(2, 0)]$

9. Do you think that I is associative under multiplication? Why or why not?

10. Find the following products:
    a. $[I(2, 0) \times I(0, 3)] \times I(0, 1)$
    b. $[I(2, 0) \times I(0, 3)] \times [I(0, 1) \times I(0, 2)]$
    c. $[I(2, 0) \times I(0, 3)] \times [I(0, 1) \times I(0, 2)] \times I(0, 1)$

11. a. What is the effect of multiplying a given number in I by a number in I^-?
    b. Look at Exercise 10. Translate the products in Exercise 10 to J.
    c. Can you formulate a rule regarding multiplication by elements of J^-?

CLOSURE FOR MULTIPLICATION ON I

E.5   Consider the definition for multiplication on I. We are given $I(a, b)$ and $I(c, d)$ in I. This means that a, b, c, and d are in W. The product is defined as $I(ad + bc, ac + bd)$. Notice that the product is an ordered pair even though its formulation looks complicated. This pair has a first entry $ad + bc$. The operations of addition and multiplication on W have been employed. More formally we can state that ad is in W and bc is in W. Why? It is also the case that $ad + bc$ is in W. Why? We have established that the first entry in any pair is in W. We could establish that the second entry in any pair is also in W. How would we do this? Once this is done, we will have shown that both the first and second entries in any pair formed by the product of two elements from I are elements of W. In this way we will have shown that the product of two pairs in I is a pair which is in I by definition. It is left to you to finish this proof by answering the questions posed above.

Closure Property for $I \times$: *For all* $I(a, b)$ *and* $I(c, d)$, $I(a, b) \times I(c, d) \in I$.

EXERCISES

1. The product of $I(a, b) \times I(c, d)$ is $I(ad + bc, ac + bd)$ for all $I(a, b)$ and $I(c, d)$ in I.
   a. Is ac in W? Why?   b. Is bd in W? Why?
   c. Is $ac + bd$ in W? Why?   d. Is ad in W? Why?
   e. Is bc in W? Why?   f. Is $ad + bc$ in W? Why?

2. From Exercise 1, can you conclude that $I(ac + bd, ad + bc)$ determines an integer?

3. Consider the product $I(c, d) \times I(a, b) = I(ca + db, cb + bc)$. How is this product different from the product in Exercise 1?

4. Compare the products in Exercises 1 and 4.
   a. Does $ac = ca$? Why?
   b. Does $bd = db$? Why?
   c. Does $ac + bd = ca + db$? Why?

5. Explain why $ad + bc = cb + da$.

6. Can you prove that $I \times$ is commutative?

COMMUTATIVE PROPERTY OF MULTIPLICATION ON I

In the last two exercise sets, evidence of commutativity for I under multiplication has been shown. You may wish to review some of these exercises. As we have stated many times before, illustrative examples do not suffice as proofs but only as evidence of the existence or possible existence of properties. Consider the example shown below as an illustration of the commutative property of $I\times$.

Example
Does $I(2, 3) \times I(3, 1) = I(3, 1) \times I(2, 3)$?

$I(2, 3) \times I(3, 1) = I(2 \times 1 + 3 \times 3, 2 \times 3 + 3 \times 1) = I(11, 9) = I(2, 0)$
$I(3, 1) \times I(2, 3) = I(3 \times 3 + 2 \times 1, 3 \times 1 + 2 \times 3) = I(11, 9) = I(2, 0)$

Since we know that the set W is commutative for both addition and multiplication, we expect the products of pairs in I to be commutative under multiplication. Here we call on properties of W to prove properties in I. Again we point out that the idea of *using prior knowledge* about a *set and operation* to *prove properties* about a *new set and operation* is one of the cornerstones of a deductive system. It often appears as though one is using a property to prove the same property for a given set and operation. It is important to make this distinction. We do this by naming properties of sets and operations very specifically.

Notice in the proof which follows how we start out by operating in the set I (steps 1 and 2), then we use definition to move into the set W (steps 3–7), and finally we translate back into I by definition. In this manner we are able to call on our knowledge of the properties of W to introduce a new property for the set I.

■ **E.6**   TO PROVE   For $I(a, b)$ and $I(c, d)$ in I, $I(a, b) \times I(c, d) = I(c, d) \times I(a, b)$.

1. $I(a, b) \times I(c, d) = I(ad + bc, ac + bd)$	1. definition of multiplication on I
2. $I(c, d) \times I(a, b) = I(cb + da, ca + db)$	2. definition of multiplication on I
3. $ad + bc = bc + ad$	3. commutative property of multiplication on W
4. $bc + ad = cb + da$	4. commutative property of multiplication on W
5. $ad + bc = cb + da$	5. transitive property using steps 3 and 4
6. $ac + bd = ca + db$	6. commutative property of multiplication on W
7. $(ad + bc, ac + bd) = (cb + da, ca + db)$	7. reflexive property of equality using steps 5 and 6
8. $I(ad + bc, ac + bd) = I(cb + da, ca + db)$	8. from step 7 using the fact that an ordered pair determines an integer
9. $I(a, b) \times I(c, d) = I(c, d) \times I(a, b)$	9. transitive property and using steps 1, 8, and 2

Commutative Property for $I \times$: For all $I(a, b)$ and $I(c, d)$ in I, $I(a, b) \times I(c, d) = I(c, d) \times I(a, b)$

EXERCISES

1. Illustrate by an example that the set I is commutative under multiplication.

2. Illustrate the associative property for I under multiplication.

3. Does I have an identity element for multiplication? If so, what is it?

4. Given the set I and an operation $*$ defined as $I(a, b) * I(c, d) = I(ac, bd)$,
   a. is the set closed under $*$?
   b. is the set commutative under $*$?
   c. prove a and b to be true or false.

ASSOCIATIVE PROPERTY OF MULTIPLICATION ON I

In this section we will not prove formally the associative property. We will give several examples for you th consider and a very brief outline of the proof.

Example 1

Show that $[I(2, 3) \times I(0, 2)] \times I(0, 1) = I(2, 3) \times [I(0, 2) \times I(0, 1)]$.

$I(2, 3) \times I(0, 2) = I(0, 2)$ and $I(0, 2) \times I(0, 1) = I(0, 2)$

$I(0, 2) \times I(0, 1) = I(0, 2)$ and $I(2, 3) \times I(0, 2) = I(0, 2)$

■ E.7 **Example 2**

Show that $[I(a, 0) \times I(b, 0)] \times I(c, 0) = I(a, 0) \times [I(b, 0) \times I(c, 0)]$.

$$[I(a, 0) \times I(b, 0)] \times I(c, 0) = [I(0, ab)] \times I(c, 0)$$
$$= I(abc, 0)$$

and

$$I(a, 0) \times [I(b, 0) \times I(c, 0)] = I(a, 0) \times [I(0, bc)]$$
$$= I(abc, 0)$$

You can justify the associative property for the integers under multiplication by showing that

■ $[I(a, b) \times I(c, d)] \times I(e, f) = I(a, b) \times [I(c, d) \times I(e, f)]$

EXERCISES

1. Find the following products:
   a. $I(2, 3) \times I(3, 3)$ b. $I(3, 5) \times I(0, 0)$
   c. $I(2, 3) \times I(a, a)$ d. $I(x, y) \times I(a, a)$

2. What element from I was represented in every problem in Exercise 1?

3. Does the number $I(a, a)$ have any special property on $I\times$?

4. Find the following products:
   a. $I(1, 5) \times I(4, 3)$ b. $I(5, 4) \times I(2, 3)$
   c. $I(2, 6) \times I(1, 0)$ d. $I(a + 1, a) \times I(9, 7)$

5. Can you see a special element illustrated in each problem in Exercise 4? What is it?

6. Can you find a property that the number $I(a, a + 1)$ has?

IDENTITY ELEMENT FOR MULTIPLICATION ON I

In several of the exercises we have alluded to the fact that there is an identity element in the set I under multiplication. It seems reasonable to consider the element $I(0, 1)$ in I or 1 in J as an identity element under multiplication. Before we prove this formally, consider the examples below.

Example 1

■ E.8 $I(2, 3) \times I(0, 1) = I(2 + 0, 0 + 3) = I(2, 3)$

Example 2

$I(0, 1) \times I(a, b) = I(a + 0, b + 0) = I(a, b)$

We can represent any element in the set whose basic pair is $(0, 1)$ as $(a, a + 1)$. If we do this and consider the product of $I(a, a + 1)$ with another integer $I(b, 0)$ or $I(0, b)$, notice the results.

Example 3

$I(a, a + 1) \times I(b, 0) = I(b, 0) = I(ab + b, ab + 0)$

Example 4

$I(a, a + 1) \times I(0, b) = I(0 + ab, ab + b) = I(0, b)$

Notice that we have not bothered to show that the identity commutes. Why?

These examples should have prepared you sufficiently to construct a formal proof of the identity property in $I \times$. The proof simply consists of showing that the number $I(0, 1)$ behaves as an identity. That is,

■ *Identity Property for $I \times$: For all $I(a, b)$ in I there exists an element $I(0, 1)$ such that $I(0, 1) \times I(a, b) = I(a, b)$ and $I(a, b) \times I(0, 1) = I(a, b)$.*

EXERCISES

1. Find a relationship between $I(1, 0)$ and $I(0, 1)$ by computing the products below.
   a. $I(1, 0) \times I(1, 0)$
   b. $I(1, 0) \times I(0, 1)$
   c. $I(0, 1) \times I(0, 1)$

2. Can you show that the identity is unique?

3. Find the following products:
   a. $I(1, 0) \times I(2, 5)$ b. $I(1, 0) \times I(3, 1)$
   c. $I(1, 0) \times I(0, 0)$ d. $I(0, 1) \times I(0, 0)$
   e. $I(0, 1) \times I(2, 4)$ f. $I(0, 1) \times I(5, 1)$

4. Can you make a generalization regarding the relationships stated below? Use examples to illustrate your generalizations.
   a. $I(1,0) \times$ a positive integer   b. $I(1, 0) \times$ a negative integer
   c. $I(1, 0) \times$ a zero integer   d. $I(0, 1) \times$ a positive integer
   e. $I(0, 1) \times$ a negative integer   f. $I(0, 1) \times$ a zero integer

5. Restate Exercise 1 with respect to the set J.

6. Find the products in Exercise 3 by translating into the set J.

7. Translate Exercise 4 into the notation for the set J.

INVERSES UNDER MULTIPLICATION ON I

In the last exercise set you showed that $I(1, 0) \times I(1, 0) = I(0, 1)$. That is, $(-1) \times (-1) = 1$. Thus, $I(1, 0)$ is the multiplicative inverse of $I(1, 0)$ since their product is the identity.

It is reasonable to ask, Does every integer have a multiplicative inverse in the set of integers? Before we attempt to answer this general question, let us see if we can answer a similar question for a specific integer, say 2. Does $I(0, 2)$ have a multiplicative inverse in the set of integers? That is, is there some integer, say $I(a, b)$, such that $I(0, 2) \times I(a, b) = I(0, 1)$?

■ E.10   Let us try to find a and b in the above equation. Using the definition of multiplication, we get $I(0b + 2a, 0a + 2b) = I(0, 1)$, that is, $I(2a, 2b) = I(0, 1)$. For this to hold we know that $(2a, 2b) \sim (0, 1)$, or

$$2a + 1 = 0 + 2b$$
$$2a + 1 = 2b$$

It is obvious that both $2a$ and $2b$ are even because they are divisible by 2. Thus, $2a + 1$ must be odd, whereas $2b$ is even. It is impossible for an even number to equal an odd number, so it is impossible to find two natural numbers, a and b, such that $2a + 1 = 2b$. Thus, there is no solution in the set of integers to the equation $I(0, 2) \times I(a, b) = I(0, 1)$.

So we see that $I(0, 2)$ has no multiplicative inverse in the integers. With this example in mind you should be able to answer the more general question of whether every integer has a multiplicative inverse in the set of integers. Remember, one counter-example is sufficient to prove a

■   statement false.

EXERCISES

1. Show why this statement is true: Some integers have a multiplicative inverse in the set of integers.

2. Show why this statement is false: Every integer has a multiplicative inverse in the set of integers.

3. Find two integers which have multiplicative inverses in the set of integers.

ZERO PROPERTY FOR MULTIPLICATION ON I ($I \times$)

In the set W, we discussed the properties of zero. Zero takes on two important functions in W, namely, the identity property in $W +$ and the zero product property in $W \times$. We have already shown that the element $I(0, 0)$ is an identity under addition for I. It would seem reasonable to expect that $I(0, 0)$ might produce $I(0, 0)$ as a product when multiplied by any element in the set I. Consider the examples below.

1. $I(0, 0) \times I(2, 5) = I(0, 0)$
2. $I(3, 6) \times I(0, 0) = I(0, 0)$
3. $I(a, b) \times I(0, 0) = I(0, 0)$
4. $I(a, a) \times I(x, y) = I(ay + ax, ax + ay) = I(0, 0)$

■ **E.11**
■   *Zero Property for* $I \times$: *For all* $I(a, b)$ *in* I, $I(a, b) \times I(0, 0)$ *and* $I(0, 0)$
     $\times I(a, b) = I(0, 0)$.

EXERCISES

■ **E.9**   **1.** Name the properties illustrated below.
     a. $I(a, b) \times I(c, d) = I(c, d) \times I(a, b)$
     b. $I(a, b) \times I(x, y) = I(a, b)$ and $I(c, d) \times I(x, y) = I(c, d)$, etc.
     c. $I(a, b) \times I(q, r) = I(q, r)$ and $I(c, d) \times I(q, r) = I(q, r)$
     d. $[I(a, b) \times I(c, d)] \times I(e, f) = [I(c, d) \times I(a, b)] \times I(e, f)$
■   e. $[I(a, 0) \times I(b, 0)] \times I(c, 0) = I(c, 0) = I(a, 0) \times [I(b, 0) \times I(c, 0)]$

2. Find the results of the mixed operations below.
     a. $I(2, 3) \times (I(3, 4) + I(6, 2))$
     b. $I(2, 3) \times I(3, 4) + I(2, 3) \times I(6, 2)$
     c. $I(2, 0) \times (I(3, 0) + I(1, 0))$
     d. $I(2, 0) \times I(3, 0) + I(2, 0) \times I(1, 0)$

3. Compare the results of *a* with *b*, and *c* with *d*, in Exercise 2. Do you think that multiplication distributes over addition?

4. Using the same examples as those in Exercise 2, replace multiplication with addition and addition with multiplication. Compare the results. Does addition distribute over multiplication?

5. Consider the set *I* and the operation & such that $I(a, b) \& I(c, d)$ means $I(a + c, bd)$. Perform this operation below.

     a. $I(2, 3) \& I(3, 5)$ b. $I(0, 0) \& I(5, 6)$
     c. $I(1, 1) \& I(2, 5)$ d. $I(0, 1) \& I(2, 3)$
     e. $I(3, 5) \& I(2, 3)$ f. $I(5, 6) \& I(0, 0)$
     g. $I(2, 5) \& I(1, 1)$ h. $I(2, 3) \& I(0, 1)$

6. From Exercise 5, do you believe that the & operation has the properties listed below? Can you prove this?
     a. closure   b. commutativity
     c. associativity   d. identity

7. Consider the set *I* and the operation @ to mean $I(a, b) @ I(c, d) = I(a \times d, b \times c)$. Perform the operation on the pairs given in Exercise 5.

8. Answer Exercise 6 for the @ operation.

9. Prove the zero property for multiplication in the integers.

E.9   **10.** For *a*, *b*, and $c \in J$, name the properties illustrated below.
     a. $a \times 0 = 0$
     b. $a \times b = b \times a$
     c. $a \times (b \times c) = (b \times c) \times a$
     d. $a \times b = b$ and $b \times a = b$
■   e. $(a \times b) \times c = a \times (b \times c)$

11. a. Translate Exercise 2 into the set J.
  b. Show that the results of the operation in I and J preserve the 1–1 correspondence.

DISTRIBUTIVE PROPERTY OF MULTIPLICATION ACROSS ADDITION ON I

The general pattern produced by a distributive property can be symbolized as $X @ (Y \& Z) = (X @ Y) \& (X @ Z)$. This pattern reads as "the operation $@$ distributes across the operation $\&$" or "the distributive property of the operation $@$ over the operation $\&$." The way in which the property is shown above is normally described as a left distributive property, because the element being distributed is on the left.

We have discussed several distributive properties thus far in the text. In some cases, such as union and intersection, it was shown that each operation distributed over the other operation. In the case of multiplication and addition on W, we stated a distributive property of multiplication over addition. This is also true for the set I. Consider the examples below as illustrations of this property.

Example 1

$$I(2,0) \times [I(3,0) + I(4,0)] \qquad = [I(2,0) \times I(7,0)] = I(0,14)$$

and

$$[I(2,0) \times I(3,0)] + [I(2,0) \times I(4,0)] = I(0,6) + I(0,8) = I(0,14)$$

Example 2

$$I(0,2) \times [I(2,0) + I(0,1)] \qquad = [I(0,2) \times I(1,0)] = I(2,0)$$

and

$$[I(0,2) \times I(2,0)] + [I(0,2) \times I(0,1)] = I(4,0) + I(0,2) = I(2,0)$$

Notice in the examples that the expressions are written in two different ways. The computational procedures depend on the way in which the examples are written, yet the final results are the same. When this happens we say that the distributive property holds for the elements involved.

Rather than prove the distributive property formally, we will state an example which is general in nature and indicate how the property affects results. Consider the expansion of each of these expressions.

■ **E.13** 1. $\begin{aligned} I(a,b) \times [I(c,d) + I(e,f)] &= I(a,b) \times I(c+e, d+f) \\ &= I(a[d+f] + b[c+e], \\ &\quad a[c+e] + b[d+f]) \\ &= I(ad + af + bc + be, \\ &\quad ac + ae + bd + bf) \end{aligned}$

2. $[I(a, b) \times I(c, d)] + [I(a, b) \times I(e, f)] = I(ad + bc, ac + bd)$
$$+ I(af + be, ae + bf)$$
$$= I(ad + bc + af + be,$$
$$ac + bd + ae + bf)$$

■ Can you show that the results in 1 and 2 are equal?

■ **E.12** *Distributive Property of Multiplication Across Addition on I*: For all $I(a, b)$, $I(c, d)$, $I(e, f)$ in I,

■ $I(a, b) \times [I(c, d) + I(e, f)] = [I(a, b) \times I(c, d)] + [I(a, b) \times I(e, f)]$

EXERCISES

1. Use the distributive property to rewrite the expressions below. Check your work by computing in two ways.
a. $I(2, 3) \times \lfloor I(5, 7) + I(6, 0) \rfloor$
b. $I(3, 0) \times I(4, 2) + I(3, 0) \times I(5, 7)$
c. $I(3, 5) \times [I(4, 0) + I(0, 6)]$
d. $I(5, 8) \times [I(4, 6) + I(6, 4)]$

2. Do Exercise 1 using the notation of the set J.

3. For the operation (∗) on I defined earlier as $(a, b) ∗ (c, d) = (ac, bd)$ and the @ operation defined as $(a, b) @ (c, d) = (a \times d, b \times c)$ does ∗ distribute over @ or does @ distribute over ∗?

4. For the ∗ operation defined above, does ∗ distribute over the operation of addition + in I? Why or why not?

5. For the operations of multiplication and subtraction, does multiplication distribute over subtraction in I? Why or why not?

6. Rewrite the following statements using the notation of this chapter. Find the integer represented by each statement.
a. $2 \times (-3) \times 4 = ?$
b. $2 \times (-3) \times (-4) = ?$
c. $(-2) \times (-3) \times 4 = ?$
d. $(-2) \times (-3) \times (-4) = ?$
e. $5 \times [(-3) + (-2)] = ?$
f. $[7 + (-2)] \times (-5) = ?$
g. $(-3) \times [(-7) + (-5)] = ?$

SUMMARY OF PROPERTIES OF I UNDER + − ×

In the previous sections, we have developed the properties of $I \times$. You should notice that several of the properties for $I +$ and $I \times$ are the same. Again, we should point out that you should make the distinction between operations and sets when stating properties. This is particularly true in cases where a property relates to several sets and several operations. It is never sufficient to state that something is true due to commutativity. Commutativity has been shown to be a property of N, W, and I under both addition and multiplication. Therefore, we have indicated a commutative property six times. The properties of $I \times$ are shown in Figure 5.6 along with the properties of $I +$ and $I -$.

PROPERTIES OF OPERATIONS ON *I*

Property	Addition	Subtraction	Multiplication
Closure	Yes	Yes	Yes
Commutativity	Yes	No	Yes
Associativity	Yes	No	Yes
Identity	Yes	No	Yes
	Zero $I(0,0)$		$I(0,1)$
Inverse	Yes	No	No
Zero Properties	Yes	Yes	Yes
	Identity	Right identity	Zero property of multiplication
Distributive	Multiplication distributes over addition		
	Yes		
	Multiplication distributes over subtraction		
	Yes		

Figure 5.6

For the most part in Figure 5.6 we listed properties of single operations. The distributive property, unlike the other properties, is not a property of a single operation. It is not correct to discuss the distributive property of multiplication. We always state the distributive property as a property of one operation over another, so we must say "the distributive property of multiplication over addition or over subtraction." We did not prove the distributive property of multiplication over subtraction in the text but left it for the exercise set in the last section.

In the section on addition, we summarized the properties of addition and showed the integers under addition to be a commutative group. Notice that multiplication on I is commutative but that the integers under multiplication do not form a group, because not every integer has a multiplicative inverse. Remember, a group consists of a set on which an operation that has the following properties is defined:

1. closure
2. associativity
3. identity
4. inverses

EXERCISES

1. Does the operation @ on I form a group? Remember $I(a, b) @ I(c, d) = I(a \times d, b \times c)$.

2. Does the operation ✱ on I form a group? Remember that $I(a, b) ✱ I(c, d) = I(ac, bd)$.

3. Does the operation & on I form a group? Remember that $I(a, b) \& I(c, d) = I(a + c, bd)$.

4. a. List the group properties that the integers have under multiplication.
   b. Give examples illustrating group properties that integers under multiplication do not have.

SECTION F   DIVISION

BEHAVIORAL OBJECTIVES

F.1 Define division for I.

F.2 Be able to perform division for I and J.

F.3 Be able to state the cancellation law for muliplication on the integers.

F.4 Be able to use the cancellation law to solve sentences in I.

DIVISION ON I

In the chapter on whole numbers, we discussed division in terms of multiplication. We will follow the same procedure here. The symbol $\div$ will be used to mean division, and $I(a, b) \div I(c, d) = I(e, f)$ will be read "$I(e,f)$ is the *quotient* (or answer to division) when $I(c, d)$ divides $I(a, b)$," or "$I(e,f)$ is $I(a, b)$ divided by $I(c, d)$." The first question we have always asked is whether an operation is closed. Consider the definition of division before we answer the closure question. As with the natural and whole numbers, if we wish to state that $I(a, b) \div I(c, d) = I(e, f)$, we must show that $I(a, b) = I(c, d) \times I(e, f)$. The examples below will illustrate this.

Examples

■ F.2   1. $I(6, 0) \div I(3, 0) = I(0, 2)$ because $I(3, 0) \times I(0, 2) = I(6, 0)$

2. $I(0, 6) \div I(3, 0) = I(2, 0)$ because $I(3, 0) \times I(2, 0) = I(0, 6)$

3. $I(0, 6) \div I(0, 3) = I(0, 2)$ because $I(0, 3) \times I(0, 2) = I(0, 6)$

Examples

If we translate the examples above into J, they read as follows:

1. $-6 \div -3 = 2$ because $-3 \times 2 = -6$

2. $6 \div -3 = -2$ because $-3 \times -2 = 6$

■   3. $6 \div 3 = 2$ because $3 \times 2 = 6$

We can see from the examples above that we can find the quotient if we can answer the question, "Is there an $I(e,f)$ such that $I(cf + de, ce + df) = I(a, b)$?" Consider the manner in which we answer this question for example 1 above.

SOLUTION $I(6, 0) \div I(3, 0) = I(x, y)$

This means that $I(3, 0) \times I(x, y) = I(6, 0)$
$$I(3y, 3x) = I(6, 0)$$
$$(3y, 3x) \sim (6, 0)$$

Thus
$$3y + 0 \;\; = 3x + 6$$
or
$$3y - 3x = 6$$
Therefore,
$$y - x = 2$$
Now
$$y = x + 2$$
or
$$y + 0 = x + 2$$
and
$$x + 2 = y + 0$$
Thus
$$(x, y) \sim (0, 2)$$
Finally,
$$I(x, y) = I(0, 2)$$

In the notation of J this problem reads $-6 \div (-3) = 2$.

In example 2 above we want to solve $I(0, 6) \div I(3, 0) = I(x, y)$. This means that we must find a solution to the equation below.

SOLUTION $I(0, 6) = I(3, 0) \times I(x, y)$
or
$$I(0, 6) = I(3y, 3x)$$

Thus $(0, 6) \sim (3y, 3x)$, so that $x = 2$, $y = 0$ gives a solution. Thus $I(0, 6) \div I(3, 0) = I(2, 0)$, or in the notation of J, $6 \div (-3) = 2$.

In Example 3, $I(0, 6) \div I(0, 3) = I(x, y)$. To solve this we write the equation below.

SOLUTION $I(0, 6) = I(0, 3) \times I(x, y)$
$$I(0, 6) = I(3x, 3y)$$

Thus $(3x, 3y) \sim (0, 6)$. If $x = 0$ and $y = 2$, we have that $(3 \cdot 0, 3 \cdot 2) = (0, 6)$ so that $I(x, y) = I(0, 2)$. Rewriting this problem in J notation, we have $6 \div 3 = 2$.

We have seen that the division of integers, $I(a, b) \div I(c, d)$, depends upon the solution of the equation $I(a, b) = I(c, d) \times I(x, y)$. The closure of the integers under multiplication implies that for any $I(c, d)$ and any $I(x, y)$ there is an $I(a, b)$, but does not imply that for all $I(a, b)$ and $I(c, d)$ there exists an $I(x, y)$ that satisfies the above equation. Consider the following example, which illustrates this fact.

Example
$$I(7, 0) \div I(2, 0) = I(x, y)$$

SOLUTION $I(7, 0) = I(2, 0) \times I(x, y)$,
or
$$I(7, 0) = I(2y + 0, 2x + 0)$$

and
$$I(2y, 2x) = I(7, 0)$$
Thus,
$$2x + 7 = 2y$$
or
$$2x - 2y = -7$$
$$x - y = -7 \div 2$$

But $-7 \div 2$ is not in W, thus, $I(x, y)$ is not in I. This means that the equation $I(7, 0) = I(2, 0) \times I(x, y)$ has no solution in I. Therefore, $I(7, 0) \div I(2, 0)$ is not defined.

The example shown above is sufficient to disprove closure for the division operation. We are now ready to define division on I.

Definition

■ F.1   *Division on I is a binary operation for which $I(a, b) \div I(c, d) = I(x, y)$*
■   *if and only if there is an integer $I(x, y)$ such that $I(a, b) = I(c, d) \times I(x, y)$.*

EXERCISES

1. a. Rewrite each of the following statements in the notation of this chapter.
   b. Find the integer that each statement represents.

1.	$2 \div 1 = ?$	2.	$4 \div 2 = ?$
3.	$-4 \div 2 = ?$	4.	$4 \div -2 = ?$
5.	$-4 \div -2 = ?$	6.	$14 \div -7 = ?$
7.	$-28 \div -7 = ?$	8.	$-28 \div 7 = ?$

2. Show that division is not closed on the set of integers.

We stated the cancellation laws for multiplication on N and W in Chapter 3. We have also stated cancellation laws for addition on N, W, and I. We will state without proof and give examples of the cancellation law for multiplication on I. Remember that the cancellation law actually contains two ideas:

1. We can multiply both sides of an equality by the same quantity without changing the equality.
2. We can divide both sides of an equality by the same quantity without changing the equality.

We state this law more formally as follows:

■ F.3 *Cancellation law for multiplication on integers: For $I(a, b)$, $I(c, d)$,*
   $I(e, f)$ in I, $I(a, b) = I(c, d)$ if and only if $I(a, b) \times I(e, f) = I(c, d) \times I(e, f)$
■   *and $I(e, f) \neq 0$.*

Notice that we restricted $I(e, f)$ to I^+ or I^- only. The reason that we excluded zero is that $2 \times 0 = 0 = 3 \times 0$ would imply that $2 = 3$, which is obviously false. When we discuss the rationals we will prove this law and give several applications. The examples which follow should be sufficient for you to understand the meaning of the statement of the law.

Example 1

■ F.4   Given $I(a, b) \times I(2, 0) = I(6, 0)$, find $I(a, b)$.

SOLUTION $[I(a, b) \times I(2, 0)] \div I(2, 0) = I(6, 0) \div I(2, 0)$

$$I(a, b) = I(0, 3)$$

Notice that $I(0, 3) \times I(2, 0) = I(6, 0)$.

Example 2

Given $I(a, b) = I(c, d)$, show that $I(a, b) \times I(0, 2) = I(c, d) \times I(0, 2)$.

SOLUTION $I(a, b) \times I(0, 2) = I(a, b) \times I(0, 2)$ by the reflexive property of equality. Since $I(a, b) = I(c, d)$, we can replace the $I(a, b)$ on the right by $I(c, d)$. Thus, $I(a, b) \times I(0, 2) = I(c, d) \times I(0, 2)$.

■

EXERCISES

1. Solve the sentences below using the cancellation law for multiplication.
   a. $I(3, 0) \times I(a, b) = I(12, 0)$ b. $I(a, b) \times I(0, 2) = I(0, 12)$
   c. $I(0, 6) = I(0, 3) \times I(a, b)$ d. $I(0, 8) = I(2, 0) \times I(a, b)$

2. Show that for $I(a, b) = I(c, d)$, $I(a, b) \times I(e, f) = I(c, d) \times I(e, f)$. (*Hint*: Start with the reflexive property of equality.)

SECTION G   ORDER

BEHAVIORAL OBJECTIVES

G.1   Define order for I.

G.2   Test for order in I.

ORDER ON I

We have already defined equivalence for pairs (a, b) and (c, d). Actually, if two pairs are equivalent, they both belong to the same set, and in this way the sets represented by $I(a, b)$ and $I(c, d)$ are equal. As we showed with the set of whole numbers, two numbers in I are equal or not. If they are not equal, then one is greater than the other, which also implies that one is less than the other.

Thus, we have a trichotomy property. Two integers $I(a, b)$ and $I(c, d)$ relate to each other in one of three ways, $I(a, b) < I(c, d)$, $I(a, b) = I(c, d)$, or $I(a, b) > I(c, d)$ Before we consider the conditions necessary for each of these relationships, we must consider a definition of "less than" on I. We will then consider the "greater than" relation in terms of "less than." For example, if $I(a, b)$ is less than $I(c, d)$, then we will consider $I(c, d)$ to be greater than $I(a, b)$.

Before formally defining "less than," let us return for a moment to the set J and the number line. We can define "less than" on the number line to mean *to the left of.* In this manner, 0 is always less than any number in J^+ and any a in J^- is less than 0 or less than any b in J^+. If we look at the number line in Figure 5.7, we can see several examples of the less than relation defined on the line.

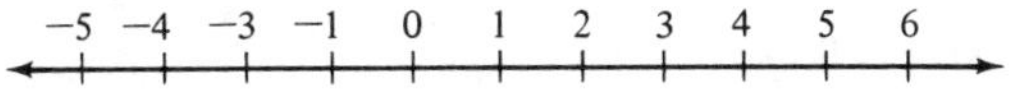

Figure 5.7

1. $-5 < -4 < -3 < -2 < -1 < 0 < 1 < 2 < 3 < 4 < 5 < 6$
2. Notice that $5 < 6$: then $6 > 5$.
3. $5 \not< 5$ and $5 \not> 5$.
4. If $-5 < -3$ and $-3 < 2$, then $-5 < 2$.

Let us translate several of the above statements directly from J to I, which should give us some insight into order in the set I. If elements in J^0 are less than those in J^+, then $I(a, a) < I(0, a)$. If elements in J^- are less than elements in J^0 and J^+, then $I(a, 0) < I(0, 0) < I(0, b)$. We can summarize these relationships below.

For $a, b \in N$,

1. $I(b, 0) < I(0, 0)$
2. $I(b, 0) < I(0, a)$
3. $I(0, 0) < I(0, a)$

We can translate some of the examples that follow Figure 5.7 from the set J to the set I.

Examples
1. If $-5 < -3$, then $I(5, 0) < I(3, 0)$. Notice that $I(5, 0) + I(0, 2) = I(3, 0)$.
2. If $-3 < 2$, then $I(3, 0) < I(0, 2)$. Notice that $I(3, 0) + I(0, 5) = I(0, 2)$.
3. If $5 < 6$, then $I(0, 5) < I(0, 6)$. Notice that $I(0, 5) + I(0, 1) = I(0, 6)$.
4. If $-1 < 0$, then $I(1, 0) < I(0, 0)$. Notice that $I(1, 0) + I(0, 1) = I(0, 0)$.

In the examples above, following the words "notice that" we always showed that there was a number from I^+ which when added to the smaller number produced the larger. This is how we have defined "less than" in the past, and we will follow this convention in the future. Therefore, if one wishes to know the relationship between $I(a, b)$ and $I(c, d)$, one must be able to find an $I(x, y)$ such that $I(x, y)$ is in I^+ and when added to one of the integers produces the other integer. Thus, if $I(a, b) + I(x, y) = I(c, d)$, then we say that $I(a, b) < I(c, d)$. If $I(c, d) + I(x, y) = I(a, b)$, then we say that $I(c, d) < I(a, b)$. Let us consider a few more examples before we define this relation.

Example 1

■ G.2   Which is the larger, $I(2, 5)$ or $I(2, 1)$?

SOLUTION   Since $I(2, 5) = I(0, 3)$ and $I(2, 1) = I(1, 0)$, we expect $I(0, 3)$ to be the larger.

We note that $I(1, 0) + I(0, 4) = I(1, 4) = I(0, 3)$ and that $I(0, 4) \in I^+$, so $I(2, 5)$ is the larger.

Example 2

Which is the larger, $I(5, 3)$ or $I(7, 3)$?

SOLUTION $I(5, 3) = I(2, 0)$ and $I(7, 3) = I(4, 0)$ The question is really, Which is the larger, $I(2, 0)$ or $I(4, 0)$? Since $I(4, 0) + I(0, 2) = I(4, 2) = I(2, 0)$ and $I(0, 2) \in I^+$, we say that $I(4, 0) < I(2, 0)$.

■

Definition

■ G.1
■   Given $I(a, b)$ and $I(c, d)$ in I, $I(a, b) < I(c, d)$ if and only if there is an $I(x, y)$ in I^+ such that $I(a,b) + I(x, y) = I(c, d)$.

EXERCISES

1. Which of each of the integers below is the larger? Why?
   a. $I(2, 3)$ and $I(5, 4)$ b. $I(6, 0)$ and $I(8, 4)$
   c. $I(3, 3)$ and $I(8, 9)$ d. $I(6, 8)$ and $I(0, 4)$
   e. $I(9, 3)$ and $I(10, 10)$

2. Name the relationship between the integers below by writing $<$, $=$, or $>$ between them.
   a. $I(2, 5)$ $I(2, 6)$ b. $I(2, 1)$ $I(2, 4)$
   c. $I(7, 5)$ $I(7, 2)$ d. $I(8, 6)$ $I(2, 0)$
   e. $I(3, 9)$ $I(2, 4)$ f. $I(a, 0)$ $I(0, b)$

3. Find an (x, y) in I^+ which will make an equality out of the integers below.
   a. $I(2, 4)$ and $I(2, 0)$ b. $I(4, 8)$ and $I(7, 5)$
   c. $I(0, 4)$ and $I(5, 2)$ d. $I(a + 2, 0)$ and $I(0, a)$
   e. $I(a + 4, a)$ and $I(a + 3, a)$ f. $I(a, b)$ and $I(a + 1, b + 2)$

CAPSULE III DISCOVERY SEQUENCES

At the end of Chapter 3, we discussed the discovery approach to teaching mathematics. We did not supply examples to clarify the discussion. We now suggest a few sequences which are examples of discovery approaches at the elementary level.

SEQUENCE 1

In the Madison Project materials, Robert Davis has suggested many excellent ideas for the classroom. Most of the approaches used can be applied at grade levels three through nine. Many of the sequences used in the Madison Project employ frames or placeholders. In this way open sentences are dealt with by students. Answers are given and checked by students. There is no attempt to formalize the material. Some examples follow below.

Examples

1. $4 + \square = 8$ solution: 4
2. $6 + \square = 8$ solution: 2

Now we jump to a more difficult problem.

Examples

1. $4 + (2 \times \square) = 12$ solution: 4
2. $6 + (2 \times \square) = 12$ solution: ?
3. $5 + (2 \times \square) = 25$ solution: 10
4. $6 + (2 \times \square) = 25$ solution: ?

In the few examples given above, the parentheses are explained simply by use. The teacher may try some number in the example $4 + (2 \times \square) = 12$ and state that $4 + (2 \times 1)$ is $4 + 2$, which does not equal 12. By doing this several times, the children learn to use parentheses without the need for rules.

Another example of sentences which can be used in the elementary grades is that of open sentences with different placeholders. Davis employs triangles and squares instead of the usual x and y of algebra. Some examples follow.

Examples

1. $\triangle = \square + 3$

Solutions can be shown in a table such as this:

$\triangle$	$\square$
4	1
5	2
6	?
?	7
?	?

2. $\triangle \times \square = 48$

Some solutions are given below. Can you find more?

$\triangle$	$\square$
2	24
4	12
6	?
?	8
?	?

If you find the sequences above to be helpful, we suggest that you go to the sources listed in the capsule for Chapter 3 and consider in greater detail the approach used by Robert Davis (1969).

SEQUENCE 2

Number machines were first introduced in programs such as the School Mathematics Study Group (SMSG). The idea of a machine which operates on numbers and transforms them into other numbers is very exciting for children in grades 1 to 6. The machines can be constructed for any level, and students can make their own machines for other children to discover. A number machine is simply a box with slots at the left and right. A number is entered at the left and another number leaves at the right. Children try to discover what the machine does. Consider a few examples below.

Example

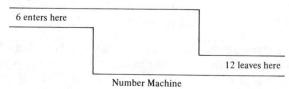

6 enters here

12 leaves here

Number Machine

Notice that the machine could be an "add 6 machine" or a "multiply by 2" machine. If we enter 5 and 11 leaves, then we might be able to see that it is an "add 6 machine," but if we enter 5 and 10 leaves, then it might be a "multiply by 2" machine.

Since we have three places to fill in the machine, we can also ask what entered the machine above if 20 leaves the machine and we have found the machine to be an "add 6 machine." By doing this we can deal with inverse operations.

If we wish to make this sequence more exciting, we might connect two machines as shown below.

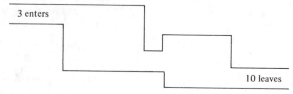

3 enters

10 leaves

Consider the machines connected above and see if you can discover the way in which they operate. Notice that 3 enters the first machine and 10 leaves. Consider that 4 enters and 12 leaves, 5 enters and 14 leaves, 6 enters and 16 leaves. Can you guess what happened?

Now suppose that you know that the first machine is a "multiply by 2" machine. Can you see that the second machine is an "add 4" machine? Suppose we know that 24 left the second machine; can you see what entered the first machine?

The examples shown above should give you some ideas about how to use number machines. Try inventing your own machines. Try using machines with young children and have them make their own machines.

The sequences shown above can be structured by the teacher, and later students can make their own sentences or machines. Each of the sequences deal with rules in a semiconcrete way. The sequence which follows is similar in nature because it deals with rules; however, it does so in a more abstract manner.

SEQUENCE 3

We call this sequence "Find My Secret." We simply employ ordered pairs which produce a result in the set with which we are dealing. We can produce results outside the set if we wish. In using this with children, one can deal with certain properties, particularly closure and commutativity. The sequence follows below.

Example

Write this statement on the blackboard: $(3, 5) \rightarrow 8$. Ask, "What is my secret? If you think you know, then tell me what this pair produces $(4, 2) \rightarrow ?$" Children might answer 6.

In the beginning, deal with simple rules like "add the entries." Later make the rules a little more complicated, as shown below.

Examples

See if you can find my secret.

$(3, 5) \rightarrow 11$
$(5, 3) \rightarrow 13$ Notice it is non-commutative.
$(1, 0) \rightarrow 2$
$(0, 1) \rightarrow 1$
$(1, 1) \rightarrow 3$ Have you discovered the secret yet?

Notice that this rule involves multiplying the first entry by 2 and then adding the second entry to the product. In using this sequence, do not ask children to verbalize the rule. By supplying the answers to the question, they indicate that they have found your secret. Notice also that this is a noncommutative rule.

Consider another secret below.

Examples

$(3, 4) \rightarrow 7$

$(5, 2) \rightarrow 7$ Does this look very simple?

$(6, 2) \rightarrow 6$ Notice that the rule here is not the add rule.

$(4, 5) \rightarrow 5$ Now you have to find another rule. Davis calls this technique *torpedoing*. The purpose is to break learning sets and make children think.

$(1, 2) \rightarrow 11$

$(2, 3) \rightarrow 9$

There are several rules that we can state for the sequence above. One is that the sum of the three numbers is 14. Another is to find the sum of the two entries and (1) *if the sum is greater than 7*, subtract the difference between the sum and 7 from 7, i.e., $7 - (\text{sum} - 7)$; or (2) if *the sum is less than 7*, subtract the difference between 7 and the sum from 7, that is, $(7 - (7 - \text{sum}))$. Notice that this is equivalent to subtracting the sum from 14; however, in looking at the pattern produced, children perceive the pattern in many ways. It is exciting to watch children formulate rules and at some point explain their rule to other children. This process helps with problem solving and in this case develops an intuitive understanding of mathematical relations.

When this sequence is used, encourage children to formulate their own secrets. Remember not to look for early verbalization, and structure a few sequences so that learning sets are broken. The sequences shown so far not only help children to discover but also allow the teacher to drill fundamentals painlessly.

The above sequences are examples of the discovery approach in action. For more about the discovery method, we again suggest the bibliography at the end of the capsule in Chapter 3.

CHAPTER 6 **STRUCTURES**

SECTION A **GROUPS**

BEHAVIORAL OBJECTIVES

A.1 Be able to state the properties of a group.

A.2 Be able to state the properties of a groupoid.

A.3 Be able to state the properties of a semigroup.

A.4 Be able to state the properties of a monoid.

A.5 Be able to classify familiar sets with operations as groups, monoids, groupoids, or semigroups.

A.6 Be able to show that a given set forms a group under the given operation.

A.7 Be able to define a subgroup.

A.8 Be able to find subgroups of a given group.

A.9 Be able to find the order of a given group.

GROUPS

In the past decade there has been a great deal of emphasis on the concept of structure in mathematics. In the chapters which preceded this, we were dealing with structures without really pointing out their importance. Essentially, we have been outlining the structure of the real number system starting with the set N and moving to larger and larger sets. We have continually mentioned that a set alone has no operation properties unless we define operations on that set. Actually, a set such as N is not completely described without the fundamental operations on the set and the properties of those operations.

In this chapter we discuss one particular type of structure, that of a *group*. We will introduce a second structure, called a *field*. We should point out that any non-empty set which has *at least one operation* defined on it can be called a structure. Therefore, we could have a name for a set which is closed and has no other properties. We could discuss a set and operation which is closed and associative only. In this manner we could build small structures. Actually, each of the above structures is named and does have meaning for the mathematician. The first is called a *groupoid*, and the second is called a *semigroup*. If we add the identity property to the semigroup (a set with an operation that is both closed and associative), we form a structure called a *monoid*. We might consider structures which have been discussed in previous chapters to see which name applies to them. Remember, our main concern is that of discussing the *group* concept.

■ A.2

▲ A.3

▲ ■

■ A.4

■

Definitions

■ A.2 ■ 1. *A groupoid is a set and operation which is closed.*

■ A.3 2. *A semigroup is a set and operation with the following properties:*
   a. closure property
   ■ *b. associative property*

■ A.4 3. *A monoid is a set and operation with the following properties:*
   a. closure property
   b. associative property
   ■ *c. identity property*

EXERCISES

■ A.5 1. What structure is formed by integers under subtraction? Why?

2. What structure is formed by even integers under addition?

3. Name the structure formed by *N* under
   a. addition
   b. subtraction
   c. multiplication

4. Name the structure formed by integers under
   a. multiplication
   ■ b. division

5. Make a table for the set $X = \{a, b, c\}$ such that it forms
   a. a semigroup
   b. a monoid
   c. a groupoid

DEFINITION OF GROUP

The concept of group had its birth in the eighteenth century. However, group concepts were used by early civilizations. The Babylonian and Egyptian cultures were the first to use the concept of a modular group, which is more familiar to us as clock arithmetic. Later in the chapter we will discuss modular arithmetic. An early application of group theory was introduced by Galois and was named Galois theory; it deals with physical interpretations and applications of groups to particle theory. Galois (1811–1832) introduced the term group and carried the concept farther than his predecessors, Euler (1707–1783), LaGrange (1736–1813), Gauss (1777–1855), and Abel (1802–1829). As we mentioned earlier, the Abelian, or commutative, group was named after Abel.

In our discussion of the integers, we pointed out that the integers form an Abelian group under addition. We defined group in that section. However, to bring you up to date, we will again define a group. Notice that the group concept did not appear until we developed the integers. This was due to the fact that we needed inverse as a property of our operation on the set.

Definition

■ **A.1** *A group is a set S and a binary operation ∘ which has the following properties:*

1. *closure: for all a, b, ∈ S, a ∘ b ∈ S*
2. *associativity: for all a, b, c ∈ S, (a ∘ b) ∘ c = a ∘ (b ∘ c)*
3. *identity: there is an element i ∈ S such that for any a ∈ S, a ∘ i = a = i ∘ a*
4. *inverses: for any a ∈ S, there is an $a^{-1} ∈ S$ such that $a ∘ a^{-1} = i$ and*
■ *$a^{-1} ∘ a = i$*

■ **A.1** We could state the definition above more simply if we had defined the
■ substructures mentioned earlier. Since a monoid has the first three
properties mentioned above, *a group is simply a monoid with inverses.*
Most of the discussion of groups in this chapter will deal with finite
groups (i.e., the group has a finite number of elements), since it is much
simpler to check the properties of a finite group. Consider next several
examples of groups.

Example 1

The set $S = \{A, B, C\}$ with an operation ∘ forms an Abelian group.

∘	A	B	C
A	B	C	A
B	C	A	B
C	A	B	C

■ **A.6** To prove that S under the operation ∘ forms a group, we must show
that S satisfies the four group properties. Let us consider these properties
in the order stated thus far (closure, associativity, identity, and inverse).

1. If we consider the given set $S = \{A, B, C\}$ and then look at the table
produced by the operation ∘ on S, we notice that all the entries in the
table are elements of S. Therefore, the operation on S is closed. We
can always show closure by inspecting the elements which are produced
under the operation. If at least one element which is produced by the
operation is *not* an element of the set, then the operation is not closed.

2. To show that the operation on the set S is associative, we must consider
all combinations of A, B, and C taken three at a time. If we list these,
we can see that there are quite a few. Therefore, we will consider only
a few examples here and leave to you the task of checking the others.
Is $(A ∘ A) ∘ A = A ∘ (A ∘ A)$? Since $(A ∘ A) ∘ A = B ∘ A$ and $A ∘ (A ∘ A)$
$= A ∘ B$, and since $A ∘ B = B ∘ A$, $A ∘ (A ∘ A) = (A ∘ A) ∘ A$. One can
readily check $(B ∘ B) ∘ B = B ∘ (B ∘ B)$ and $(C ∘ C) ∘ C = C ∘ (C ∘ C)$. Consider
now $(A ∘ B) ∘ B$ and $A ∘ (B ∘ B)$. $(A ∘ B) ∘ B = C ∘ B$ because
$A ∘ B = C$. $A ∘ (B ∘ B) = A ∘ A$ because $B ∘ B = A$. Now $C ∘ B = B$
and $A ∘ A = B$. Therefore, $A ∘ (B ∘ B) = (A ∘ B) ∘ B$. We will show the
last few examples in a shorter form.

a. $(A \circ C) \circ B = A \circ (C \circ B)$ because $(A \circ C) = A$ and $(C \circ B) = B$. Thus, $(A \circ C) \circ B = A \circ B$ and $A \circ (C \circ B) = A \circ B$.

b. $(B \circ A) \circ C = B \circ (A \circ C)$ because $(B \circ A) = C$ and $(A \circ C) = A$. Thus, $(B \circ A) \circ C = C \circ C$, or C, and $B \circ (A \circ C) = B \circ A$, or C.

c. $B \circ (A \circ B) = (B \circ A) \circ B$ because $(B \circ A)$ and $(A \circ B)$ each equal C; $B \circ (A \circ B) = B \circ C$; and $(B \circ A) \circ B = C \circ B$. Since $C \circ B = B \circ C$ from the operation table, we know that $B \circ (A \circ B) = (B \circ A) \circ B$.

We have shown that five examples hold true for the associative property. We have also implied that two others are true. By checking you can see that the associative property holds. Without further examples, we will state that this operation on S is associative. We must point out that we have not proven associativity. To prove beyond a doubt that S under $\circ$ is associative, we must show that associativity holds for all combinations of elements from the set taken three at a time.

Let us now ask the question: How many combinations of three elements are there? To answer this question we must consider that elements can be repeated and that a change in order also changes the configuration. If we place A in first position, there are three elements which can be placed in second position. For each pair of elements there are three choices for the third element. For example:

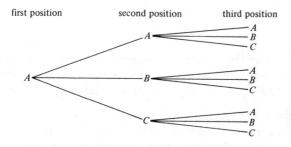

Notice that if A is first, there are 9 possible combinations. If we do the same for B and C, there will be 27 possible combinations of the elements A, B, and C. Can you find the number of combinations of two elements taken three at a time? Notice that for A and B only, the pattern looks like this.

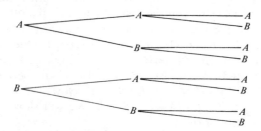

Therefore we have 8 possibilities. Consider the pattern below.

2 elements → 8 possibilities
3 elements → 27 possibilities
4 elements → ? possibilities

If you guessed 64 possibilities, then you see the emerging rule!

3. Consider now the identity element in S. Since $C \circ A = A = A \circ C$, $C \circ B = B = B \circ C$, $C \circ C = C = C \circ C$. C acts as an identity. In general, we can state that for all $x \in S$, $C \circ x = x$ and $x \circ C = x$. Therefore, C is the identity element in S under $\circ$.

4. The last property needed to form a group is that of inverse. We must show that for every element in the set there is another element such that the two operating together will produce the identity C. The two elements need not be different from each other. We can see that C appears in three places in the operation table: $A \circ B = C$, $B \circ A = C$, and $C \circ C = C$. This means that A and B are inverses and C is its own inverse. Since each element has an inverse, S forms a group under $\circ$.

If we consider the table which defines S under $\circ$, we can see that S is commutative under the $\circ$ operation. Therefore, S forms an Abelian group under $\circ$. We leave the proof of this to you.

Example 2

A.6   The set $A = \{I(1, 0), I(0, 1)\}$ with the operation $\times$ forms an Abelian group. Consider the operation table first.

$\times$	$I(1, 0)$	$I(0, 1)$
$I(1, 0)$	$I(1, 0)$	$I(0, 1)$
$I(0, 1)$	$I(0, 1)$	$I(1, 0)$

Since the set A is a subset of I, we know that it is both commutative and associative under $\times$. From the table we can see that $I(1, 0)$ is an identity and that $I(1, 0)$ and $I(0, 1)$ are their own inverses, because $I(1, 0) \times I(1, 0) = I(1, 0)$, and $I(0, 1) \times I(0, 1) = I(1, 0)$.

Notice this is a group and yet the set I under $\times$ does not form a group. Why?

Example 3

The following table defines the operation $*$ on the set $\{I, A, B, C, D, E\}$. Notice that the operation on the set is not commutative, yet all the requirements needed for a group under $*$ are satisfied.

A.9   The groups shown are called groups of order two and order six respectively. *The order of a finite group refers to the number of elements in the set.* If you look at the group of order six, you will notice that the subset $\{I, A, B\}$ forms a group of order three. In this case the group $\{I, A, B\}$ is commutative even though the group which contains it is

*	I	A	B	C	D	E
I	I	A	B	C	D	E
A	A	B	I	D	E	C
B	B	I	A	E	C	D
C	C	E	D	I	B	A
D	D	C	E	A	I	B
E	E	D	C	B	A	I

non-commutative. We showed earlier that the sets $S = \{A, B, C\}$ and $A = \{I(1, 0), I(0, 1)\}$ formed groups under the operations of $\circ$ and $\times$.

■ **A.6** We have not satisfied you that the set $S = \{I, A, B, C, D, E\}$ forms a group. To do so we must show that the four group properties are satisfied. We will not do this in order because the associative property is a long one to satisfy. For a finite set under an operation, we satisfy associativity by listing all examples or by calling on properties which the set already possesses. For example, if we know that a set is associative under $\times$, then its subsets, if they are closed, are also associative under $\times$. In this case, we know no properties of S which will help us; therefore we must list results. The set S is closed under *, as can be seen from the operation table. The set S has an identity under *, because I operating on any element on the right or left produces that element. We can see from the table that I, C, D, and E are their own inverses and that A and B are inverses. Remember, two elements are inverses if the operation on the elements produces the identity. *An easy way to check for inverses is to see if the identity element appears in each row and column of the table.* If it does not appear, then there cannot be inverses on this structure. If the identity does appear in this manner, then see if it follows the definition. Remember, if $ab = i$ where i is the identity, then ba must also equal i. In other words, inverses must commute.

So far we have shown that three of the four properties of a group are exhibited by S and the operation $\times$. To prove that this set operation is associative, we must list all ways in which the six objects can be taken three at a time. In this case there are $6^3 = 216$ possible ways of writing these elements three at a time. We will simply indicate what some of these triples are as an indication of associativity. This is not a proof. In the case of smaller sets, it can be demonstrated more easily.

$$A \times (A \times A) = I \quad \text{and} \quad (A \times A) \times A = I$$
$$A \times (A \times B) = A \quad \text{and} \quad (A \times A) \times B = A$$
$$A \times (A \times C) = E \quad \text{and} \quad (A \times A) \times C = E$$
$$\cdot \qquad\qquad\qquad \cdot \qquad\qquad\qquad \cdot$$
$$\cdot \qquad\qquad\qquad \cdot \qquad\qquad\qquad \cdot$$
$$\cdot \qquad\qquad\qquad \cdot \qquad\qquad\qquad \cdot$$

$$E \times (I \times A) = D \quad \text{and} \quad (E \times I) \times A = D$$
■ $$E \times (I \times I) = E \quad \text{and} \quad (E \times I) \times I = E$$

We suggest that you satisfy yourself that the examples above are correct by using the table. You may wish to explore other combinations on your own as an exercise. We do not suggest it be done for every case because it is such a tedious job. The exercises which follow will allow you to carefully show whether a structure is or is not a group.

EXERCISES

1. Which of the tables below define groups? Why?

□	A	B
A	A	B
B	B	A

*	A	B
A	A	A
B	A	B

○	A	B
A	A	B
B	B	B

2. Does the larger number operation on the set $\{2, 4, 6\}$ form a group? Why or why not? Remember $2 \triangle 4 = 4$ or $(2, 4) \xrightarrow{\triangle} 4$

3. Does the first number operation on the set $\{2, 4, 6\}$ form a group? Why or why not? Remember $(2, 4) \xrightarrow{*} 2$ but $(4, 2) \xrightarrow{*} 4$

Note: The operations $*$ and $\triangle$ were defined in Chapter 3.

4. From Example 3 of this section, show that the set $S = \{I, A, B\}$ is a commutative group.

SUBGROUPS

In the previous section, we showed a table for a group of order six. Within the table, we mentioned that the subset of S which consisted of the elements I, A, B formed a group. Let us consider what was needed to satisfy the conditions for a subset to form a group under an operation. We repeat the part of the table under discussion below.

*	I	A	B
I	I	A	B
A	A	B	I
B	B	I	A

A.6 If this set forms a group, it must satisfy all the properties listed in the definition. We can see from the table that it is closed. The identity element I is present in the subset; therefore, the operation has the identity property. Since A and B are inverses and I is its own inverse, then we have inverses in this operation on the set. We need only show that the operation is associative on $\{I, A, B\}$. Since the set $\{I, A, B\}$ is closed and the elements of S satisfy the associative property, the set containing I, A, B is associative under the operation defined on S. In this way we have shown that the subset of S is a group.

To show that a subset of a given group forms a group, we need only satisfy three conditions—closure, existence of an identity, and existence of inverses. If these conditions hold, associativity is satisfied automatically. Why?

When a subset of a given group also forms a group, we call it a subgroup.
■ We can now define subgroup.

Definition
■ A.7   *A is a subgroup of S (where S is a group) if*

1. *A is a subset of S*
2. *A has the following properties under the operation on S:*
   a. closure
   b. identity
   ■ *c. inverse*

Consider the example below to illustrate the definition of a subgroup.

Example
■ A.8   Assume set $S = \{a, b, c\}$ and the operation on S in the table below forms a group. Which of the following sets are subgroups of S: $A = \{a, b\}$, $B = \{a, c\}$, $C = \{c, b\}$, $D = \{a\}$? Why?

○	a	b	c
a	a	b	c
b	b	a	c
c	c	c	a

1. The set A is a subgroup because it contains the identity, is closed, and forms inverses.
2. Similarly, the set B forms a subgroup of S.
3. The set C is not a subgroup because it is not closed under the operation. Why?
4. The set D forms a subgroup because it is the identity, it is its own inverse, ■ and, of course, it is closed under the operation.

Can you find any other subgroups?

EXERCISES

1. Given a group, can you show that the identity always forms a subgroup?
2. The set $\{a, b, c, d\}$ under the operation defined by the table below is a group. How many subgroups are there? Name the subgroups and explain why they are subgroups. Try all subsets which contain the identity element.

○	a	b	c	d
a	a	b	c	d
b	b	c	d	a
c	c	d	a	b
d	d	a	b	c

3. Let G be a group and S a subset of G with the following properties:
   a. S is closed.
   b. For each element in S, its inverse is also in S.
   Show that S is a subgroup.

4. If S is a subset of a group G and is closed under the operation of G, show that S is associative under the operation of G.

SECTION B MODULAR SYSTEMS

BEHAVIORAL OBJECTIVES

B.1 Be able to describe a modular system.

B.2 Be able to add in a modular system.

B.3 Be able to show that addition mod 3 forms a commutative group.

B.4 Be able to form an addition table for a modular system.

B.5 Be able to subtract in a modular system.

B.6 Be able to construct a subtraction table for a modular system.

B.7 Be able to find the properties of subtraction in a modular system.

B.8 Be able to partition the natural numbers into equivalence classes mod m.

B.9 Be able to construct a multiplication table for a modular system.

B.10 Be able to find the properties of multiplication in a modular system.

B.11 Be able to define the distributive property in mod 3.

B.12 Be able to define a field.

B.13 Be able to show that mod 3 under addition and multiplication forms a field.

B.14 Be able to find solutions to sentences in mod 3.

B.15 Be able to construct a division table for mod 3.

B.16 Be able to find the properties of division mod 3.

B.17 Be able to translate base ten numerals to mod m.

MODULAR SYSTEMS

B.1   The clock that we depend on daily is an example of an instrument that uses a modular system of counting. The system is referred to as the mod 12 system and operates much like base 12 except that there is a numeral 12 and there is no numeral 0. Notice that 12 actually acts like a zero, because three hours past 12 o'clock is $3 + 12 \pmod{12}$, or $3 \pmod{12}$.

In Figure 6.1 you will notice the similarity in counting in base twelve and
■ mod 12.

Mod 12	Base Twelve
1	1
2	2
.	.
.	.
.	.
10	T
11	E
12	10
1	11
2	12

Mod 12	Base Twelve
3	13
4	14
.	.
.	.
.	.
10	1T
11	1E
12	20
1	21
2	22
3	23

Figure 6.1

You can see that if we replace T, E, and 10 in base twelve with 10, 11, and
12 (mod 12), and only write the first digit of the numeral in base twelve,
we can correspond the base twelve system to the modular system easily.
Consider a few examples.

Examples

1. 25_{twelve} corresponds to 5 (mod 12).
2. $3T_{\text{twelve}}$ corresponds to 10 (mod 12).
3. 7 (mod 12) corresponds to 7_{twelve}, 17_{twelve}, 27_{twelve}, ...

■ B.1   Notice in the third example above that 7 (mod 12) is equivalent to a
set of objects in base twelve, that is, $7_{\text{twelve}} \sim 17_{\text{twelve}} \sim 27_{\text{twelve}}$. . . . Thus,
7_{twelve}, 17_{twelve}, 27_{twelve}, . . . are all members of the equivalence class called
7 (mod 12). Again we see the idea of equivalence classes or sets of equivalent
objects. This points out another property of a modulus system, that of
equivalence classes. A modulus system, mod a, will partition the natural
numbers into a equivalence classes, mod 5 forms five equivalence classes,
and so on. A modulus system with three elements is called a modulus-3
■ system, or mod 3.

ADDITIONS AND SUBTRACTION MOD 3

■ B.1   Consider now the mod 3 system. We will discuss this in terms of a
three-hour clock. Figure 6.2 shows the clock with the numeral 3 where
the twelve is on our regular clock. Some authors use a 0, but this notation
is not consistent with clocks as we know them; therefore, our system
will always have a set of natural numbers $S = \{1, 2, \ldots, m\}$, where m is
■ the modulus.

We will define addition in terms of movement on the clock. For example,
1 (mod 3) + 1 (mod 3) will mean move from 3 (the starting position) to

Figure 6.2

B.2 1 on the clock and then move one more space to 2. In this way 1 (mod 3) + 1 (mod 3) is 2 (mod 3). Consider the examples below.

$$2 \text{ (mod 3)} + 2 \text{ (mod 3)} = 1 \text{ (mod 3)}$$
$$1 \text{ (mod 3)} + 2 \text{ (mod 3)} = 3 \text{ (mod 3)}$$
$$2 \text{ (mod 3)} + 3 \text{ (mod 3)} = 2 \text{ (mod 3)}$$

We have constructed an addition table in Figure 6.3 for the mod 3 system. We will name the operations in mod 3 circle addition, circle subtraction, etc. The symbols then will be those of our operations enclosed in a circle. Thus, addition is symbolized as $\oplus$. In this way we need not write 3 (mod 3) + 2 (mod 3) unless we are discussing more than one modular system.

B.4

$\oplus$	1	2	3
1	2	3	1
2	3	1	2
3	1	2	3

Figure 6.3

B.3 Before we consider the subtraction operation in mod 3, let us check the properties which belong to addition mod 3. Most of the properties follow from the table; therefore, we will look at them first.

1. The set $\{1, 2, 3\}$ is closed under addition.
2. Symmetry in the table indicates commutativity. This can be shown by listing all possible $a \oplus b$ combinations with $b \oplus a$.
3. The element 3 acts as an identity.
4. The element 3 is its own inverse, and 1 and 2 are inverses of each other.

$$1 \oplus 2 = 3 = 2 \oplus 1 \quad \text{and} \quad 3 \oplus 3 = 3$$

This leaves the associative property to consider. Since there are only three elements in the set, there are 3^3, or 27, possible ways to write them

taken three at a time. We will test only the first nine ways. You can test the other combinations as an exercise.

$$1 \oplus (1 \oplus 1) = 3 \quad \text{and} \quad (1 \oplus 1) \oplus 1 = 3$$
$$1 \oplus (1 \oplus 2) = 1 \quad \text{and} \quad (1 \oplus 1) \oplus 2 = 1$$
$$1 \oplus (1 \oplus 3) = 2 \quad \text{and} \quad (1 \oplus 1) \oplus 3 = 2$$
$$1 \oplus (2 \oplus 1) = 1 \quad \text{and} \quad (1 \oplus 2) \oplus 1 = 1$$
$$1 \oplus (2 \oplus 2) = 2 \quad \text{and} \quad (1 \oplus 2) \oplus 2 = 2$$
$$1 \oplus (2 \oplus 3) = 3 \quad \text{and} \quad (1 \oplus 2) \oplus 3 = 3$$
$$1 \oplus (3 \oplus 1) = 2 \quad \text{and} \quad (1 \oplus 3) \oplus 1 = 2$$
$$1 \oplus (3 \oplus 2) = 3 \quad \text{and} \quad (1 \oplus 3) \oplus 2 = 3$$
$$1 \oplus (3 \oplus 3) = 1 \quad \text{and} \quad (1 \oplus 3) \oplus 3 = 1$$

After checking all possibilities, you will realize that the set and operation is associative.

We have just shown that mod 3 under circle addition is a commutative group. Consider next the operation of circle subtraction mod 3. We will again consider circle subtraction as the inverse of circle addition so that $a \ominus b = c$ (mod 3) will mean that $b \oplus c = a$ (mod 3). We can employ the addition table for this operation as we did in Chapter 3 (on natural and whole numbers), or we might produce a circle subtraction table in order to consider the properties under subtraction. Consider Figure 6.4. Can you construct such a table from the addition table given earlier? To use Figure 6.4, we enter the left-hand column first and move to the column headed by the second number. In this way we find $2 \ominus 3$ by entering on the left at 2 and moving across the row to the column headed by 3. Thus, $2 \ominus 3$ is 2. Notice that $3 \ominus 2$ is 1.

B.6

$\ominus$	1	2	3
1	3	2	1
2	1	3	2
3	2	1	3

Figure 6.4

B.7 The subtraction table is closed, since we can see all entries belong to the set given for mod 3. We might think of the subtraction operation here as moving in a counterclockwise direction. Thus, $2 \ominus 3$ would mean move counterclockwise from 2 three units to 2, $1 \ominus 2$ would mean move counterclockwise two units starting at 1, which would produce 2, and $2 \ominus 2$ would mean move counterclockwise two units starting from 2, which would produce 3.

If we consider the subtraction table in Figure 6.4, we will notice that there is no evidence of commutativity, or of an identity. Since there is no identity, there is no inverse. If we consider one example of associativity, we will find that the set is non-associative. Notice that $(1 \ominus 2) - 2$

$= 3$ and $1 - (2 \ominus 2) = 1$. So far, subtraction has closure as its only property in mod 3. Notice that 3 acts as a right identity, because

$$1 \ominus 3 = 1$$
$$2 \ominus 3 = 2$$
▲ $\quad 3 \ominus 3 = 3$

We can now state the properties of subtraction in mod 3 as closure and
■ right identity.

EQUIVALENCE CLASSES MOD 3

We mentioned earlier that equivalence classes are formed by the modulus m. Let us consider what this means. When we count in the set of natural numbers, the counting process produces a set which is infinite. Modulus 3, however, is finite in that there are only three objects in the set. We have shown that we can add in modulus 3 as well as in the set N. Let us look at some particular natural numbers and rewrite them using the numbers in mod 3.

1 can be written 1 (mod 3)
$4 = 1 + 3$ can be written 1 (mod 3) $+ 3$ (mod 3) $= 1$ (mod 3)
$7 = 4 + 3$ can be written 1 (mod 3) $+ 3$ (mod 3) $= 1$ (mod 3)
$10 = 7 + 3$ can be written 1 (mod 3) $+ 3$ (mod 3) $= 1$ (mod 3)

$\quad \cdot \qquad \cdot \qquad \qquad \cdot$
$\quad \cdot \qquad \cdot \qquad \qquad \cdot$
$\quad \cdot \qquad \cdot \qquad \qquad \cdot$

$n = 1 + 3k$ can be written 1 (mod 3) $+ 3$ (mod 3) $= 1$ (mod 3)

■ B.8   In this way we can see that 1 added to any multiple of 3 is always 1 mod 3. We can also show that $2 + 3, 2 + 6, 2 + 9, \ldots$ are all 2 (mod 3) and that $3 + 3k$ is equivalent to 3 (mod 3). This produces a partition in the set N as follows:

$$A = \{1, 4, 7, 10, \ldots, 1 + 3k, \ldots\} \sim 1 \ (\text{mod } 3)$$
$$B = \{2, 5, 8, 11, \ldots, 2 + 3k, \ldots\} \sim 2 \ (\text{mod } 3)$$
■ $\quad C = \{3, 6, 9, 12, \ldots, 3 + 3k, \ldots\} \sim 3 \ (\text{mod } 3)$

We use the symbol $\sim$ to mean "is equivalent to." Notice that $A \cup B \cup C = N$ and $A \cap B \cap C = \varnothing$.

From the explanation above, you should be able to translate from base ten to mod 3. The operation simply involves dividing by 3. When we divide a base ten number by 3, the only possible remainders are 1, 2, or 3. Consider the examples listed below.

Examples

◄ B.17   1. $25_{\text{ten}} = (8 \times 3) + 1 \sim 1 \ (\text{mod } 3)$
2. $18_{\text{ten}} = (6 \times 3) + 0 \sim 3 \ (\text{mod } 3)$
■ $\quad$ 3. $20_{\text{ten}} = (6 \times 3) + 2 \sim 2 \ (\text{mod } 3)$

EXERCISES

1. Translate the following base-ten numerals to mod 3.
  a. 21   b. 16   c. 14   d. 18

2. Construct equivalence classes for mod 4 and mod 5.

3. Translate the following base-ten numerals to mod 4 and mod 5.
  a. 20   b. 24   c. 25   d. 13   e. 14

4. Find the following sums in mod 3. Find the differences in mod 3 by replacing + with −.
  a. 2 + 3   b. 3 + (2 + 2)
  c. 3 + (1 + 1)   d. 2 + (1 + 2)

5. Construct an addition table for mod 5.

6. Find the following sums and differences in mod 5 for the given ordered pairs.
  a. (2, 3)   b. (3, 4)   c. (5, 3)
  d. (3, 5)   e. (2, 4)   f. (4, 2)

7. What properties has mod 5 under addition?

8. Does mod 5 form a group under addition?

9. Do all modular systems form groups under addition?

10. Are all modular systems closed under subtraction?

11. What is the name of the structure formed by the mod-3 system under circle subtraction?

12. Does mod 4 form an Abelian group under addition?

MULTIPLICATION AND DIVISION MOD 3

We defined addition modulus 3 in terms of movement on the clock. We will define multiplication in terms of repeated additions. In this way $3 \otimes 2$ will mean $2 \oplus 2 \oplus 2 = 3$ (mod 3). The multiplication table for mod 3 is seen in Figure 6.5. See if you can construct it without looking at Figure 6.5, then check your work with the figure.

■ B.9

$\otimes$	1	2	3
1	1	2	3
2	2	1	3
3	3	3	3

■

Figure 6.5

■ B.10 Let us now list some of the properties under circle multiplication. You should be able to verify these properties from the table.

1. The set $\{1, 2, 3\}$ is closed under $\otimes$.
2. The set $\{1, 2, 3\}$ is commutative under $\otimes$. (Recall that this is not necessary to produce a group.)
3. The set $\{1, 2, 3\}$ has an identity element 1.
4. Except for the element 3, all the elements have inverses. That is, all the elements except the additive identity 3 have multiplicative inverses.

5. Again, to show associativity, we must explore all 27 cases. We will simply state that mod 3 is associative under $\otimes$. It is left to you to prove this.

6. The set $\{1, 2\}$ forms a group under $\otimes$. That is, the set of elements other than the additive identity forms a group under circle multiplication.

We would like now to introduce a new structure called a *field*, using mod 3 under the operations of circle addition and circle multiplication as an example. A field is a set upon which two operations are defined where the operations are linked by a distributive property. We now state a distributive property for mod 3.

B.11 *Distributive property of $\otimes$ over $\oplus$ mod 3: For all a, b, c in mod 3, a $\otimes$ (b $\oplus$ c) = (a $\otimes$ b) $\oplus$ (a $\otimes$ c).*

B.13 The field described by the set $\{1, 2, 3\}$ mod 3 has the following properties:

1. closure under $\oplus$ and $\otimes$
2. commutativity under $\oplus$ and $\otimes$
3. associativity under $\oplus$ and $\otimes$
4. identities under $\oplus$ and $\otimes$
5. inverses under $\oplus$ for all elements and under $\otimes$ for all elements except 3.
6. distributive property of $\otimes$ over $\oplus$

It can be shown that all prime modular systems produce fields under the operations of $\oplus$ and $\otimes$. You have already found that mod 4 and mod 5 form Abelian groups under $\oplus$. Do the elements other than the $\oplus$ identity (i.e., the "non-zero" elements) form an Abelian group under the operation $\otimes$? If so, and if the distributive property holds, then the natural numbers mod 4 form a field. This is left as an exercise at the end of this section.

We now give a formal definition of a field.

Definition

B.12   *A set S with two operations + and × is called a field if*
1. *S is a commutative group under +*
2. *the elements of S, not including the + identity, form a commutative group under ×*
3. *for all a, b, c $\in$ S, a × (b + c) = (a × b) + (a × c)*

Although the operations + and × in the definition of a field do not have to be addition and multiplication, the terminology associated with these operations is often used so that the identity with respect to the first operation + is called the "zero". Using this vocabulary in the above definition, point 2 becomes "the non-zero element of S forms a commutative group under × ".

Consider now the operation of division mod 3. We will again define division as the inverse of multiplication. From Figure 6.5, we see that

1 and 2 have inverses, but 3 does not. Thus, to find $a \oplus b$ we need only find a c so that $a = b \otimes c$. For any choice of b other than 3, we have

$$b^{-1} \otimes a = b^{-1} \otimes (b \oplus c) = (b^{-1} \otimes b) \otimes c = c$$

Thus

$$c = b^{-1} \otimes a$$
$$= a \otimes b^{-1}$$

This shows how division can be defined as the inverse of multiplication. Namely, $a \oplus b = c$ means $c = a \otimes b^{-1}$. That is, $a \oplus b = a \otimes b^{-1}$. Consider how this works by observing the following examples in mod 3.

Examples

B.14
1. Solve the sentence $1 \oplus 2 = c$. The multiplicative inverse of 2 is 2; therefore, if $1 = 2 \otimes c$, $2 \otimes 1 = (2 \otimes 2) \otimes c$, or $2 = c$. In this way, $1 \oplus 2 = 2$. (*Note*: From Figure 6.6, $2 \otimes 2 = 1$; thus $2 = c$).
2. Solve the sentence $3 \oplus 2 = c$. $3 = 2 \otimes c$; then $2 \otimes 3 = (2 \otimes 2) \otimes c$, so $3 = c$.
3. Solve the sentence $2 \oplus 3 = c$. If $2 = 3c$, then $2 = 3$ since $3c = 3$, but this is not true. *Notice that we cannot divide by 3 in this system. Does this seem familiar?*

From the examples above, we can now construct a table. Notice in Figure 6.6 that entries are not included for division by 3. This is much like the whole numbers, where we showed that division by 0 is not possible. We still call this operation closed, but we restrict division to all $a \neq 0$.

B.15

$\oplus$	1	2	3
1	1	2	–
2	2	1	–
3	3	3	–

Figure 6.6

B.16 From the figure, we can state the following properties.

1. The system is closed under division except for division by 3.
2. The system is not commutative under $\oplus$, because $3 \oplus 1 = 3$, and $1 \oplus 3 \neq 3$.
3. There is no identity.
4. There are no inverses.
5. The operation is non-associative.

EXERCISES

1. Find the products and quotients for the following pairs mod 3.
  a. (2, 3)   b. (3, 2)   c. (1, 2)   d. (2, 1)

2. Construct multiplication and division tables for mod 4 and mod 5.

3. a. Does mod 4 produce an Abelian group under $\otimes$? Why or why not?
   b. Does the set of "non-zero" elements in mod 4 form a commutative group under $\otimes$?
   c. Does the mod 4 form a field under $\oplus$ and $\otimes$?

4. Does mod 5 produce a field under $\oplus$ and $\otimes$?

5. Find the following products and quotients mod 5.
  a. (4, 5)   b. (3, 2)   c. (3, 4)   d. (2, 1)

6. Find the following products and quotients mod 4.
  a. (4, 3)   b. (3, 2)   c. (2, 3)   d. (1, 3)   e. (4, 2)

7. Which of the following systems form groups? Find their subgroups.
  a. mod 3 under $\oplus$ b. mod 5 under $\oplus$
  c. mod 4 under $\oplus$ d. mod 5 under $\otimes$

8. Restate the formal definition of a field in terms of properties and operations on a set.

SECTION C   CYCLIC GROUPS

BEHAVIORAL OBJECTIVES

C.1 Be able to produce the operation tables for the group of rotations of an equilateral triangle.

C.2 Be able to define an isomorphism between groups.

C.3 Be able to show that the group of rotations of an equilateral triangle is isomorphic to the mod-3 group.

C.4 Be able to describe a cyclic group.

C.5 Be able to generate a cyclic group.

C.6 Be able to produce the group of rotations of an equilateral triangle as a cyclic group.

C.7 Be able to show using the operation table that the symmetries of a rectangle form a group.

CYCLIC GROUPS AND ISOMORPHISM

Let us start with an equilateral triangle and rotate clockwise about its center point so that after the rotation the triangle coincides with itself (i.e., fits exactly upon itself). As we can see, there are only three rotations that we need consider: the first, R_1, a clockwise rotation of 120°; the second, R_2, a clockwise rotation through 240°, and the third, R_3, a rotation through 360°. We give examples of these rotations in Figure 6.7, where one vertex of the triangle is marked so that the particular rotation

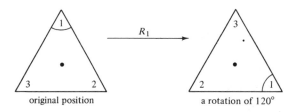

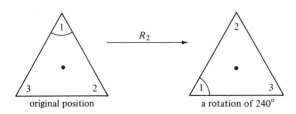

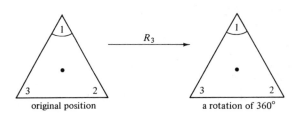

Figure 6.7

can be emphasized. It should be obvious that rotation R_3 results in the position of the triangle remaining unchanged.

We will consider the set of these rotations $\{R_1, R_2, R_3\}$ and on this set define an operation. Under this operation this set of rotations will form a group, called a *rotation group*. After some investigation we will see that this rotation group is in some sense an already familiar group.

On the set of rotations $\{R_1, R_2, R_3\}$ let us define an operation, $*$; $R_1 * R_2$ will mean first rotate 120 degrees, then rotate 240 degrees. Thus $X * Y$ will be the rotation that results when we perform the rotation X followed by the rotation Y.

Figure 6.8 shows the results of this operation. Consider the properties of this set and operation and you will see that the structure is an Abelian group. How does this group differ from the group formed by addition mod 3?

■ C.1

*	R_1	R_2	R_3
R_1	R_2	R_3	R_1
R_2	R_3	R_1	R_2
R_3	R_1	R_2	R_3

rotations

+	1	2	3
1	2	3	1
2	3	1	2
3	1	2	3

mod 3

■

Figure 6.8

Compare the operation table for addition mod 3 with the operation table of the group shown above. You may see that there is a 1–1 correspondence between the elements of the set $\{R_1, R_2, R_3\}$ and $\{1, 2, 3\}$ and that this correspondence holds even when operations are performed on the sets. This correspondence,

$$R_3 \leftrightarrow 3$$
$$R_1 \leftrightarrow 1$$
$$R_2 \leftrightarrow 2$$

has the property that whenever we perform an operation between two elements of one group and then perform the operation between the corresponding elements of the other group, the results are corresponding elements. Consider the following examples.

Examples

■ C.3   1. $R_1 \leftrightarrow 1$ and $R_2 \leftrightarrow 2$. Operating in one group we have $R_1 * R_2 = R_3$, and adding corresponding elements in mod 3 we have $1 \oplus 2 = 3$ and $R_3 \leftrightarrow 3$; so $R_1 * R_2 \leftrightarrow 1 \oplus 2$.

2. In mod 3, $2 \oplus 2 = 1$. Using the corresponding element of our group of rotation we have $R_2 * R_2 = R_1$, and since $R_1 \leftrightarrow 1$, $2 \oplus 2 \leftrightarrow R_2 * R_2$.

3. In mod 3, $2 \oplus 3 = 2$. In group of rotations $R_2 * R_3 = R_2$.

You should satisfy yourself that this property of correspondence holds for all nine combinations. The phenomenon described here is called an *isomorphism*, and the two sets with their respective operations are called isomorphs. We say that mod 3 under addition is isomorphic to the set
■ of rotations of an equilateral triangle under *.

Definition

◀ C.2   *The set S is isomorphic to the set T if and only if S and T are in a 1–1 correspondence and the correspondence is preserved under the operations*
■ *defined on S and T.*

We pointed out that the groups formed under addition in mod 3 and the rotation group of the equilateral triangle are isomorphic. Actually, they are the same group in an abstract sense; even though they are physically different and appear to be different groups, they are actually the cyclic group of order 3. This group is often described as C_3.

■ C.4   A cyclic group can be produced by repeating the operation on a single element. In this case, we say that the element generates the group. For example, we choose some element, x, other than the identity and get a second element by operating upon x by itself. That is, $x * x = x^2$, the second element. $x^2 * x$, or x^3, is the third, etc. We continue in this fashion until x^n gives the identity element. Can you see why no new elements will be generated by continued operations with x? Thus, the set of elements
■   that comprise the cyclic group is $\{x, x^2, x^3, \ldots, x^n = I\}$.
■ C.6   Let us now show that the set of rotations of an equilateral triangle,
▲ C.5   or C_3, is a cyclic group. If we choose R_1 as our initial or generating element, then let

$$x = R_1$$
$$x^2 = R_1 * R_1 = R_2$$
$$x^3 = (R_1 * R_1) * R_1 = R_2 * R_1 = R_3$$

We have produced or generated the set $\{R_1, R_2, R_3\}$, which forms the
■   group discussed in this section. Mod 3 can be generated in the same fashion. Notice that these groups differ only in the symbols used. We may find several interpretations for the symbols but essentially we are
▲   dealing with the same group. Consider the tables in Figure 6.9 which represent the rotation group, the mod 3 group under $\oplus$, and the group C_3. The groups are written in an order which allows us to compare them.

*	R_1	R_2	R_3
R_1	R_2	R_3	R_1
R_2	R_3	R_1	R_2
R_3	R_1	R_2	R_3

rotation group

$\oplus$	1	2	3
1	2	3	1
2	3	1	2
3	1	2	3

mod 3

	x	x^2	x^3
x	x^2	x^3	x
x^2	x^3	x	x^2
x^3	x	x^2	x^3

cyclic group, C_3

Figure 6.9

You should note from the figure that $R_1 \leftrightarrow 1 \leftrightarrow x$, $R_2 \leftrightarrow 2 \leftrightarrow x^2$, and $R_3 \leftrightarrow 3 \leftrightarrow x^3$. Careful consideration of each table will satisfy you that these three groups are isomorphic.

EXERCISES

1. Show that the sets under the operations defined in the table below are isomorphic.

×	1	−1
1	1	−1
−1	−1	1

$\oplus$	1	2
1	2	1
2	1	2

$\oplus$	A	A^2
A	A^2	A
A^2	A	A^2

2. Name the identity element in each set above.

3. Produce a table for the group formed by rotating a square around its center. Define the elements as follows: R_4 = initial position, R_1 = 90° rotation, R_2 = 180° rotation, R_3 = 270° rotation. Define the operation as we did in the rotation of the equilateral triangle.

4. Produce the group C_4. Make a table to illustrate the operations using R_1 from Exercise 3 as the generator.

5. Are the groups formed by C_4 and the rotation of the square isomorphic? How do they compare to the group formed by addition in mod 4?

6. Take a piece of paper marked $ABCD$ as shown below. By holding it horizontally and flipping it over you produce the second diagram shown. Let F_1 be a flip around axis 1. Name the initial position R_2. Can you define the operation produced by flipping and make a table? Does this produce a group? (*Hint*: Does $F_1 * F_1 = R_2$?

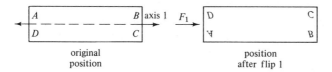

original
position

position
after flip 1

7. Hold the paper in the same manner only this time flip it around a vertical axis. Call this operation F_2. Does F_2 generate a group? (*Hint*: Does $F_2 * F_2 = R_2$?)

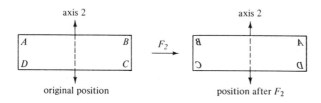

original position

position after F_2

8. Hold the paper in the same position again. This time rotate the paper around its center 180°. Call this operation R_1. Make a table and see if R_1 generates a group. (*Hint*: Does $R_1 * R_1 = R_2$)

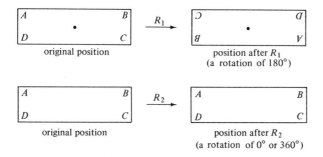

original position

position after R_1
(a rotation of 180°)

original position

position after R_2
(a rotation of 0° or 360°)

9. Do the groups in Exercises 6–8 have anything in common? If so, what?

10. Consider what happens when you perform the operations shown below using the elements $R_1, R_2, F_1,$ and F_2 from Exercises 6–8. Find the results indicated below by rotating as you did in Exercises 6–8.

a. $F_1 * F_2$ b. $F_2 * F_1$ c. $F * F_1$

d. $F_1 * R_1$ e. $R_1 * F_2$ f. $R_1 * F_1$

11. Compare the operations $F_1{}^2, F_2{}^2,$ and $R_1{}^2$ from Exercises 6–8. What common characteristics do they have?

12. In generating a cyclic group, why do we not pick the identity as our generating element?

13. a. Generate mod 3 as a cyclic group using 1 as the generating element.

b. Generate mod 3 as a cyclic group using 2 as the generating element.

SYMMETRY GROUPS

The operation described in Exercises 6, 7, 8, and 10 of the last exercise set form a group under the operation of "performing successive motions." The elements of the set are motions that result in the rectangle fitting exactly on itself. We will call this set S_4, where the elements $R_1, R_2, F_1,$ and F_2 can be described as follows:

R_2 : a motion that leaves the rectangle unchanged (a rotation of 0° or 360°)

F_1 : a flip around a horizontal axle

F_2 : a flip around a verticle axle

R_1 : a rotation of 180° around the center of the rectangle

On S_4 we define an operation ◦: $X \circ Y$ will be the motion that results when we perform the motion X followed by the motion Y.

Figure 6.10 shows the results of this operation. This table describes a group of order 4. Is it isomorphic to any of the groups of order 4 which we described earlier?

■ **C.7**

◦	F_1	F_2	R_1	R_2
F_1	R_2	R_1	F_2	F_1
F_2	R_1	R_2	F_1	F_2
R_1	F_2	F_1	R_2	R_1
R_2	F_1	F_2	R_1	R_2

Figure 6.10

EXERCISES

1. A symmetry group can be formed by reflecting and rotating an equilateral triangle. The elements of this group are $R_1, R_2,$ and R_3 (the rotation through 120°, 240°, and 360°, respectively), and $F_1, F_2,$ and F_3 where F_1 is a flip in line l_1, F_2 is a flip in line l_2, and F_3 is a flip in line l_3. Produce the table for this group where the operation $*$ is the familiar one, namely, perform the first operation and then the second.

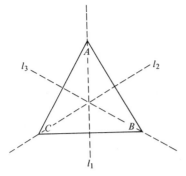

2. How many rotations will an isosceles triangle have? Remember that the rotation must bring the triangle back into coincidence with itself.

3. Define an operation on the isosceles triangle. Will the operation produce a group?

4. Consider the square and its movements. How many rotations are there; how many flips are there?

5. Do the rotations around the center alone produce a group?

6. Do flips around the axes produce a group?

7. Does rotation around the center and the flips around the axes produce a group?

8. Find the subgroups of the group of symmetries of a square.

CAPSULE IV **BRAIDS**

It is possible to teach elementary school children concepts which are abstract, particularly when the presentation is at the concrete level. We have discussed the group concept in this chapter. Consider how this structure can be introduced in the upper elementary grades.

We can define a structure and produce the operation on a *geoboard*. A geoboard is a board which has nails driven into it at intervals of 1 or $1\frac{1}{2}$ inches. The board is usually 12 inches square and looks like a 6 × 6 array of points. It is an excellent material to be used in elementary school as well as secondary school. The materials needed for this discussion are a geoboard and rubber bands or a pencil and paper.

We start by defining a braid. We could define a braid as a way to connect a set of given points to an equal set of given points. Consider a set of n points spaced equally along a horizontal line, with an equal number of points along a parallel line below it. A braid is a 1–1 correspondence between the first set of points and the second set. We describe the braid by listing, in order, the points that correspond to the points of the first set, $1, 2, 3, \ldots, n$.

Consider Figure 6.11. Imagine this diagram on a geoboard in which the lines are represented by two rubberbands. The braids shown in Figure 6.11 are named (1, 2) and (2, 1), respectively, since the first braid describes a correspondence of *point 1 to point 1* and *point 2 to point 2*, whereas the second braid describes a correspondence of *point 1 to point 2* and *point 2 to point 1*.

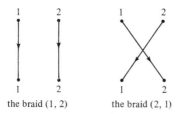

the braid (1, 2)   the braid (2, 1)

Figure 6.11

This can be further explained if we have two people, 1 and 2, and two positions into which the people can move. The first person moves either into the first or the second position; this, in turn, determines the position into which the second person can move. The final position in which these people stand, or the final place at which the points rest, determines the name of the braid.

Now we are ready to define an operation. We may wish to write braids as shown in Figure 6.11, or we may wish to place them on a pegboard or geoboard. In either case this operation works equally well. To perform the operation ∘ on (1, 2) and (2, 1), we simply place the first and second braid in a position where the endpoints of the first braid become the starting points of the second braid. The operation shown in Figure 6.12 is written (1, 2) ∘ (2, 1).

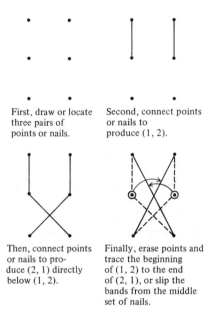

First, draw or locate three pairs of points or nails.

Second, connect points or nails to produce (1, 2).

Then, connect points or nails to produce (2, 1) directly below (1, 2).

Finally, erase points and trace the beginning of (1, 2) to the end of (2, 1), or slip the bands from the middle set of nails.

Figure 6.12

Notice in Figure 6.12 that (1, 2) ∘ (2, 1) = (2, 1). Notice that the only change is that the new braid formed is stretched. Consider the operations in Figure 6.13.

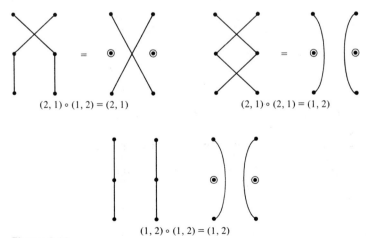

(2, 1) ∘ (1, 2) = (2, 1)

(2, 1) ∘ (2, 1) = (1, 2)

(1, 2) ∘ (1, 2) = (1, 2)

Figure 6.13

See the table in Figure 6.14, which describes this structure.

○	(1, 2)	(2, 1)
(1, 2)	(1, 2)	(2, 1)
(2, 1)	(2, 1)	(1, 2)

Figure 6.14

If we call the element (1, 2) a zero, and if we name the element (2, 1) the first element, or 1, the table translates as in Figure 6.15.

⊕	0	1
0	0	1
1	1	0

Figure 6.15

This operation looks like that of modular addition in mod 2. Actually these two systems are abstractly the same even though they are different in physical appearance. Both systems are Abelian groups.

After working with 2–2 braids, the transition to 3–3 braids is easy. Find out how many positions three people can occupy in a line. Students should easily list them:

```
1  2  3
2  1  3
3  1  2
1  3  2
2  3  1
3  2  1
```

We might call this the possible arrangements of three things taken three at a time. If we draw braids to illustrate this, they look like Figure 6.16.

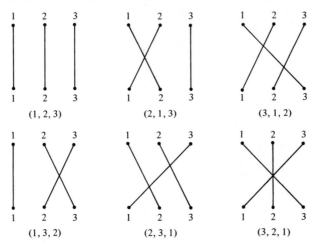

Figure 6.16

If we apply the operation already described to the set of braids B_3 = {(1, 2, 3), (2, 3, 1), (3, 1, 2)}, the operations appear as in Figure 6.17.

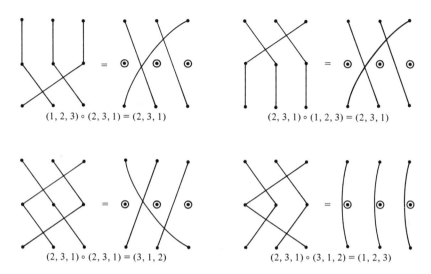

$$(1, 2, 3) \circ (2, 3, 1) = (2, 3, 1)$$

$$(2, 3, 1) \circ (1, 2, 3) = (2, 3, 1)$$

$$(2, 3, 1) \circ (2, 3, 1) = (3, 1, 2)$$

$$(2, 3, 1) \circ (3, 1, 2) = (1, 2, 3)$$

Figure 6.17

From the examples shown in Figure 6.17, you should try to complete the following operations:

$(1, 2, 3) \circ (1, 2, 3) =$
$(1, 2, 3) \circ (3, 1, 2) =$
$(3, 1, 2) \circ (1, 2, 3) =$
$(3, 1, 2) \circ (3, 1, 2) =$
$(3, 1, 2) \circ (2, 3, 1) =$.

Now compare your results with those shown in Figure 6.18.

$\circ$	(1, 2, 3)	(2, 3, 1)	(3, 1, 2)
(1, 2, 3)	(1, 2, 3)	(2, 3, 1)	(3, 1, 2)
(2, 3, 1)	(2, 3, 1)	(3, 1, 2)	(1, 2, 3)
(3, 1, 2)	(3, 1, 2)	(1, 2, 3)	(2, 3, 1)

Figure 6.18

Notice again that we can rename the braids as follows:

$(1, 2, 3) \leftrightarrow 3$
$(2, 3, 1) \leftrightarrow 1$
$(3, 1, 2) \leftrightarrow 2$

In this way we produce the modular group formed under addition mod 3. See Figure 6.19.

⊕	3	1	2
3	3	1	2
1	1	2	3
2	2	3	1

Figure 6.19

It is easy to show that one structure can be translated into another one, even though the two structures seem to differ physically. Consider now an equilateral triangle. Label the vertices as shown below.

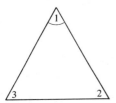

Now consider what happens when we rotate the triangle around a center point until it coincides with itself. See Figure 6.20.

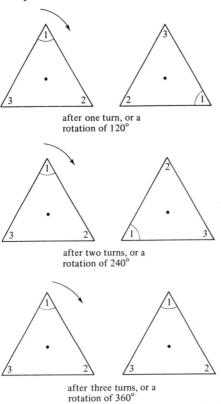

after one turn, or a
rotation of 120°

after two turns, or a
rotation of 240°

after three turns, or a
rotation of 360°

Figure 6.20

In Figure 6.20 we defined a set of operations on an equilateral triangle. The operation is that of rotating or turning it around its center. If we name the triangle $(1, 2, 3)$ when it is in the beginning position, then we name the first operation $(3, 1, 2)$, because the triangle is transformed from position $(1, 2, 3)$ to position $(3, 1, 2)$. This would allow us to name the $240°$ rotation $(2, 3, 1)$ and the $360°$ rotation $(1, 2, 3)$. This gives us three elements in the set of turns or rotations. If we perform these turns two at a time, they take on the following meaning: $(1, 2, 3) \circ (2, 3, 1)$ directs us to take the triangle in position $(1, 2, 3)$, rotate it $360°$, then rotate it $240°$. In this way we produce $(2, 3, 1)$.

If you review rotation of the equilateral triangle, you may compare the three structures discussed above. It would be of value to compare Figures 6.18 and 6.19 with the table for rotation of the equilateral triangle. By using the notation described above, you can see that $\oplus$ mod 3, rotation of the equilateral triangle, and the operation defined on B_3 are all isomorphic.

Each of the operations discussed in this capsule can be used easily with children in intermediate and upper elementary grades. The operation on braids is most effective when a geoboard is employed.

CHAPTER 7 RATIONAL NUMBERS

SECTION A OPERATIONS ON THE RATIONALS

BEHAVIORAL OBJECTIVES

A.1 Be able to write equivalence classes for a given rational number.

A.2 Be able to define the set of rational numbers.

A.3 For two given pairs of integers, decide whether or not they belong to the same equivalence class.

A.4 State the definition of equivalent pairs of integers.

A.5 State a definition of addition for rational numbers.

A.6 Be able to add any two rational numbers.

A.7 Be able to add rational numbers of the form $R[a, b]$, $R[c, d]$.

A.8 State the definition of multiplication for rational numbers.

A.9 Be able to multiply rational numbers.

A.10 Be able to show that the operations on the rational numbers are independent of the ordered pairs chosen to represent the numbers.

A.11 State the definition of equality for rational numbers.

RATIONAL NUMBERS DEFINED

We have developed several sets of numbers and considered their properties under the basic operations. You should be aware of the fact that as we introduced a larger set we produced a structure with more properties. For example, the natural numbers had no identity for addition, but the whole numbers did. Subtraction was not closed for N and W, but it was closed in I. With the introduction of I, we also were made cognizant of a new concept, that of inverses under addition. In this chapter we develop a new set of numbers called the *rationals*.

The rational numbers are a set which will contain N, W, and I as subsets. They will possess many properties not exhibited by the subsets. Later in this chapter we will show that the rational numbers form a field.

The fact that new properties will emerge from the development of the rational numbers is not the only reason for introducing them here. The rational numbers were developed historically out of need. One of the earliest developments of positive rationals or common fractions can be traced to the Babylonian civilization. In that civilization, fractions with denominators of 60 were used primarily as a result of time measure. Other civilizations developed the concept of common fractions for measurement and for barter. Most well-developed cultures used money, and as a result fractional units were developed to simplify transactions.

We will define the rational numbers in much the same way as we defined integers. But first we will consider a few examples of positive rationals of fractions. You will notice that the fractions shown below have an unlimited number of names. In other words, every rational fraction generates a set of equivalent fractions or pairs which have the same value but are named differently. Using this fact we will consider rational numbers to be a *set* of *equivalence classes*. This means that each element in the set of rational numbers is itself a set. We will use the notation $R[\![a, b]\!]$ to mean a rational number and the notation $[\![a, b]\!]$ to represent an ordered pair in the set $R[\![a, b]\!]$.

Let us first give some examples of what these equivalence classes will look like.

Examples

■ **A.1** 1. the equivalence class corresponding to the basic fraction $\frac{1}{2}$:

$$R_{\frac{1}{2}} = \{\tfrac{1}{2}, \tfrac{2}{4}, \tfrac{3}{6}, \tfrac{4}{8}, \ldots, n/2n, \ldots\}$$

2. the equivalence class whose basic pair is $[\![1, 2]\!]$:

$$R[\![1, 2]\!] = \{[\![1, 2]\!], [\![2, 4]\!], [\![4, 8]\!], \ldots, [\![n, 2n]\!], \ldots\}$$

3. the equivalence class whose basic pair is $[\![3, 5]\!]$:

$$R[\![3, 5]\!] = \{[\![3, 5]\!], [\![6, 10]\!], [\![9, 15]\!], \ldots, [\![3n, 5n]\!], \ldots\}$$

4. the equivalence class for the rational number $2/3$ or $R[\![2, 3]\!]$ using both fractional and ordered pair notation.

$$R_{\frac{2}{3}} = \{\tfrac{2}{3}, \tfrac{4}{6}, \tfrac{6}{9}, \ldots, 2n/3n, \ldots\}$$
■
$$R[\![2, 3]\!] = \{[\![2, 3]\!], [\![4, 6]\!], [\![6, 9]\!], \ldots, [\![2n, 3n]\!], \ldots\}$$

You will notice in the examples that we used two notations. The first is probably more familiar and is often used for computation. The second will be used here for the mathematical development of the set. But you can see that the notation can easily be translated. We will supply examples of this translation later. Consider now how the equivalence class of a rational number can be found. In the first example, notice that the numerator of the fraction given, 1, appears first in the set and the numerators which follow are simply 2, 3, 4, ...

In the second example, the first entries in the pairs correspond to the numerators in Example 1, namely, 1, 2, 3, 4, ... In Example 3, the first entries in each pair are multiples of 3 because the first entry in the basic pair is 3. In the last example, the numerators and the first entries in the set of ordered pairs shown are multiples of 2. If we consider the denominators or second entries in the pairs in each example, we will find that a set of multiples is produced. In Examples 1 and 2, the set produced can be named $M_2 = \{2, 4, 6, \ldots, 2n\}$ and $n \in N$. In Example 3, the second entries can be noted as $M_5 = \{5, 10, 15, \ldots, 5n\}$ and $n \in N$. In the last example, the second entries or denominators of the basic fraction can be written as the set $M_3 = \{3, 6, 9, \ldots, 3n\}$ and $n \in N$. In this manner we employ two sets of multiples which when paired produce a set which represents a given rational number. Therefore, we suggest that if the rational in question is $R[\![4, 7]\!]$, we can produce its equivalence class by considering the set of multiples of 4 and the set of multiples of 7. Then we form ordered pairs by taking pairs of corresponding multiples, as shown below.

$$M_4 = \{4, 8, 12, \ldots, 4n\}$$
$$M_7 = \{7, 14, 21, \ldots, 7n\}$$
$$R[\![4, 7]\!] = \{[\![4, 7]\!], [\![8, 14]\!], [\![12, 21]\!], \ldots, [\![4n, 7n]\!], \ldots\}$$
$$= \{[\![a, b]\!] \mid a = 4k \text{ and } b = 7k \text{ for all } k \in N\}$$

We have shown in the previous example how we might generate the equivalence class of a rational number. We have not yet formally defined a rational number, nor have we defined equivalence for elements of the equivalence class. We will use the set I to develop our definition for the set of rational numbers, which we will abbreviate as R. Let us first see how we will define the set of equivalence classes that will be our rational number.

Example 1

■ A.3   Which of the fractions listed belong in the same equivalence class: $\frac{3}{4}$, $\frac{8}{12}$, and $\frac{9}{12}$?

SOLUTION   Notice that $\frac{3}{4}$ is a pair such that $(3, 4)$ are relatively prime. The pair $\frac{8}{12}$ is such that it can be written $\frac{2 \times 4}{3 \times 4}$. This means that the numerator and denominator are multiples of 2 and 3, respectively, so $\frac{8}{12}$ is in the equivalence class of $\frac{2}{3}$. The pair $\frac{9}{12}$ can be written $\frac{3 \times 3}{4 \times 3}$. Thus, the numerator and denominator are multiples of 3 and 4, respectively. This places the pair in the equivalence class $\frac{3}{4}$. Thus, $\frac{9}{12}$ is equivalent to $\frac{3}{4}$. Also, note that $3 \times 12 = 4 \times 9$.

Example 2

The pairs $[\![3, 5]\!]$ and $[\![9, 15]\!]$ are equivalent. How are they related other than in the manner indicated in Example 1?

SOLUTION   Notice for the pairs $[\![3, 5]\!]$ and $[\![9, 15]\!]$ that the products 3×15 and 5×9 are equal.

Example 3

Using the relationship from Example 2, see if $\frac{3}{4}$ and $\frac{9}{12}$ are equivalent.

SOLUTION   We know that $\frac{3}{4}$ and $\frac{9}{12}$ are in the same equivalence class.
■   Notice that the products 3×12 and 4×9 are equal.

In the examples shown above, we have indicated two ways of finding equivalent pairs. The first example, which requires factorization, was probably more familiar to you. The second method requires multiplication only. We will define equivalence by using the second method.

Definition

■ A.4   *Two pairs of integers $[\![a, b]\!]$ and $[\![c, d]\!]$ are members of the same equivalence class if and only if $a \times d = b \times c$.*

■   Notation: *If $[\![a, b]\!]$ and $[\![c, d]\!]$ belong to the same equivalence class, we write $[\![a, b]\!] \sim [\![c, d]\!]$ and say that $[\![a, b]\!] \in R[\![c, d]\!]$ or that $[\![c, d]\!] \in R[\![a, b]\!]$.*

Since $[\![a, b]\!]$ and $[\![c, d]\!]$ belong to the same equivalence class, we can say that $R[\![c, d]\!] = R[\![a, b]\!]$ because $R[\![c, d]\!]$ and $R[\![a, b]\!]$ are simply the equivalence classes to which $[\![c, d]\!]$ and $[\![a, b]\!]$ belong, namely, the same class.

Definition

■ A.11   *Two rational numbers $R[\![a, b]\!]$ and $R[\![c, d]\!]$ are equal if and only if $[\![a, b]\!]$*
■   *$\sim [\![c, d]\!]$.*

Example 1

■ A.3   Do $[\![5, 8]\!]$ and $[\![25, 40]\!]$ belong to the same equivalence class?

SOLUTION   Since 5×40 equals 8×25, $[\![5, 8]\!] \sim [\![25, 40]\!]$.

Example 2

Do the fractions $\frac{3}{12}$ and $\frac{5}{20}$ belong to the same equivalence class?

SOLUTION $3 \times 20 = 12 \times 5$; then $\frac{3}{12} \sim \frac{5}{20}$.

Example 3

What value must a have such that $[\![a, 6]\!]$ and $[\![3, 9]\!]$ are equivalent fractions?

SOLUTION   If $a \times 9 = 6 \times 3$, then $a = 2$, and $[\![2, 6]\!] \sim [\![3, 9]\!]$.

Example 4

If $[\![3, 4]\!] \sim [\![15, 25]\!]$, then 3×25 must equal 4×15.

SOLUTION   Since $3 \times 25 \neq 4 \times 15$, we conclude that $[\![3, 4]\!]$ is not equiv-
■   alent to $[\![15, 25]\!]$, and we write $[\![3, 4]\!] \not\sim [\![15, 25]\!]$.

Now that we have a definition of equivalence classes as sets of ordered pairs of integers, we will formally define a rational number as one of these classes.

Definition

■ A.2   *The set of rational numbers R is the set of equivalence classes $R[\![a, b]\!]$*
■   *such that a and b are in I and $b > 0$.*

We know what an equivalence class of a fraction looks like; what would an equivalence class of the "fraction" $-\frac{1}{2}$ look like? That is, what is $R[\![-1,2]\!]$?

$$R[\![-1,2]\!] = \{[\![-1,2]\!], [\![-2,4]\!], [\![-3,6]\!], \ldots, [\![-n,2n]\!], \ldots\}$$

EXERCISES

1. Write the equivalence class for each of the following fractions written in standard form.
  a. $\frac{1}{3}$ b. $\frac{2}{5}$ c. $\frac{3}{9}$ d. $\frac{5}{6}$

2. Write the equivalence class for the following pairs of integers.
  a. $[\![5,8]\!]$ b. $[\![1,5]\!]$ c. $[\![2,6]\!]$
  d. $[\![5,6]\!]$ e. $[\![-5,6]\!]$

3. Do the pairs $[\![2,6]\!]$ and $[\![1,3]\!]$ generate the same set?

4. Are the pairs $[\![2,6]\!]$ and $[\![1,3]\!]$ elements of the same equivalence class?

5. Which of the following pairs are equivalent?
  a. $[\![3,9]\!]$ and $[\![15,45]\!]$ b. $[\![6,10]\!]$ and $[\![66,120]\!]$
  c. $[\![3,4]\!]$ and $[\![26,35]\!]$ d. $[\![7,14]\!]$ and $[\![12,24]\!]$

6. Find a value for a which makes the given pairs equivalent.
  a. $[\![a,3]\!]$ and $[\![4,6]\!]$ b. $[\![2,a]\!]$ and $[\![5,10]\!]$
  c. $[\![2,5a]\!]$ and $[\![8,40]\!]$ d. $[\![5,a]\!]$ and $[\![2,6]\!]$

7. Write the equivalence class for each of the following rational numbers.
  a. $R[\![5,2]\!]$ b. $R[\![-3,1]\!]$ c. $R[\![-4,3]\!]$ d. $R[\![3,1]\!]$

ADDITION ON THE RATIONAL NUMBERS
We mentioned earlier in the text that a mathematical system is a set and the operations on the set. We will define the four arithmetic operations on the set of rational numbers and then investigate the properties of these operations.

Let us informally examine why addition on the rationals requires that we add rational numbers whose denominators are alike. It seems reasonable to talk about *numerators* as describing the number of things under discussion and *denominators* as describing the size of the things being discussed. In this way, the rational number three-sevenths means three parts each of which is one-seventh of a unit. If we wish to add three-sevenths and two-sevenths, the sum becomes five-sevenths. A few simple examples follow below.

Examples
■ A.6 1. $\frac{3}{5} + \frac{2}{5} = \frac{5}{5}$, or 1
  2. $\frac{2}{9} + \frac{5}{9} = \frac{7}{9}$
■ 3. $\frac{3}{4} + \frac{2}{4} = \frac{5}{4}$, or $\frac{4}{4} + \frac{1}{4} = 1\frac{1}{4}$

In each of the examples above, we had no difficulty performing the operation, because the denominators were the same size. We could tentatively define addition in this manner: When the denominators describe units of the same size, the sum of two rationals is found by adding numerators. A definition based on this concept is stated below.

Definition

■ **A.5** ■ $R[\![a, b]\!] + R[\![c, b]\!] = R[\![a + c, b]\!]$

Some examples of this definition follow.

Examples

■ **A.6**   1. $R[\![3, 5]\!] + R[\![1, 5]\!] = R[\![3 + 1, 5]\!]$, or $R[\![4, 5]\!]$

2. $R[\![-4, 7]\!] + R[\![-2, 7]\!] = R[\![-4 + -2, 7]\!]$, or $R[\![-6, 7]\!]$

3. $R[\![-2, 8]\!] + R[\![4, 8]\!] = R[\![-2 + 4, 8]\!]$, or $R[\![2, 8]\!]$ Notice that $[\![2, 8]\!]$

■   is in the same class as $[\![1, 4]\!]$. Therefore, we can write $R[\![2, 8]\!]$ as $R[\![1, 4]\!]$ if we wish.

The definition of addition used above will be sufficient for the set of rationals so long as we can express two rationals in such a manner as to find pairs which have the same denominators. This can always be done. Let us see why. Suppose we want to add the rational numbers $R[\![a, b]\!]$ and $R[\![c, d]\!]$. Since a rational number is a set of ordered pairs,

$$R[\![a, b]\!] = \{ [\![a, b]\!], [\![2a, 2b]\!], [\![3a, 3b]\!], \ldots, [\![da, db]\!], \ldots, [\![na, nb]\!], \ldots \}$$

and

$$R[\![c, d]\!] = \{ [\![c, d]\!], [\![2c, 2d]\!], [\![3c, 3d]\!], \ldots, [\![bc, bd]\!], \ldots, [\![nc, nd]\!], \ldots \}$$

Since all the pairs that comprise a rational number are equivalent, it does not matter which pair is chosen to represent the rational number. (This statement is important and will be discussed later.) That is,

$$R[\![a, b]\!] = \{ [\![a, b]\!], [\![2a, 2b]\!], \ldots, [\![da, db]\!], \ldots \}$$

and

$$R[\![da, db]\!] = \{ [\![a, b]\!], [\![2a, 2b]\!], \ldots, [\![da, db]\!], \ldots \}$$

so

$$R[\![a, b]\!] = R[\![da, db]\!]$$

and in a similar fashion

$$R[\![c, d]\!] = R[\![bc, bd]\!]$$

Now we can use the definition of addition, because the second entries in $R[\![da, db]\!]$ and $R[\![bc, bd]\!]$ are equal, $db = bd$.

$$R[\![a, b]\!] + R[\![c, d]\!] = R[\![da, db]\!] + R[\![bc, bd]\!]$$
$$= R[\![da + bc, db]\!]$$

Thus, we can always express two rationals in such a way as to find pairs that have the same denominators, and then we can use our definition to add.

Definition

■ A.5   *For rational numbers* $R[a, b]$ *and* $R[c, d]$, $R[a, b] + R[c, d]$
■ $= R[ad + bc, bd]$.

Several examples of this definition follow below.

Examples

■ A.7   1. $R[2, 3] + R[3, 7] = R[2 \times 7 + 3 \times 3, 3 \times 7]$, or $R[23, 21]$
2. $R[-3, 5] + R[3, 4] = R[-3 \times 4 + 5 \times 3, 5 \times 4]$, or $R[3, 20]$
3. $R[-2, 3] + R[-4, 7] = R[-2 \times 7 + 3 \times -4, 3 \times 7] = R[-26, 21]$
■   4. $R[x, y] + R[q, r] = R[xr + yq, yr]$

EXERCISES

1. Find the following sums.
   a. $R[2, 3] + R[5, 6]$ b. $R[3, 4] + R[-2, 3]$
   c. $R[-3, 5] + R[6, 10]$ d. $\frac{2}{7} + \frac{3}{8}$
   e. $(-\frac{3}{4}) + \frac{3}{8}$ f. $R[a, x] + R[2, b]$

2. Perform the addition operation below, and change order of addends to check for commutativity.
   a. $\frac{2}{3} + \frac{2}{5}$ b. $R[3, 5] + R[-3, 5]$
   c. $R[-5, 6] + R[-7, 8]$ d. $R[a, b] + R[c, d]$

3. Do you think that R is commutative under addition? Justify your reason.

4. Find the sums below, and change the grouping to check for associativity.
   a. $(R[2, 3] + R[3, 4]) + R[5, 6]$
   b. $(R[3, 5] + R[2, 7]) + R[1, 2]$
   c. $(R[a, b] + R[c, d]) + R[e, f]$

5. Do you think that addition on R is associative? Justify your reason.

6. Find the following sums.
   a. $R[2, 3] + R[0, 3]$ b. $R[-3, 5] + R[0, 5]$
   c. $R[0, 2] + R[1, 2]$ d. $R[0, b] + R[a, b]$

7. Do you believe that there is an identity in R for addition? Why or why not?

8. Find the following sums.
   a. $R[2, 3] + R[-2, 3]$
   b. $R[-3, 5] + R[3, 5]$
   c. $R[a, b] + R[-a, b]$

9. What concept is illustrated in Exercise 8?

10. Write equivalence classes for the following rational numbers.
    a. $R[2, 3]$ b. $R[3, 4]$ c. $R[-1, 2]$ d. $R[-2, 3]$

11. From the classes in Exercise 10, choose pairs that have the same second entry.

12. From Exercises 10 and 11, can you devise a method for writing numbers with like denominators given two ordered pairs? Cite one example.

13. For each of the sums of fractions below, write the sum using the notation of this section. Perform the addition and then translate back to fraction notation.
    a. $\frac{2}{3} + \frac{1}{2}$ b. $(-\frac{5}{2}) + \frac{2}{3}$
    c. $(-\frac{2}{3}) + (-\frac{7}{3})$ d. $\frac{1}{2} + (-\frac{1}{2})$

MULTIPLICATION ON THE RATIONAL NUMBERS

To multiply rational numbers in fraction form, we simply find the product of the numerators and the product of the denominators. In a later section on computation, we will discuss models which may clarify the operation. We will supply some examples below before we define the operation of multiplication on the rational numbers.

Examples

■ A.9
1. $\frac{2}{3} \times \frac{3}{5} = \frac{6}{15}$ or $\frac{2}{5}$
2. $-\frac{3}{4} \times \frac{2}{3} = -\frac{6}{12}$ or $-\frac{1}{2}$
3. $-\frac{5}{8} \times -\frac{1}{2} = \frac{5}{16}$
4. $\frac{a}{b} \times \frac{c}{d} = \frac{ac}{bd}$

The examples shown above can be written in the form of ordered pairs rather than fractions, as shown below.

Examples

■ A.9
1. $R[\![2,3]\!] \times R[\![3,5]\!] = R[\![6,15]\!]$, or $R[\![2,5]\!]$
2. $R[\![-3,4]\!] \times R[\![2,3]\!] = R[\![-6,12]\!]$, or $R[\![-1,2]\!]$
3. $R[\![-5,8]\!] \times R[\![-1,2]\!] = R[\![5,16]\!]$
4. $R[\![a,b]\!] \times R[\![c,d]\!] = R[\![ac,bd]\!]$

Notice that Example 4 implies a definition, because a, b, c, and d can stand for any integer. If we place the restriction on b and d which our definition requires, then we define multiplication as follows.

Definition

A.8 ■ *For rational numbers $R[\![a,b]\!]$ and $R[\![c,d]\!]$, $R[\![a,b]\!] \times R[\![c,d]\!] = R[\![ac,bd]\!]$.*

Before we supply an exercise set, we will consider a few more examples to illustrate multiplication.

Examples

■ A.9
1. $R[\![2,3]\!] \times R[\![0,6]\!] = R[\![0,18]\!]$
2. $R[\![2,3]\!] \times R[\![2,2]\!] = R[\![4,6]\!]$, or $R[\![2,3]\!]$
3. $R[\![-3,5]\!] \times R[\![-5,3]\!] = R[\![15,15]\!]$, or $R[\![1,1]\!]$
4. $R[\![-3,2]\!] \times R[\![2,3]\!] = R[\![-6,6]\!]$, or $R[\![-1,1]\!]$

In this chapter, as in the chapter on integers, we have been differentiating between an object $[\![a,b]\!]$, which is an element in an equivalence class, and $R[\![a,b]\!]$, which represents a rational number. Remember, a rational number is a set or equivalence class whose members are objects of the form $[\![a,b]\!]$. Because we have made this distinction, the question arises as to whether the choice of ordered pairs used to represent the set will affect operations such as multiplication and addition on the set.

Before we deal with this question, let us consider some rational number $R[\![a, b]\!]$. We indicated earlier that the set or equivalence class of a rational number can be generated by writing first the set of multiples of a and b, which we called M_a and M_b. Suppose that the pair $[\![a, b]\!]$ is not the simplest pair. How would we generate the set in this instance? Let us consider an example.

Example

$R[\![2, 3]\!]$ generates $\{[\![2, 3]\!], [\![4, 6]\!], [\![6, 9]\!], \ldots, [\![2n, 3n]\!], \ldots\}$ since $M_2 = \{2, 4, 6, \ldots, 2n, \ldots\}$ and $M_3 = \{3, 6, 9, \ldots, 3n, \ldots\}$. Now we can show that $[\![2, 3]\!] \sim [\![4, 6]\!]$ by definition or by observing that they belong to $R[\![2, 3]\!]$. We would expect $R[\![4, 6]\!]$ to generate the same set as $R[\![2, 3]\!]$.

If we use the method shown above to generate the set $R[\![4, 6]\!]$, we produce the set shown below. $R[\![4, 6]\!]$ generates $\{[\![4, 6]\!], [\![8, 12]\!], [\![12, 18]\!], \ldots, [\![4n, 6n]\!], \ldots\}$ since $M_4 = \{4, 8, 12, \ldots, 4n, \ldots\}$ and $M_6 = \{6, 12, 18, \ldots, 6n, \ldots\}$. The entries in the pair $[\![4, 6]\!]$ have a common factor of 2. That is, $[\![4, 6]\!]$ can be written $[\![2 \times 2, 2 \times 3]\!]$. This tells us that $[\![4, 6]\!] \sim [\![2, 3]\!]$ and that $R[\![4, 6]\!]$ is generated by the multiples of $[\![2, 3]\!]$. Thus,

$$R[\![4, 6]\!] = \{[\![2, 3]\!], [\![4, 6]\!], [\![6, 9]\!], \ldots\}$$

We are now ready to show that operations on R are independent of our choice of ordered pairs to represent the rational number. Since we represent all pairs in the equivalence class $R[\![a, b]\!]$ as $[\![na, nb]\!]$, we know that $[\![a, b]\!]$ and $[\![na, nb]\!]$ are equivalent. In the same way, $[\![c, d]\!]$ and $[\![mc, md]\!]$ are equivalent. Since m and n are any elements of N, $[\![na, nb]\!]$ and $[\![mc, md]\!]$ represent any pair in the classes $R[\![a, b]\!]$ and $R[\![c, d]\!]$, respectively. Consider what happens when we add the rational numbers $R[\![a, b]\!]$ and $R[\![c, d]\!]$ using first the pairs $[\![a, b]\!]$ and $[\![c, d]\!]$ to represent the numbers and then using $[\![na, nb]\!]$ and $[\![mc, md]\!]$ to represent the numbers. We hope or should expect that the equivalence class produced by $R[\![a, b]\!] + R[\![c, d]\!]$ is the same as that produced by $R[\![na, nb]\!] + R[\![mc, md]\!]$. That is, the sum of two rationals does not depend upon the pair of integers of the equivalence class used to represent the rational numbers. We now show a proof of this.

Theorem

■ A.10   *If $R[\![a, b]\!]$ and $R[\![na, nb]\!]$ are used to represent a rational number, and if $R[\![c, d]\!]$ and $R[\![mc, md]\!]$ are used to represent another rational number, then $R[\![a, b]\!] + R[\![c, d]\!] = R[\![na, nb]\!] + R[\![mc, md]\!]$.*

PROOF   We know by definition that $R[\![a, b]\!] + R[\![c, d]\!] = R[\![ad + bc, bd]\!]$. We will replace $ad + bc$ with x and bd with y. The sum will be called $R[\![x, y]\!]$. This is done to simplify our notation; that is, $R[\![a, b]\!] + R[\![c, d]\!] = R[\![x, y]\!]$. Now $R[\![na, nb]\!] + R[\![mc, md]\!] = R[\![namd + nbmc, nbmd]\!]$. By commutative, associative, and distributive properties for I under addition and multiplication, we can rewrite the sum as $R[\![nm(ad + bc), nm(bd)]\!]$. Replacing $ab + bc$ by x and bd by y, we have $R[\![nmx, nmy]\!] = R[\![na, nb]\!]$

$+ R[\![mc, md]\!]$. But $[\![x, y]\!] \sim [\![nmx, nmy]\!]$, because $xnmy = ynmx$. Therefore, $[\![x, y]\!]$ and $[\![nmx, nmy]\!]$ are elements of the same equivalence class and can be employed to represent the same rational number. Thus, $R[\![x, y]\!] = R[\![nmx, nmy]\!]$! That is, the sum of two rational numbers does not depend upon the pairs of integers chosen to represent the rational numbers.

Consider now a similar argument for the operation of multiplication on R. Let $R[\![a, b]\!]$, $R[\![na, nb]\!]$, $R[\![c, d]\!]$, and $R[\![mc, md]\!]$ be used again. This time we must show that the products $R[\![a, b]\!] \times R[\![c, d]\!]$ is the same rational number as $R[\![na, nb]\!] \times R[\![mc, md]\!]$.

PROOF The first product is $R[\![ac, bd]\!]$ and the second is $R[\![mnac, mnbd]\!]$, and since $mnbdac = mnacbd$, we see that $[\![ac, bd]\!] \sim [\![mnac, mnbd]\!]$. Therefore, $R[\![ac, bd]\!] = R[\![mnac, mnbd]\!]$, or $R[\![a, c]\!] \times R[\![b, d]\!]$ $= R[\![na, nb]\!] \times R[\![mc, md]\!]$, and the operation of multiplication does not depend upon the pairs chosen to represent the rational numbers being multiplied.

EXERCISES

1. Find the products below.
  a. $R[\![1, 2]\!] \times R[\![1, 4]\!]$
  b. $R[\![1, 3]\!] \times R[\![-1, 2]\!]$
  c. $R[\![-2, 1]\!] \times R[\![-3, 1]\!]$

2. Do you think that this operation is closed? Why or why not?

3. Find the products below, and change the order of the factors to check for commutativity.
  a. $R[\![1, 2]\!] \times R[\![2, 1]\!]$
  b. $R[\![-2, 3]\!] \times R[\![1, 3]\!]$
  c. $R[\![x, y]\!] \times R[\![a, b]\!]$

4. Do you think that this operation is commutative? Why or why not?

5. Find the products below, and change the grouping to check for associativity.
  a. $(R[\![2, 3]\!] \times R[\![1, 2]\!]) \times R[\![2, 1]\!]$
  b. $(R[\![3, 2]\!] \times R[\![-2, 3]\!]) \times R[\![-2, 1]\!]$
  c. $R[\![a, b]\!] \times (R[\![c, d]\!] \times R[\![e, f]\!])$

6. Can you show that this operation is associative?

7. Find the products below.
  a. $R[\![2, 3]\!] \times R[\![1, 1]\!]$ 　　　 b. $R[\![2, 2]\!] \times R[\![3, 4]\!]$
  c. $R[\![a, a]\!] \times R[\![b, c]\!]$ 　　　 d. $R[\![0, 2]\!] \times R[\![1, 2]\!]$
  e. $R[\![-3, 4]\!] \times R[\![0, 4]\!]$ 　　　 f. $R[\![0, a]\!] \times R[\![b, c]\!]$
  g. $R[\![1, 2]\!] \times R[\![2, 1]\!]$ 　　　 h. $R[\![-2, 3]\!] \times R[\![-3, 2]\!]$
  i. $R[\![a, b]\!] \times R[\![b, a]\!]$

8. Three properties are illustrated in Exercise 7. Can you name them?

9. Show that
  a. $R[\![1, 2]\!] + R[\![3, 4]\!] = R[\![2, 4]\!] + R[\![6, 8]\!]$
  b. $R[\![1, 2]\!] \times R[\![3, 4]\!] = R[\![2, 4]\!] \times R[\![6, 8]\!]$
  c. $R[\![-1, 5]\!] \times R[\![2, 3]\!] = R[\![-4, 20]\!] \times R[\![6, 9]\!]$
  d. $R[\![-1, 5]\!] + R[\![2, 3]\!] = R[\![-4, 20]\!] + R[\![6, 9]\!]$

SECTION B   **PROPERTIES OF ADDITION AND MULTIPLICATION**

BEHAVIORAL OBJECTIVES

B.1   Show that the rationals are closed under addition.

B.2   Be able to state the commutative property of addition for rationals.

B.3   Be able to illustrate the commutative property of addition for rationals.

B.4   Be able to state the associative property of addition for rationals.

B.5   Identify the rational number that is the additive identity.

B.6   Be able to show that $R\,[\![0, 1]\!]$ is the additive identity.

B.7   Be able to state the relationship between a number and its inverse.

B.8   Be able to find the additive inverse of a rational number.

B.9   Be able to define the operation of multiplication on the rational numbers.

B.10 Be able to show that the rationals are closed under multiplication.

B.11 Be able to show that multiplication is commutative for the rational numbers.

B.12 State the commutative property for rational numbers under multiplication.

B.13 Be able to show that the associative property holds for the rationals under multiplication.

B.14 Be able to state the associative property of multiplication for the rationals.

B.15 Be able to state the identity property for multiplication.

B.16 Be able to identify an identity element for multiplication on the set of rationals.

B.17 Be able to describe inverses for multiplication in R.

B.18 Given an arbitrary rational number, find the multiplicative inverse.

B.19 Be able to give an argument why $R\,[\![0, 1]\!]$ acts as zero.

B.20 Be able to state the zero property for multiplication of rationals.

CLOSURE PROPERTY OF ADDITION ON R

In this section we will show that the rational numbers produce a field under addition and multiplication. We will not prove every property, but we will demonstrate proofs for several properties. Those properties which are not proven are left as an exercise.

Since we know that the integers are closed under addition and multiplication, and that the definition of addition on R involves only addition and multiplication of integers, the rationals must be closed under addition. To see why, it is necessary merely to observe the definition of addition of rational numbers with some understanding of what the objects involved really are:

$$R[\![a, b]\!] + R[\![c, d]\!] = R[\![ad + bc, bd]\!]$$

This says that the sum of the two equivalence classes corresponding to $R[\![a, b]\!]$ and $R[\![c, d]\!]$, respectively, is the equivalence class corresponding to $R[\![ad + bc, bd]\!]$. For *closure under addition* we need only convince ourselves that there is an equivalence class corresponding to $R[\![ad + bc, bd]\!]$. This follows because $ad + bc$ is an integer (the integers are closed under addition and multiplication) and bd is a positive integer (the product of two positive integers is again a positive integer). Thus, by the definition of equivalence class, there is an equivalence class corresponding to $R[\![ad + bc, bd]\!]$. Therefore, the sum of two equivalence classes is an equivalence class, so the rationals are closed under addition.

Closure Property of Addition on R: For all $R[\![a, b]\!]$ and $R[\![c, d]\!]$ in R, $R[\![a, b]\!] + R[\![c, d]\!] = R[\![ad + bc, bd]\!] \in R$.

COMMUTATIVE AND ASSOCIATIVE PROPERTY OF ADDITION ON R

We will state the commutative and associative properties without proof. You should be able to construct a proof of these properties by considering our proof for the operation of multiplication. We will simply indicate by example why the commutative and associative properties hold for addition. These examples do not constitute a proof.

Example
B.3 Show that $R[\![a, b]\!] + R[\![c, d]\!] = R[\![c, d]\!] + R[\![a, b]\!]$.

SOLUTION $R[\![a, b]\!] + R[\![c, d]\!] = R[\![ad + bc, bd]\!]$
$R[\![c, d]\!] + R[\![a, b]\!] = R[\![cb + da, db]\!]$
It can be shown that $ad + bc = cb + da$. Why? Also, $bd = db$. Why? Therefore, we can show that $R[\![a, b]\!] + R[\![c, d]\!]$
$= R[\![c, d]\!] + R[\![a, b]\!]$.

B.2 *Commutative Property for Addition of Rational Numbers: For rational numbers $R[\![a, b]\!]$ and $R[\![c, d]\!]$, $R[\![a, b]\!] + R[\![c, d]\!] = R[\![c, d]\!] + R[\![a, b]\!]$.*

We will illustrate the associative property below. Again, this is simply an example and does not satisfy the proof; the method for proof is implied.

Example
Show that $(R[\![a, b]\!] + R[\![c, d]\!]) + R[\![e, f]\!] = R[\![a, b]\!] + (R[\![c, d]\!] + R[\![e, f]\!])$.

SOLUTION $(R[\![a, b]\!] + R[\![c, d]\!]) + R[\![e, f]\!]$

$$= R[\![ad + bc, bd]\!] + R[\![e, f]\!]$$
$$= R[\![\{ad\} f + \{bc\} f + \{bd\} e, \{bd\} f]\!]$$

Why? Also,

$R[\![a, b]\!] + (R[\![c, d]\!] + R[\![e, f]\!])$

$$= R[\![a, b]\!] + R[\![cf + de, df]\!]$$
$$= R[\![a\{df\} + b\{cf\} + b\{de\}, b\{df\}]\!]$$

It can be shown that $\{ad\} f = a\{df\}$, $\{bc\} f = b\{cf\}$, $\{bd\} e = b\{de\}$, and $\{bd\} f = b\{df\}$. Why? It can also be shown that $\{ad\}f + \{bc\}f + \{bd\} e = a\{df\} + b\{cf\} + b\{de\}$. It follows that

$(R[\![a, b]\!] + R[\![c, d]\!]) + R[\![e, f]\!]$

$$= R[\![a, b]\!] + (R[\![c, d]\!] + R[\![e, f]\!]).$$

■ **B.4**   *Associative Property for Addition of Rational Numbers: For rational*
■   *numbers* $R[\![a, b]\!], R[\![c, d]\!], R[\![e, f]\!], (R[\![a, b]\!] + R[\![c, d]\!]) + R[\![e, f]\!] = R[\![a, b]\!]$
   $+ (R[\![c, d]\!] + R[\![e, f]\!])$.

EXERCISES

1. Illustrate the commutative property for R under addition.

2. Prove the commutative property for R under addition.

3. Illustrate the associative property for R under addition.

4. Prove the associative property for R under addition.

5. Which properties are illustrated in the examples below?
   a. $(R[\![2, 3]\!] + R[\![3, 4]\!]) + R[\![5, 6]\!] = R[\![5, 6]\!] + (R[\![2, 3]\!] + R[\![3, 4]\!])$
   b. $(R[\![2, 3]\!] + R[\![3, 4]\!]) + R[\![5, 6]\!] = (R[\![3, 4]\!] + R[\![2, 3]\!]) + R[\![5, 6]\!]$
   c. $R[\![2, 3]\!] + (R[\![3, 4]\!] + R[\![5, 6]\!]) = (R[\![2, 3]\!] + R[\![3, 4]\!]) + R[\![5, 6]\!]$
   d. $(R[\![2, 3]\!] + R[\![3, 4]\!]) + R[\![5, 6]\!] = (R[\![5, 6]\!] + R[\![3, 4]\!]) + R[\![2, 3]\!]$

6. Can you prove the following to be true?
   a. $R[\![a, b]\!] + (R[\![c, d]\!] + R[\![e, f]\!]) = (R[\![a, b]\!] + R[\![e, f]\!]) + R[\![c, d]\!]$
   b. $(R[\![a, b]\!] + R[\![c, d]\!]) + R[\![e, f]\!] = (R[\![e, f]\!] + R[\![c, d]\!]) + R[\![a, b]\!]$
   c. $(R[\![a, b]\!] + R[\![e, f]\!]) + R[\![c, d]\!] = (R[\![a, b]\!] + R[\![c, d]\!]) + R[\![e, f]\!]$

IDENTITY AND INVERSE ON R UNDER ADDITION

Ever since we developed the set of whole numbers, we have had an identity element for the operation of addition, namely, a zero element. We showed under integers that such an element existed, and we called it $I(a, a)$. We will show now that the rationals have an identity, which is not surprising, and that it relates directly to the zero of the whole numbers. Several examples below will give you the idea of the identity in R under addition.

Examples
■ **B.5**   1. $R[\![2, 3]\!] + R[\![0, 3]\!] = R[\![2, 3]\!]$

2. $R[\![3,5]\!] + R[\![0,2]\!] = R[\![6,10]\!]$. Notice that since $[\![6,10]\!] \sim [\![3,5]\!]$, $R[\![6,10]\!]$ represents the set $R[\![3,5]\!]$, and $R[\![6,10]\!] = R[\![3,5]\!]$.

3. $R[\![a,b]\!] + R[\![0,4]\!] = R[\![4a,4b]\!] = R[\![a,b]\!]$

4. $R[\![x,y]\!] + R[\![0,a]\!] = R[\![xa,ya]\!] = R[\![x,y]\!]$

You will notice that any rational number whose first entry in the pair ■ $[\![a,b]\!]$ is 0 acts as an additive identity. That is, we will call the identity $R[\![0,a]\!]$, where a represents any element in N. Why not allow a to be an element of W? or I? Remember the restriction on $R[\![a,b]\!]$! We defined $R[\![a,b]\!]$ so that a and b are in I with $b > 0$. This is equivalent to stating that $a \in I$ and $b \in N$.

■ **B.6** *Identity Element for R Under Addition: For any $R[\![b,c]\!] \in R$, $R[\![0,a]\!]$ $+ R[\![b,c]\!] = R[\![b,c]\!]$ and $R[\![b,c]\!] + R[\![0,a]\!] = R[\![b,c]\!]$.*

In our definition of identity, we again point out that the identity must commute. Notice that we continually do this even though the operation on the set is commutative. We can generate the set which represents the rational number $R(0,a)$ as shown:

$$R[\![0,a]\!] = \{[\![0,1]\!], [\![0,2]\!], [\![0,3]\!], \ldots, [\![0,a]\!], \ldots\} = R[\![0,1]\!]$$

■ Notice that the additive identity, $R[\![0,1]\!]$, corresponds to the fraction $\frac{0}{1}$, or 0, the identity for addition in the whole numbers.

■ **B.7** To have inverses we must first have an identity element. *Remember, the inverse of an element is an element that operates on the given element* ■ *to produce the identity,* which in this case is $R[\![0,a]\!]$. This means that for a given $R[\![x,y]\!]$ there must be another number $R[\![x,y]\!]^*$, called the inverse of $R[\![x,y]\!]$, such that the sum of $R[\![x,y]\!]$ and $R[\![x,y]\!]^*$ is $R[\![0,a]\!]$. Notice that we are using the * notation to read "additive inverse of $R[\![x,y]\!]$." Before we formally show that $R[\![x,y]\!]$ has an additive inverse, we will consider some examples of sums in R which are in fact $R[\![0,a]\!]$.

Examples

1. $R[\![2,3]\!] + R[\![-2,3]\!] = R[\![0,3]\!]$
2. $R[\![-3,5]\!] + R[\![3,5]\!] = R[\![0,5]\!]$
3. $R[\![a,b]\!] + R[\![-a,b]\!] = R[\![0,b]\!]$

From the examples we would expect that the additive inverse of $R[\![x,y]\!]$ would be $R[\![-x,y]\!]$. If $x = 0$, then $R[\![0,y]\!] + R[\![0,y]\!] = R[\![0,y]\!]$. Now we are ready to formalize this concept. Consider a sentence which requires that the sum of two rational numbers produce the identity. In order to discover exactly how a number is related to its inverse, the sentence would read

◀ **B.8** $R[\![x,y]\!] + R[\![b,c]\!] = R[\![0,a]\!]$

Then $R[\![xc+yb,yc]\!] = R[\![0,a]\!]$ by definition of addition in R. Continuing with this equation,

1. $[\![xc + yb, yc]\!] \sim [\![0, a]\!]$ why?
2. $(xc + yb)\, a = (yc)\, 0$ definition of $\sim$
3. $xca + yba = 0$ zero property of multiplication on I
4. $yba = -\, xca$ cancellation law of addition on I
5. $yb = -\, xc$ cancellation law of multiplication on I
6. $by = c(-\, x)$ why?
7. $[\![b, c]\!] \sim [\![-\, x, y]\!]$ definition of $\sim$
8. $R[\![b, c]\!] = R[\![-\, x, y]\!]$ why?

Thus, if $R[\![b, c]\!]$ is an additive inverse of $R[\![x, y]\!]$, then $R[\![b, c]\!]$ = $R[\![-\, x, y]\!]$. This is summarized by saying that $R[\![x, y]\!]^* = R[\![-\, x, y]\!]$. The statements shown above could serve as a proof that there exists an additive inverse for all $R[\![x, y]\!]$, and we name that inverse $R[\![-\, x, y]\!]$. The statements as given only suggest a proof and are given as an indication of how a proof might be constructed. Notice that we forced the inverse to be written $R[\![-\, x, y]\!]$ rather than $R[\![x, -\, y]\!]$. This was done to continue our convention of writing any negative notation in the first entry. *Notice that $-\, x$ does not actually indicate a negative integer*, because if x is a negative integer, $-\, x$ is its additive inverse under addition on I. In this way, if x is $-\, 2$, $-\, x$ is 2.

Before we define inverse for R under addition, let us consider several examples of additive inverse.

Examples

B.8   1. Find the additive inverse of $R[\![2, 3]\!]$. Since the additive inverse of 2 in the set I is $-\, 2$, $R[\![2, 3]\!] + R[\![-\, 2, 3]\!] = R[\![0, 3]\!]$.

2. Find the additive inverse of $R[\![-\, 3, 5]\!]$. Since the additive inverse of $-\, 3$ in the set I is 3, $R[\![-\, 3, 5]\!] + R[\![3, 5]\!] = R[\![0, 5]\!]$.

From the examples you can see that to find the additive inverse of a number in R, we find the additive inverse of the leading element in the pair which represents the rational number.

B.7   *Additive Inverse for Rational Numbers: For all $R[\![x, y]\!]$ in R, the additive inverse of $R[\![x, y]\!]$ is $R[\![-\, x, y]\!]$, sometimes written $R[\![x, y]\!]^*$.*

EXERCISES

1. Which of the following pairs are in the equivalence class $R[\![0, a]\!]$?
   a. $[\![0, 2]\!]$ b. $[\![2, 0]\!]$ c. $[\![0, 0]\!]$
   d. $[\![a, a]\!]$ e. $[\![0, 5]\!]$ f. $[\![5, 0]\!]$

2. Prove using the definition of addition that $R[\![x, y]\!] + R[\![0, a]\!] = R[\![0, a]\!] + R[\![x, y]\!]$.

3. Find the additive inverse of the following rational numbers.
   a. $R[\![-\, 2, 3]\!]$ b. $R[\![3, a]\!]$ c. $R[\![3, 7]\!]$
   d. $R[\![a, b]\!]$ e. $R[\![-\, a, b]\!]$

4. For the rational numbers given below:

a. Generate the set.

b. Find the inverse.

c. Generate the set for the inverse.

d. Prove that the elements in the set of the inverse can be used to represent the inverse.

1. $R[\![2, 5]\!]$ 2. $R[\![-3, 5]\!]$ 3. $R[\![a, b]\!]$ 4. $R[\![-a, b]\!]$

5. a. What fraction corresponds to $R[\![x, y]\!]$?

b. What fraction corresponds to $R[\![-x, y]\!]$?

c. How are $R[\![-x, y]\!]$ and $R[\![x, y]\!]^*$ related?

d. Use a, b, and c to perform the following additions.

1. $\frac{1}{3} + (-\frac{1}{3})$ 2. $\frac{2}{5} + (-\frac{2}{5})$ 3. $(-\frac{8}{3}) + \frac{8}{3}$

CLOSURE PROPERTY FOR MULTIPLICATION ON *R*

We will review the definition of multiplication on the set of rational numbers using the notation of this chapter. We will then look at some of the properties of *R* under multiplication.

Definition

■ **B.9**   *For $R[\![a, b]\!]$ and $R[\![c, d]\!] \in R$, we define the operation $\times$ by $R[\![a, b]\!]$*
■   *$\times R[\![c, d]\!] = R[\![ac, bd]\!]$.*

In the set of rational numbers, the definition of an operation implies closure, because the product of two rationals is defined to be an equivalence class. Since the entries in the pair defining the equivalence class are formed by multiplication on the set *I*, the entries in the pair are integers.

B.10   Let us now consider a more formal proof of this property. Given that $R[\![a, b]\!]$ and $R[\![c, d]\!] \in R$, prove that $R[\![a, b]\!] \times R[\![c, d]\!] = R[\![ac, bd]\!] \in R$.

1. $R[\![a, b]\!] \times R[\![c, d]\!] = R[\![ac, bd]\!]$ 1. definition of $\times$ *R*

2. a, b, c, d in *I* and $b, d > 0$ 2. definition of *R*

3. ac and bd in *I*   3. closure $\times$ *I*

4. $bd > 0$ 4. multiplication of positive integers

■ 5. $R[\![ac, bd]\!]$ in *R*   5. definition of *R*

Closure Property for Multiplication on R: For all $R[\![a, b]\!]$ and $R[\![c, d]\!]$ in *R*, $R[\![a, b]\!] \times R[\![c, d]\!] = R[\![ac, bd]\!] \in R$.

Before we continue to the next property for *R* under multiplication, we should point out that we are operating in the set *R*, which means that the objects we use are sets, not ordered pairs. Also, since operations on *R* are performed by operating on entries in ordered pairs of integers, we are using sets from *I* to perform these operations. It may often seem that we are operating on ordered pairs. Actually, the pairs used are only vehicles. Each time we add or multiply two elements in *R*, we are operating on sets which have been built from sets. That is, our objects are defined recursively and relate back directly to the first set developed, namely *N*. That we do all this by simply writing $R[\![a, b]\!]$ instead of tabulating the set each time is simply a convenience. Furthermore, *we write a and b as entries in $R[\![a, b]\!]$ rather than write $R[\![I(0, a), I(0, b)]\!]$.* You should see that to minimize the proliferation of parentheses and symbols we use the convention of writing $R[\![a, b]\!]$ and hope that it does not obscure the real objects with which we are operating.

COMMUTATIVE AND ASSOCIATIVE PROPERTIES FOR MULTIPLICATION OF RATIONALS

Before we prove the commutative property for multiplication on the set of rational numbers, it is instructive to consider several examples of commutativity. The examples should give you some feeling for the proof which follows.

Example 1

Does $R[\![2, 3]\!] \times R[\![3, 2]\!] = R[\![3, 2]\!] \times R[\![2, 3]\!]$?

SOLUTION $R[\![2, 3]\!] \times R[\![3, 2]\!] = R[\![6, 6]\!]$, or $R[\![1, 1]\!]$, and $R[\![3, 2]\!] \times R[\![2, 3]\!]$
$= R[\![6, 6]\!]$, or $R[\![1, 1]\!]$. The answer to the question is Yes.

Example 2

Show that $R[\![2, a]\!] \times R[\![c, 3]\!] = R[\![c, 3]\!] \times R[\![2, a]\!]$.

SOLUTION $R[\![2, a]\!] \times R[\![c, 3]\!] = R[\![2c, a3]\!]$, and $R[\![c, 3]\!] \times R[\![2, a]\!]$
$= R[\![c2, 3a]\!]$. Since $c2$ and $2c$ are equal, and since $a3 = 3a$,
$R[\![2, a]\!] \times R[\![c, 3]\!] = R[\![c, 3]\!] \times R[\![2, a]\!]$. Why?

Example 3

Does $R[\![0, 1]\!] \times R[\![5, 6]\!] = R[\![5, 6]\!] \times R[\![0, 1]\!]$?

SOLUTION $R[\![0, 1]\!] \times R[\![5, 6]\!] = R[\![0, 6]\!]$, and $R[\![5, 6]\!] \times R[\![0, 1]\!]$
$= R[\![0, 6]\!]$. Therefore, the answer to the question is Yes.

Example 4

Show that $R[\![a, b]\!] \times R[\![x, y]\!] = R[\![x, y]\!] \times R[\![a, b]\!]$.

SOLUTION $R[\![a, b]\!] \times R[\![x, y]\!] = R[\![ax, by]\!]$, and $R[\![x, y]\!] \times R[\![a, b]\!]$
$= R[\![xa, yb]\!]$. Since $xa = ax$ and $yb = by$, $R[\![a, b]\!] \times R[\![x, y]\!]$
$= R[\![x, y]\!] \times R[\![a, b]\!]$. Why?

Example 4 gives us an idea of how the entries in pairs are affected by changing order of the pairs. Notice that the change in order of pairs under multiplication produces a change in the order of the integers multiplied, for example, ax to xa and by to yb. Since we already know that I is commutative under multiplication, we will use this fact to complete our proof. It is often instructive for a student to do several problems of the type listed above before considering the proof which follows.

Example

■ B.11   Given that $R[\![a, b]\!]$ and $R[\![c, d]\!] \in R$, prove that $R[\![a, b]\!] \times R[\![c, d]\!]$
$= R[\![c, d]\!] \times R[\![a, b]\!]$.

1. $R[\![a, b]\!] \times R[\![c, d]\!] = R[\![ac, bd]\!]$	1. definition of multiplication on R
2. $ac = ca$ and $bd = db$	2. commutative property of multiplication on I
3. $R[\![ac, bd]\!] = R[\![ca, db]\!]$	3. substituting step 2 into step 1
4. $R[\![ca, db]\!] = R[\![c, d]\!] \times R[\![a, b]\!]$	4. definition of multiplication on R
5. $R[\![a, b]\!] \times R[\![c, d]\!] = R[\![c, d]\!]$ $\times R[\![a, b]\!]$	5. transitive property of equality, steps 1, 3, and 4

We can now state a commutative property for R under multiplication.

■ **B.12** *Commutative Property for R Under Multiplication*: For all $R[\![a, b]\!]$ and $R[\![c, d]\!]$ in R, $R[\![a, b]\!] \times R[\![c, d]\!] = R[\![c, d]\!] \times R[\![a, b]\!]$.

The proof of the associative property of multiplication on R is much the same as that of the commutative property in that it depends directly on the properties of I under multiplication. We will again supply several examples before proving that the property holds. The examples should help clarify the proof. Notice that we use parentheses to indicate the grouping of integers within the pairs which represent rational numbers. Since we know that I is associative, we could compute without regard for the parentheses, but we use them here to emphasize how the operation is affected by this change in grouping. We should remind you that even though the grouping changes, the order of the integers employed is preserved. We note that *commutativity does not imply associativity* and that *associativity does not imply commutativity*. These properties are completely independent from each other.

Example 1

Does grouping affect the product $R[\![2, 3]\!] \times (R[\![2, 2]\!] \times R[\![3, 2]\!])$?

SOLUTION $\quad R[\![2, 3]\!] \times (R[\![2, 2]\!] \times R[\![3, 2]\!]) = R[\![2, 3]\!] \times R[\![6, 4]\!]$
$$= R[\![12, 12]\!]$$
$$= R[\![1, 1]\!]$$
$$(R[\![2, 3]\!] \times R[\![2, 2]\!]) \times R[\![3, 2]\!] = R[\![4, 6]\!] \times R[\![3, 2]\!]$$
$$= R[\![12, 12]\!]$$
$$= R[\![1, 1]\!]$$

The answer to the question is that grouping does not seem to affect the product in this instance.

Example 2

Are the following products in R equal? $(R[\![0, 1]\!] \times R[\![1, 2]\!]) \times R[\![2, 3]\!]$ and $R[\![0, 1]\!] \times (R[\![1, 2]\!] \times R[\![2, 3]\!])$.

SOLUTION $\quad$ Since $(R[\![0, 1]\!] \times R[\![1, 2]\!]) \times R[\![2, 3]\!] = R[\![0, 2]\!] \times R[\![2, 3]\!]$
$$= R[\![0, 6]\!]$$

and

$$R[\![0, 1]\!] \times (R[\![1, 2]\!] \times R[\![2, 3]\!]) = R[\![0, 1]\!] \times R[\![2, 6]\!]$$
$$= R[\![0, 6]\!]$$

in this example, the answer is Yes.

Example 3

Find the products of $(R[\![a, 2]\!] \times R[\![2, a]\!]) \times R[\![b, c]\!]$ and $R[\![a, 2]\!] \times (R[\![2, a]\!] \times R[\![b, c]\!])$. Compare the results. Are the products the same?

SOLUTION
$$(R[\![a, 2]\!] \times R[\![2, a]\!]) \times R[\![b, c]\!] = R[\![a2, 2a]\!] \times R[\![b, c]\!]$$
$$= R[\![(a2)\,b, (2a)\,c]\!]$$
and
$$R[\![a, 2]\!] \times (R[\![2, a]\!] \times R[\![b, c]\!]) = R[\![a, 2]\!] \times R[\![2b, ac]\!]$$
$$= R[\![a(2b), 2(ac)]\!]$$

Notice that $(a2)\,b$ and $a(2b)$ are equal, as are $(2a)\,c$ and $2(ac)$. Why? This means that the two products are equal.

Example 4

Compare the products $(R[\![a, b]\!] \times R[\![c, d]\!]) \times R[\![e, f]\!]$ and $R[\![a, b]\!] \times (R[\![c, d]\!] \times R[\![e, f]\!])$.

SOLUTION
$$(R[\![a, b]\!] \times R[\![c, d]\!]) \times R[\![e, f]\!] = R[\![ac, bd]\!] \times R[\![e, f]\!]$$
$$= R[\![(ac)\,e, (bd)\,f]\!]$$
$$R[\![a, b]\!] \times (R[\![c, d]\!] \times R[\![e, f]\!]) = R[\![a, b]\!] \times R[\![ce, df]\!]$$
$$= R[\![a(ce), b(df)]\!]$$

Notice again that the grouping of integers is different in the products above. But this does not affect the products. Why? Note that $(ac)\,e$ and $a(ce)$ are equal, as are $(bd)\,f$ and $b(df)$.

■ B.13   Therefore, given that $R[\![a, b]\!]$, $R[\![c, d]\!]$, and $R[\![e, f]\!]$ are rational numbers, prove that

$$(R[\![a, b]\!] \times R[\![c, d]\!]) \times R[\![e, f]\!] = R[\![a, b]\!] \times (R[\![c, d]\!] \times R[\![e, f]\!]).$$

1. $(R[\![a, b]\!] \times R[\![c, d]\!]) \times R[\![e, f]\!]$ $= R[\![ac, bd]\!] \times R[\![e, f]\!]$
   1. definition of multiplication on R

2. $R[\![ac, bd]\!] \times R[\![e, f]\!]$ $= R[\![(ac)\,e, (bd)\,f]\!]$
   2. definition of multiplication on R

3. $(ac)\,e = a(ce)$ and $(bd)\,f = b(df)$
   3. associative property of multiplication on I

4. $R[\![(ac)\,e, (bd)\,f]\!]$ $= R[\![a(ce), b(df)]\!]$
   4. reflexive property of equality and step 3

5. $R[\![a(ce), b(df)]\!]$ $= R[\![a, b]\!] \times R[\![ce, df]\!]$
   5. definition of multiplication on R

6. $R[\![a, b]\!] \times R[\![ce, df]\!]$ $= R[\![a, b]\!] \times (R[\![c, d]\!] \times R[\![e, f]\!])$
   6. definition of multiplication on R

7. $(R[\![a, b]\!] \times R[\![c, d]\!]) \times R[\![e, f]\!]$ $= R[\![a, b]\!] \times (R[\![c, d]\!] \times R[\![e, f]\!])$
   7. transitive property of equality and steps 1, 2, 4, 5, and 6

■

We have proven the associativity property for multiplication on R and can now state the property below. These proofs should serve as models for the proofs under addition on R.

■ **B.14** *Associative Property of Multiplication on R: For all R(a, b), R(c, d), R(e, f)*
■ *in R, $[\![R(a, b) \times R(c, d)]\!] \times R[\![e, f]\!] = R[\![a, b]\!] \times (R[\![c, d]\!] \times R[\![e, f]\!])$.*

EXERCISES

1. Generate the set $R[\![3, 5]\!]$ and $R[\![2, 1]\!]$, and show by example that commutativity holds no matter what pair is used to represent the set.

2. Find the following products.
a. $R[\![0, 1]\!] \times R[\![3, 5]\!]$ b. $R[\![0, 2]\!] \times R[\![2, 1]\!]$
c. $R[\![0, 3]\!] \times R[\![a, 2]\!]$ d. $R[\![0, a]\!] \times R[\![c, b]\!]$

3. How would you describe the property illustrated in Exercise 2?

4. State three examples that illustrate the associative property for R under multiplication.

5. Is associativity independent of the pairs used to represent the rational numbers? Why or why not?

6. Find the following products. Compare the products to the given rational numbers.
a. $R[\![1, 1]\!] \times R[\![2, 3]\!]$ b. $R[\![3, 4]\!] \times R[\![2, 3]\!]$
c. $R[\![3, 3]\!] \times R[\![1, 2]\!]$ d. $R[\![a, b]\!] \times R[\![c, c]\!]$

7. How would you name the property illustrated in Exercise 6?

8. Perform the indicated operations on the following fractions.
a. $\frac{1}{2} \times \frac{0}{3}$ b. $\frac{2}{3} \times \frac{2}{2}$ c. $\frac{3}{4} \times \frac{4}{3}$ d. $\frac{2}{2} \times \frac{3}{2}$

9. Find the products indicated below. How would you describe the products in each case?
a. $R[\![3, 5]\!] \times R[\![5, 3]\!]$ b. $R[\![2, 1]\!] \times R[\![1, 2]\!]$
c. $R[\![3, 1]\!] \times R[\![1, 3]\!]$ d. $R[\![a, b]\!] \times R[\![b, a]\!]$

10. Name the property illustrated in Exercise 9.

11. Can you show by example that rational numbers, written as the fraction (a/b), have the properties of rational numbers under multiplication?

IDENTITY, INVERSE, AND ZERO PROPERTIES
OF MULTIPLICATION ON *R*

Let us now try to find an identity element for R under multiplication. The method we will use to find this element is fairly standard. We will suppose that $R[\![x, y]\!]$ acts as an identity and then discover enough properties of x and y to be able to identify the particular rational number that is the identity.

If $R[\![x, y]\!]$ is the multiplicative identity, then $R[\![x, y]\!] \times R[\![a, b]\!] = R[\![a, b]\!]$ for any $R[\![a, b]\!]$ and $R[\![a, b]\!] \times R[\![x, y]\!] = R[\![a, b]\!]$. Using the definition of multiplication, we have that $R[\![ax, by]\!] = R[\![a, b]\!]$

What does this tell us about x and y? Since the pair $[\![ax, by]\!]$ is in the same rational number as $[\![a, b]\!]$, we must have that $x = y$.

Thus, the identity for multiplication must be of the form $R[\![a, a]\!]$. We
B.16 generally choose $R[\![1, 1]\!]$ to represent this integer. Consider some examples

below to see if our rational number 1 acts as an identity element. Remember that $R[\![1, 1]\!]$ is $\{[\![1, 1]\!], [\![2, 2]\!], [\![3, 3]\!], \ldots\}$.

Example 1

Does $R[\![1, 1]\!] \times R[\![3, 5]\!] = R[\![3, 5]\!]$?

SOLUTION $R[\![1, 1]\!] \times R[\![3, 5]\!] = R[\![3, 5]\!]$ by definition of multiplication.

Example 2

Does $R[\![6, 6]\!] \times R[\![2, 5]\!] = R[\![2, 5]\!]$?

SOLUTION $R[\![6, 6]\!] \times R[\![2, 5]\!] = R[\![12, 30]\!]$, and since $2 \times 30 = 5 \times 12$, then $[\![2, 5]\!] \sim [\![12, 30]\!]$ and they can be used to represent the same rational number.

Example 3

Does $R[\![3, 2]\!] \times R[\![a, a]\!] = R[\![3, 2]\!]$?

SOLUTION $R[\![3, 2]\!] \times R[\![a, a]\!] = R[\![3a, 2a]\!]$ and $3 \times 2a = 2 \times 3a$. Therefore, $[\![3, 2]\!] \sim [\![3a, 2a]\!]$, which means that $R[\![3, 2]\!]$ and $R[\![3a, 2a]\!]$ represent the same rational number.

Example 4

Does $R[\![x, y]\!] \times R[\![a, a]\!] = R[\![x, y]\!]$?

SOLUTION $R[\![x, y]\!] \times R[\![a, a]\!] = R[\![xa, ya]\!]$, and $xya = yxa$. Therefore, $[\![x, y]\!] \sim [\![xa, ya]\!]$, which means that $R[\![x, y]\!]$ and $R[\![xa, ya]\!]$ are names for the same rational number.

We will not prove that $R[\![1, 1]\!]$ is the identity, nor will we prove that the identity is unique. We simply state the property below.

■ **B.15** *Identity Element for Multiplication on R: For all a, b, c in I with a and* ■ *c > 0, $R[\![a, a]\!] \times R[\![b, c]\!] = R[\![b, c]\!]$ and $R[\![b, c]\!] \times R[\![a, a]\!] = R[\![b, c]\!]$.*

To this point in the text, we have not found inverses for multiplication. As the sets of numbers have grown larger, the number of properties associated with the operation has increased. The first inverse we encountered was in I under addition, then in R under addition. We are now ready to introduce inverses in R under multiplication. Remember, *inverse depends upon the existence of identities.* We define inverse as that element which operates on another element to produce the identity. In this case, we will use the convention of calling the multiplicative inverse of $R[\![a, b]\!]$, $R[\![a, b]\!]^{-1}$. Sometimes this element is called the *reciprocal* of $R[\![a, b]\!]$. We also call it the *multiplicative inverse* to differentiate between $R[\![a, b]\!]^{-1}$ and $R[\![a, b]\!]^*$, the additive inverse.

Consider a few numerical examples which may lead you to an understanding of the definition which will follow.

Examples

1. $R[2, 3] \times R[3, 2] = R[6, 6]$, or $R[1, 1]$. Therefore, $R[2, 3]$ has as its multiplicative inverse $R[3, 2]$.

2. $R[-3, 4] \times R[-4, 3] = R[12, 12]$, or $R[1, 1]$. Therefore, $R[-3, 4]$ has $R[-4, 3]$ as its multiplicative inverse.

We should next consider the solution of an equation in which the multiplicative inverse is given as $R[a, b]$.

Example 1

B.18   Find the reciprocal of $R[3, 2]$ by solving the equation $R[3, 2] \times R[a, b] = R[1, 1]$.

SOLUTION   Now $R[3, 2] \times R[a, b] = R[1, 1]$ and $R[3, 2] \times R[a, b] = R[3a, 2b]$. Therefore, $[3a, 2b] \sim [1, 1]$, which implies that $3a = 2b$. We would like to find $R[a, b]$, and we know that $[a, b] \sim [x, y]$ iff $ay = bx$. Therefore, we can write $a3 = b2$, which implies $[a, b] \sim [2, 3]$; $R[2, 3]$ is the reciprocal of $R[3, 2]$. ∎

The next example is even more general. If we can solve the equation $R[a, b] \times R[x, y] = R[1, 1]$, then we can write a statement describing the multiplicative inverse which is based upon our solution set. This should give you a deeper understanding of the statement which will follow. In the equation below, we indicate that $R[x, y]$ is the reciprocal of $R[a, b]$ since the product of the two rational numbers produces $R[1, 1]$, which is the identity element for R under multiplication.

Example 2

B.18
1. $R[a, b] \times R[x, y] = R[1, 1]$ 1. given
2. $R[ax, by] = R[1, 1]$ 2. definition of multiplication under R
3. $[ax, by] \sim [1, 1]$ 3. definition of R
4. $ax = by$ 4. definition of $\sim$
5. $xa = yb$ 5. commutative property of multiplication in I

In order to change this equivalence into a statement about rational numbers, we must note that our definition of rational number required that the second term of a pair of integers representing the rationals be positive; that is, since $R[a, b]$ is rational, b is positive. But a is an integer and can be either positive or negative. If a is positive, we have from step 5 above that $[x, y] \sim [b, a]$, so $[b, a] \in R[b, a]$ and $R[b, a]$ is the inverse of $R[a, b]$.

If, on the other hand, a is negative, then $-a$ is positive and $[x, y] \sim [-b, -a]$, since if $xa = yb$, then $x(-a) = y(-b)$. Thus, $[-b, -a]$

$\in R[\![-b, -a]\!]$. So $R[\![x, y]\!] = R[\![-b, -a]\!]$; and $R[\![-b, -a]\!]$ is the multiplicative inverse of $R[\![a, b]\!]$. For example, the multiplicative inverse of $R[\![-1, 3]\!]$ is *not* $R[\![3, -1]\!]$, since $R[\![3, -1]\!]$ is not really defined. The multiplicative inverse of $R[\![-1, 3]\!]$ is $R[\![-3, 1]\!]$. Let us show this:

$$R[\![-1, 3]\!] \times R[\![-3, 1]\!] = R[\![3, 3]\!]$$
$$= R[\![1, 1]\!]$$

■ **B.17** Thus, the multiplicative inverse of $R[\![a, b]\!]$, symbolized by $R[\![a, b]\!]^{-1}$, is
■ $R[\![b, a]\!]$ if a is positive and $R[\![-b, -a]\!]$ if a is negative. Can you see what happens if $a = 0$?

Before we state this property, let us consider some examples.

Example 1
■ **B.18** Find the reciprocal of $R[\![2, 3]\!]$.

SOLUTION The reciprocal of $R[\![2, 3]\!]$ is $R[\![3, 2]\!]$, and $R[\![2, 3]\!] \times R[\![3, 2]\!]$ $= R[\![6, 6]\!]$, or $R[\![1, 1]\!]$.

Example 2
Find the reciprocal of $R[\![0, 3]\!]$.

SOLUTION Since there is no rational $R[\![3, 0]\!]$, $R[\![0, 3]\!]$ has no reciprocal.

Example 3
If $R[\![b, a]\!]$ is the multiplicative inverse of $R[\![a, b]\!]$, what is the inverse of $R[\![b, a]\!]$?

SOLUTION Notice that since R is commutative under $\times$, $R[\![a, b]\!]$
$\times R[\![b, a]\!] = R[\![ab, ba]\!] = R[\![ba, ab]\!] = R[\![b, a]\!] \times R[\![a, b]\!]$.
■ Therefore, $R[\![a, b]\!]$ is the multiplicative inverse of $R[\![b, a]\!]$.

■ **B.17** *The Multiplicative Inverse for Rational Numbers: If a is positive, the multiplicative inverse of $R[\![a, b]\!]$ is $R[\![b, a]\!]$; if a is negative, the multiplicative inverse of $R[\![a, b]\!]$ is $R[\![-b, -a]\!]$.*

Consider the set $R[\![0, 1]\!]$:

$$R[\![0, 1]\!] = \{[\![0, 1]\!], [\![0, 2]\!], [\![0, 3]\!], \ldots, [\![0, n]\!], \ldots\}$$

■ **B.19** As we mentioned in Example 2 above, this rational number has no inverse under multiplication. It is its own inverse under addition. It represents the identity element for addition on R. You can see that this element has some interesting properties. In addition to the other properties, when we multiply any rational by $R[\![0, n]\!]$, we always produce $R[\![0, n]\!]$. You should be careful not to confuse the three properties listed in this section. Make sure that each is clear in your mind. There is a tendency to confuse the identity property with this zero property. Consider the following examples
■ before reading our statement of the zero property.

Examples

1. $R[2, 4] \times R[0, 1] = R[0, 4] = R[0, 1]$
2. $R[0, 8] \times R[5, 3] = R[0, 24] = R[0, 1]$
3. $R[a, b] \times R[0, 3] = R[0, b3] = R[0, 1]$
4. $R[0, a] \times R[x, y] = R[0, ay] = R[0, 1]$

B.20 *Zero Property for Multiplication of Rational Numbers: For all $R[b, c]$ in R, $R[0, a] \times R[b, c] = R[0, a]$ and $R[b, c] \times R[0, a] = R[0, a]$.*

We leave the proof of this property to you. The examples may help you construct a proof for the zero property.

EXERCISES

1. Can you generate the sets below by using the multiplicative identity and the operation of multiplication? (*Hint:* $R[2, 3] \times R[1, 1] = R[2, 3]$, $R[2, 3] \times R[2, 2] = R[4, 6]$, ...)
   a. $R[2, 3]$ b. $R[-3, 5]$ c. $R[-2, 4]$ d. $R[1, 3]$

2. Prove that the identity element is unique.

3. a. If a is positive, prove that the inverse of $R[a, b]$ is $R[b, a]$.
   b. If a is negative, prove that the inverse of $R[a, b]$ is $R[-b, -a]$.

4. Generate the following sets.
   a. identity for $R \times$
   b. inverse of $R[3, 2]$
   c. $R[0, a]$

5. Find the following products.
   a. $R[2, 3] \times R[3, 2]$
   b. $R[-2, 5] \times R[0, 2]$
   c. $R[3, 3] \times R[2, 1]$

6. Explain why any element in the set can be used to represent the rational numbers $R[a, a]$ and $R[0, a]$.

7. What kind of structure is formed by the following?
   a. R under $+$
   b. R under $\times$

8. Find the following results as indicated.
   a. $R[2, 1] \times (R[1, 2] + R[2, 3])$
   b. $R[1, 5] \times (R[3, 5] + R[0, 2])$
   c. $(R[2, 1] \times R[1, 2]) + (R[2, 1] \times R[2, 3])$
   d. $(R[1, 5] \times R[3, 5]) + (R[1, 5] \times R[0, 2])$

9. From Exercise 8, compare a and c as well as b and d. What would you name the property illustrated in this problem?

10. Can you define the property illustrated by Exercises 8 and 9 for R?

SECTION C   SUBTRACTION

BEHAVIORAL OBJECTIVES

C.1   Write a correspondence between the integers and a subset of the rational numbers.

C.2   Show that there is a subset of the rational numbers that acts as the integers.

C.3   Be able to state the field properties for the rational numbers.

C.4   Be able to state and prove the cancellation law for addition in the rationals.

C.5   Be able to state and prove the cancellation law for multiplication in the rationals.

C.6   Be able to define subtraction for rational numbers.

C.7   Be able to subtract two given rational numbers.

C.8   Be able to prove that the difference of two rationals is equal to the sum of the first and the inverse of the second.

C.9   Be able to show that the rationals are closed under subtraction.

C.10   Be able to state the zero property of subtraction.

C.11   Be able to give examples of the zero property for subtraction.

C.12   Be able to state the subtraction property for the rationals.

C.13   Be able to give examples of the subtraction property for the rationals.

A RELATIONSHIP BETWEEN THE INTEGERS AND THE RATIONAL NUMBERS

We have seen that the rational number $R[\![1, 1]\!]$ is a multiplicative identity. What happens when we add $R[\![1, 1]\!]$ to itself successively?

$$R[\![1, 1]\!] + R[\![1, 1]\!] = R[\![1 + 1, 1]\!] = R[\![2, 1]\!]$$
$$R[\![2, 1]\!] + R[\![1, 1]\!] = R[\![2 + 1, 1]\!] = R[\![3, 1]\!]$$
$$R[\![3, 1]\!] + R[\![1, 1]\!] = R[\![3 + 1, 1]\!] = R[\![4, 1]\!]$$
$$\vdots \qquad \vdots \qquad \vdots \qquad \vdots$$

$$R[\![k, 1]\!] + R[\![1, 1]\!] = R[\![k + 1, 1]\!]$$

Let us compare this list with successive additions of 1, the multiplicative identity of the integers.

$$1 + 1 = 2$$
$$2 + 1 = 3$$
$$3 + 1 = 4$$
$$\vdots \quad \vdots \quad \vdots$$

$$k + 1 = k + 1$$

There is an almost obvious correspondence between the elements in these two lists, namely

■ C.1
$$R[\![1, 1]\!] \leftrightarrow 1$$
$$R[\![2, 1]\!] \leftrightarrow 2$$
$$R[\![3, 1]\!] \leftrightarrow 3$$
$$\vdots \qquad \vdots$$

■
$$R[\![k + 1, 1]\!] \leftrightarrow k + 1$$

The existence of this correspondence raises two questions: First, can this correspondence be expanded so that *every* integer corresponds to some rational number? Second, does this correspondence hold under addition and multiplication? That is, if we add or multiply two rational numbers, does the result correspond to the sum of the corresponding integers?

It is easy to answer the first question. We simply generalize the correspondence. Let the rational number $R(n, 1)$ correspond with the integer n. Let us write out some of the corresponding terms.

■ C.1

$$
\begin{aligned}
R[\![-3, 1]\!] &\leftrightarrow -3 = I\ (3, 0)\\
R[\![-2, 1]\!] &\leftrightarrow -2 = I\ (2, 0)\\
R[\![-1, 1]\!] &\leftrightarrow -1 = I\ (1, 0)\\
R[\![0, 1]\!] &\leftrightarrow 0 = I\ (0, 0)\\
R[\![1, 1]\!] &\leftrightarrow 1 = I\ (0, 1)\\
R[\![2, 1]\!] &\leftrightarrow 2 = I\ (0, 2)\\
R[\![3, 1]\!] &\leftrightarrow 3 = I\ (0, k)
\end{aligned}
$$

■

It should be obvious that every integer corresponds to some rational number.

■ C.2 To answer the second question, let us add two integers, say a and b. Their sum is $a + b$ and corresponds to $R[\![a + b, 1]\!]$. The rational number $R[\![a, 1]\!]$ and $R[\![b, 1]\!]$ corresponds to a and b, respectively, and their sum $R[\![a, 1]\!] + R[\![b, 1]\!]$ is $R[\![a + b, 1]\!]$. Thus, the correspondence holds under addition, and we can state that there is an isomorphism between the set I and this subset of the rationals. We leave as an exercise the verification

■ that the correspondence holds under multiplication.

It should not be surprising that we found $R[\![0, 1]\!]$ to be the identity under addition or that $R[\![1, 1]\!]$ is the identity under multiplication, because $I(1, 1)$, or 0, is the additive identity for I, and 1, or $I(0, 1)$, is the multiplicative identity for I, and $0 \leftrightarrow R[\![0, 1]\!]$ and $1 \leftrightarrow R[\![1, 1]\!]$.

Let us note that in the integers $I(0, k)$ and $I(k, 0)$ are additive inverses; that is, $k + (-k) = 0$. Looking at the corresponding rational numbers $R[\![k, 1]\!]$ and $R[\![-k, 1]\!]$, we see that $R[\![k, 1]\!] + R[\![-k, 1]\!] = R[\![k - k, 1]\!] = R[\![0, 1]\!]$ and that $R[\![0, 1]\!]$ corresponds to 0 in I.

■ **C.2**   It is common to give the set of rational numbers that correspond to
the integers a special name. That is, the set $\{R[\![k, 1]\!] \mid k \text{ is an integer}\}$ is commonly called the integers because of the correspondence between this
set and the integers. Because the set $\{R[\![k, 1]\!] \mid k \text{ is an integer}\}$ acts like the
■ set of integers, we quite reasonably call it the integers.

EXERCISES

1. Show that the correspondence between integers and the subset of rationals $\{R[\![k, 1]\!] \mid k \text{ is an}$
integer$\}$ is preserved under multiplication. That is, show that $\{R[\![k, 1]\!] \mid k \text{ is an integer}\}$ is
isomorphic to I and J under multiplication.

2. Find the rational number that corresponds to the following integers.
  a. 7   b. -9 c. $I(0, 12)$ d. $I(19, 0)$

3. For every integer, is there a rational number that corresponds to it under the correspondence
of this section?

4. Is the opposite question of Exercise 3 true? That is, for every rational number is there an
integer that corresponds to it under the correspondence of this section?

THE FIELD OF RATIONAL NUMBERS
We have shown that the set R forms an Abelian group under addition
and that the non-zero elements of R form an Abelian group under multiplication. For R to satisfy the conditions of a field, it must also satisfy
the distributive property of $\times$ over $+$. We will not prove that this property
exists; however, the examples below illustrate the property.

Example 1
  Find the result of $R[\![2, 1]\!] \times (R[\![1, 1]\!] + R[\![1, 2]\!])$ in two ways, and
compare the results.

SOLUTION $R[\![2, 1]\!] \times (R[\![1, 1]\!] + R[\![1, 2]\!]) = R[\![2, 1]\!] \times R[\![3, 2]\!]$
$$= R[\![6, 2]\!]$$

and

$$R[\![2, 1]\!] \times (R[\![1, 1]\!] + R[\![1, 2]\!]) = (R[\![2, 1]\!] \times R[\![1, 1]\!])$$
$$+ (R[\![2, 1]\!] \times R[\![1, 2]\!])$$
$$= R[\![2, 1]\!] + R[\![2, 2]\!]$$
$$= R[\![6, 2]\!]$$

The results of the operations performed are the same whether
we add and then multiply or whether we distribute $R[\![2, 1]\!]$
and then add.

Example 2
  Use the distributive property to find the results of the mixed operation
$R[\![a, b]\!] \times (R[\![c, d]\!] + R[\![e, f]\!])$. Perform the operations as indicated first.

SOLUTION $R[\![a, b]\!] \times (R[\![c, d]\!] + R[\![e, f]\!]) = R[\![a, b]\!] \times R[\![cf + de, df]\!]$
$$= R[\![acf + ade, bdf]\!]$$
and, using the distributive property, we have
$$R[\![a, b]\!] \times (R[\![c, d]\!] + R[\![e, f]\!]) = (R[\![a, b]\!] \times R[\![c, d]\!])$$
$$+ (R[\![a, b]\!] \times R[\![e, f]\!])$$
$$= R[\![ac, bd]\!] + R[\![ae, bf]\!]$$
$$= R[\![acbf + bdae, bdbf]\!]$$
$$= R[\![acf + dae, dbf]\!]$$

The two examples will give you some idea of a proof for the distributive property. We simply state the property below and leave this proof as an exercise.

Distributive Property of × over + for Rational Numbers: For all $R[\![a, b]\!]$, $R[\![c, d]\!]$, $R[\![e, f]\!]$ *in* R, $R[\![a, b]\!] \times (R[\![c, d]\!] + R[\![e, f]\!]) = (R[\![a, b]\!]$ $\times R[\![c, d]\!]) + (R[\![a, b]\!] \times R[\![e, f]\!])$.

We will now summarize the properties needed for the field of rational numbers.

■ **C.3**
1. The operations of addition and multiplication on R are closed.
2. The operations of addition and multiplication on R are commutative.
3. The operations of addition and multiplication on R are associative.
4. The identity for addition is $R[\![0, 1]\!]$ and for multiplication the identity is $R[\![1, 1]\!]$.
5. The operation of addition on R has inverses symbolized by $R[\![a, b]\!]^*$, where $R[\![a, b]\!] = R[\![-a, b]\!]$.
6. Every element of R except $R[\![0, 1]\!]$ has an inverse under multiplication, symbolized by $R[\![a, b]\!]^{-1}$, where $R[\![a, b]\!]^{-1} = R[\![b, a]\!]$ if a is positive and $R[\![a, b]\!]^{-1} = R[\![-b, -a]\!]$ if a is negative.
7. The operation of multiplication distributes over the addition operation in R.

■

EXERCISES

1. Consider the subset of R in which all second elements in $R[\![a, b]\!]$ are of the form $2n$, that is, $R[\![a, 2n]\!]$.
   a. Does this form an Abelian group under $+$? Why or why not?
   b. Does this form an Abelian group under $\times$? Why or why not?
   c. Does this form a field under $+$ and $\times$? Why or why not?

2. Which of the following statements are true? Why?
   a. $R[\![2, 3]\!] \times (R[\![3, 2]\!] + R[\![4, 2]\!]) = (R[\![2, 3]\!] \times R[\![3, 2]\!]) + (R[\![2, 3]\!] \times R[\![4, 2]\!])$
   b. $R[\![2, 3]\!] + (R[\![3, 2]\!] \times R[\![4, 2]\!]) = (R[\![2, 3]\!] + R[\![3, 2]\!]) \times (R[\![2, 3]\!] + R[\![4, 2]\!])$
   c. $R[\![2, 3]\!] \times (R[\![3, 2]\!] + R[\![4, 2]\!]) = (R[\![3, 2]\!] + R[\![4, 2]\!]) \times R[\![2, 3]\!]$
   d. $R[\![2, 3]\!] \times (R[\![3, 2]\!] + R[\![4, 2]\!]) = (R[\![2, 3]\!] \times R[\![3, 2]\!]) + R[\![4, 2]\!]$
   e. $R[\![2, 3]\!] \times (R[\![3, 2]\!] + R[\![4, 2]\!]) = R[\![3, 2]\!] \times (R[\![4, 2]\!] + R[\![3, 2]\!])$

3. In the set R, does addition distribute over multiplication?

CANCELLATION LAWS FOR ADDITION
AND MULTIPLICATION ON R

The cancellation laws developed in previous chapters can be translated to read as follows:

1. Given an equality, we can add equal numbers to both sides of the equality without affecting the equality.
2. Given an equality, we can subtract equal numbers from both sides of the equality without affecting the equality.
3. Given an equality, we can multiply equal numbers on both sides of the equality without affecting the equality.
4. Given an equality, we can divide equal numbers on both sides of the equality by a non-zero number without affecting the equality.

These laws are often stated in this form as rules for solving equations. In this manner, open sentences or equations can be solved for the sets N, W, and I. Before we state these laws for R, we will indicate how they are used in solving equations. Notice that we have combined 1 and 2 and called them the *cancellation law for addition*. We have also combined 3 and 4 and named them the *cancellation law for multiplication*. We will continue this practice in this section.

For x in W, find the solution set for the sentences listed.

Example 1

$x - 6 = 12$ using rule 1.

SOLUTION $(x - 6) + 6 = 12 + 6$
$$x = 18$$

Example 2

$x + 6 = 18$ using rule 2.

SOLUTION $(x + 6) - 6 = 18 - 6$
$$x = 12$$

Example 3

$x \div 2 = 5$ using rule 3.

SOLUTIONS $(x \div 2) \times 2 = 5 \times 2$
$$x = 10$$

Example 4

$4x = 12$ using rule 4.

SOLUTION $4x \div 4 = 12 \div 4$
$$x = 3$$

If you consider the sentences above, you will notice that we used the inverse operation to find the solution or solution set in each case. We considered what operation was performed on × and then "undid" the operation by using the opposite operation. In this way, we "undid" "addition of 6" by "subtracting 6" in Example 1; we undid "subtraction of 6" by "adding 6" in Example 2. In a later chapter, we will devote more time to *open sentences*. Actually, we have been solving such sentences throughout the text although we have not yet formally treated the subject.

Consider now a statement of the cancellation law for the set R. Remember, this statement will actually say two things:

1. We can add equal numbers to both sides of an equality.

2. We can subtract equal numbers from both sides of an equality.

We often describe the sides of an equality as the right-hand side and the left-hand side. Notice also in the statement of this law that we use the phrase "if and only if," which indicates that given the first statement, we can produce the second statement, or given the second statement, we can produce the first statement.

■ C.4   *Cancellation Law of = for Addition in R: For $R[\![a, b]\!], R[\![c, d]\!], R[\![x, y]\!]$ in R, $R[\![a, b]\!] + R[\![x, y]\!] = R[\![c, d]\!] + R[\![x, y]\!]$ if and only if $R[\![a, b]\!] = R[\![c, d]\!]$.*

This can be restated as follows:

1. If $R[\![a, b]\!] = R[\![c, d]\!]$, then $R[\![a, b]\!] + R[\![x, y]\!] = R[\![c, d]\!] + R[\![x, y]\!]$. and

■   2. If $R[\![a, b]\!] + R[\![x, y]\!] = R[\![c, d]\!] + R[\![x, y]\!]$, then $R[\![a, b]\!] = R[\![c, d]\!]$.

Consider some examples of this law as it is applied on the set R.

Example 1

Find the solution set for the open sentence $R[\![2, 3]\!] + R[\![x, y]\!] = R[\![5, 6]\!]$.

SOLUTION   Adding the inverse of $R[\![2, 3]\!]$ to both sides, we produce
$$R[\![-2, 3]\!] + R[\![2, 3]\!] + R[\![x, y]\!] = R[\![-2, 3]\!] + R[\![5, 6]\!]$$
$$R[\![0, 3]\!] + R[\![x, y]\!] = R[\![3, 18]\!]$$
$$R[\![x, y]\!] = R[\![3, 18]\!], \text{ or } R[\![1, 6]\!]$$
Thus, the rational number $R[\![3, 18]\!]$ makes the sentence $R[\![2, 3]\!] + R[\![x, y]\!] = R[\![5, 6]\!]$ a true statement.

Example 2

Find the solution set $R[\![x, y]\!]$ for the sentence $R[\![x, y]\!] - R[\![2, 3]\!]$ $= R[\![5, 6]\!]$.

SOLUTION   By adding $R[\![2, 3]\!]$ to both sides, we produce
$$R[\![x, y]\!] - R[\![2, 3]\!] + R[\![2, 3]\!] = R[\![5, 6]\!] + R[\![2, 3]\!]$$
$$R[\![x, y]\!] = R[\![27, 18]\!], \text{ or } R[\![3, 2]\!]$$

We should point out that in Example 1 that we subtracted $R[\![2, 3]\!]$ from both sides by adding the inverse $R[\![-2, 3]\!]$ to both sides. We did not explain it in this way because we have not yet defined subtraction on R. Therefore, we could only add the inverse. You will see in the next section that subtraction of $R[a, b]$ is defined to mean adding its inverse $R[a, b]^*$.

Consider now the cancellation law for multiplication. We offer first a proof of one part of this law prior to the statement of the law. This proof serves as a model for proof of both laws stated here:

■ **C.4** Given that $R[a, b] \times R[x, y] = R[c, d] \times R[x, y]$ and $R[x, y] \neq R[0, a]$, prove that $R[a, b] = R[c, d]$.

1. $R[a, b] \times R[x, y] = R[c, d] \times R[x, y]$	1. given
2. $R[ay + bx, by] = R[cy + dx, dy]$	2. definition $\times R$
3. $[ay + bx, by] \sim [cy + dx, dy]$	3. definition of R
4. $(ay + bx) dy = (cy + dx) by$	4. definition $\sim R$
5. $(ay + bx) d = (cy + dx) b$	5. cancellation of $\times I$
6. $ayd + bxd = cyb + dxb$	6. distributive property of $\times$ over $+ I$
7. $ayd = cyb$	7. cancellation for $+ I$
8. $ad = cb$	8. cancellation for $\times I$
9. $ad = bc$	9. commutative for $\times I$
■ 10. $[a, b] \sim [c, d]$	10. definition $\sim R$
11. $R[a, b] = R[c, d]$	11. definition of R

Notice that this proves the statement "if $R[a, b] \times R[x, y] = R[c, d] \times R[x, y]$, then $R[a, b] = R[c, d]$." We have not proven the statement "if $R[a, b] = R[c, d]$, then $R[a, b] \times R[x, y] = R[c, d] \times R[x, y]$. This is left as an exercise.

■ **C.5** *Cancellation Law of $=$ for Multiplication on R: For $R[a, b], R[c, d],$ $R[x, y]$ in R, $R[a, b] \times R[x, y] = R[c, d] \times R[x, y]$ if and only if $R[a, b] = R[c, d]$.*

Again this can be translated into two statements:

1. If $R[a, b] = R[c, d]$, then $R[a, b] \times R[x, y] = R[c, d] \times R[x, y]$. and
■ 2. If $R[a, b] \times R[x, y] = R[c, d] \times R[x, y]$, then $R[a, b] = R[c, d]$.

Consider the following examples.

Example 1

Find $R[x, y]$ for the sentence $R[x, y] \times R[1, 2] = R[5, 6]$.

SOLUTION Multiplying both sides by $R[2, 1]$ so as to get on the left side $R[x, y]$ times the multiplicative identity,

$$R[x, y] \times R[1, 2] \times R[2, 1] = R[5, 6] \times R[2, 1]$$
$$R[x, y] \times R[1, 1] = R[10, 6]$$
$$R[x, y] = R[10, 6]$$

Notice that $R[\![x, y]\!] = R[\![10, 6]\!]$ satisfies the given sentence: $R[\![10, 6]\!] \times R[\![1, 2]\!] = R[\![10, 12]\!]$. Since $[\![10, 12]\!] \sim [\![5, 6]\!]$, $R[\![10, 12]\!] = R[\![5, 6]\!]$, and $R[\![10, 6]\!]$ makes our original sentence true.

Example 2

If $R[\![2, 3]\!] \times R[\![-1, 5]\!] = R[\![2, 3]\!] \times R[\![x, y]\!]$, then $R[\![x, y]\!] = R[\![-1, 5]\!]$.

EXERCISES

1. State the reason for the statements below.
   a. If $R[\![2, 3]\!] + R[\![x, y]\!] = R[\![2, 3]\!]$, then $R[\![-2, 3]\!] + R[\![2, 3]\!] + R[\![x, y]\!] = R[\![-2, 3]\!] + R[\![2, 3]\!]$
   b. If $R[\![x, y]\!] - R[\![1, 4]\!] = R[\![2, 5]\!]$, then $R[\![x, y]\!] - R[\![1, 4]\!] + R[\![1, 4]\!] = R[\![2, 5]\!] + R[\![1, 4]\!]$.
   c. If $R[\![x, y]\!] \times R[\![2, 3]\!] = R[\![3, 5]\!]$, then $R[\![x, y]\!] \times R[\![2, 3]\!] \times R[\![3, 2]\!] = R[\![3, 5]\!] \times R[\![3, 2]\!]$.

2. Solve the sentences below for all x in W.
   a. $x + 7 = 12$ b. $2x = 16$ c. $3x - 5 = 17$
   d. $2x + 6 = 12$

3. Solve the sentences below for $R(x, y)$.
   a. $R[\![x, y]\!] \times R[\![2, 1]\!] = R[\![6, 2]\!]$
   b. $R[\![x, y]\!] + R[\![3, 5]\!] = R[\![6, 15]\!]$
   c. $R[\![x, y]\!] + R[\![2, 3]\!] = R[\![3, 5]\!]$
   d. $R[\![x, y]\!] \times R[\![a, b]\!] = R[\![c, d]\!]$
   e. $R[\![x, y]\!] + R[\![a, b]\!] = R[\![c, d]\!]$

SUBTRACTION AND ITS PROPERTIES ON R

We have been defining subtraction as the inverse operation to addition. In this way the difference between $R[\![a, b]\!]$ and $R[\![c, d]\!]$, $R[\![a, b]\!] - R[\![c, d]\!]$, is the rational number which when added to $R[\![c, d]\!]$ produces $R[\![a, b]\!]$. That is, if $R[\![a, b]\!] - R[\![c, d]\!]$ is $R[\![e, f]\!]$, then $R[\![e, f]\!] + R[\![c, d]\!] = R[\![a, b]\!]$. This is the same method that we used for defining subtraction in the integers, which is not surprising, because the integers are a subset of the rationals in the sense of our previous correspondence. Just as in the integers, we have two ways of performing the operation of subtraction, as is illustrated below.

Example 1

■ C.7 $R[\![2, 3]\!] - R[\![3, 5]\!] = R[\![x, y]\!]$ means
$R[\![2, 3]\!] = R[\![3, 5]\!] + R[\![x, y]\!]$
$R[\![2, 3]\!] = R[\![3y + 5x, 5y]\!]$

$[\![2, 3]\!] \sim [\![3y + 5x, 5y]\!]$ if $10y = 9y + 15x$. Why? Or $y = 15x$, so $[\![x, y]\!] \sim [\![x, 15x]\!]$, or $[\![x, y]\!] \sim [\![1, 15]\!]$ and $R[\![x, y]\!] = R[\![1, 15]\!]$

Example 2

$R[\![2, 3]\!] - R[\![3, 5]\!] = R[\![x, y]\!]$ means that $R[\![3, 5]\!] + R[\![x, y]\!] = R[\![2, 3]\!]$. Now add $R[\![-3, 5]\!]$ to both sides, noting that $R[\![-3, 5]\!] = R[\![3, 5]\!]^*$. That is, we add the additive inverse of $R[\![3, 5]\!]$.

$$R[\![-3,5]\!] + R[\![3,5]\!] + R[\![x,y]\!] = R[\![2,3]\!] + R[\![-3,5]\!]$$
$$R[\![0,5]\!] + R[\![x,y]\!] = R[\![10-9,15]\!]$$
$$R[\![x,y]\!] = R[\![1,15]\!]$$

Definition

■ **C.6** *For all $R[\![a,b]\!]$ and $R[\![c,d]\!]$ in R there is an $R[\![x,y]\!]$ in R such that*
■ $R[\![a,b]\!] - R[\![c,d]\!] = R[\![x,y]\!]$ *if and only if* $R[\![a,b]\!] = R[\![x,y]\!] + R[\![c,d]\!]$.

Notice in the examples above that translating the subtraction example to addition by using the additive inverse is a much simpler operation than applying the usual definition for subtraction. We justify this method with the following theorem.

Theorem

■ **C.8** $R[\![a,b]\!] - R[\![c,d]\!] = R[\![a,b]\!] + R[\![-c,d]\!]$.

PROOF $R[\![a,b]\!] - R[\![c,d]\!] = R[\![x,y]\!]$ means that $R[\![x,y]\!] + R[\![c,d]\!]$
= $R[\![a,b]\!]$. Now add $R[\![c,d]\!]^*$ to both sides.

$$R[\![x,y]\!] + R[\![c,d]\!] + R[\![c,d]\!]^* = R[\![a,b]\!] + R[\![c,d]\!]^* \quad \text{Why?}$$
$$R[\![x,y]\!] + R[\![0,1]\!] = R[\![a,b]\!] + R[\![c,d]\!]^* \quad \text{Why?}$$
$$R[\![x,y]\!] = R[\![a,b]\!] + R[\![c,d]\!]^* \quad \text{Why?}$$

But we know that $R[\![c,d]\!]^* = R[\![-c,d]\!]$. Then, replacing $R[\![c,d]\!]^*$ by $R[\![-c,d]\!]$ in the last sentence, we have $R[\![x,y]\!]$ = $R[\![a,b]\!] + R[\![-c,d]\!]$. That is, $R[\![a,b]\!] - R[\![c,d]\!] = R[\![x,y]\!]$, or $R[\![a,b]\!] - R[\![c,d]\!] = R[\![a,b]\!] + R[\![-c,d]\!]$.

Notice, to perform subtraction in R, we follow these steps:

1. Change the operation from $-$ to $+$.
2. Replace the number subtracted by its additive inverse.

Consider now a few examples of subtraction using the above method.

Examples

■ **C.7** 1. $R[\![1,2]\!] - R[\![-2,3]\!] = R[\![1,2]\!] + R[\![2,3]\!] = R[\![7,6]\!]$
2. $R[\![5,4]\!] - R[\![3,2]\!] = R[\![5,4]\!] + R[\![-3,2]\!] = R[\![-2,8]\!]$
■ 3. $R[\![-2,3]\!] - R[\![-4,1]\!] = R[\![-2,3]\!] + R[\![4,1]\!] = R[\![10,3]\!]$

We would like to consider next the question of closure for subtraction.
■ **C.9** You can readily see that if we can always translate subtraction to addition by using our theorem, subtraction must be closed because addition is closed. We will not construct a formal proof but simply state the closure
■ property.

Closure Property for Subtraction on R: For all $R[\![a,b]\!]$ and $R[\![c,d]\!]$ in R, the difference $R[\![a,b]\!] - R[\![c,d]\!] \in R$.

We again wish to point out that the operation of subtraction is not commutative or associative. We leave to you the task of finding counter-examples to illustrate this statement.

Notice also that there is no identity for subtraction. However, the identity for addition, $R[0, a]$, acts as a right identity for subtraction. We call this the zero property for subtraction. Notice below that substituting the element zero on the right of a number produces a correspondence of the number with itself.

Examples

■ **C.11** 1. $R[2, 3] - R[0, 3] = R[2, 3] + R[0, 3]$
$$= R[6, 9], \text{ or } R[2, 3]$$
Why?

■ 2. $R[a, b] - R[0, c] = R[a, b] + R[0, c]^*$
  $R[a, b] + R[0, c] = R[ac, bc], \text{ or } R[a, b]$

■ **C.10** *Zero Property for Subtraction on R: For all $R[a, b]$ in R, $R[a, b] - R[0, c]$*
■ $= R[a, b]$.

There is one last property of subtraction that is almost trivial. Namely, when we subtract a number from itself, the result is zero. This is easily shown, because

$$R[a, b] - R[a, b] = R[a, b] + R[-a, b] \qquad \text{by our theorem}$$
$$= R[0, b]$$
$$= R[0, 1]$$

We call this the subtraction property for R and state it as follows:

■ **C.12** *Subtraction Property for R: For all $R[a, b]$ in R, $R[a, b] - R[a, b]$*
■ $= R[0, b]$.

Notice that with the zero property as defined, we could call the element zero a right identity for subtraction. Every $R[a, b]$ in R can be subtracted from itself to produce the right identity, $R[0, b]$. Consider some examples.

Examples

◄ **C.13** 1. $R[1, 2] - R[1, 2] = R[1, 2] + R[-1, 2] = R[0, 2]$
  2. $R[3, 5] - R[6, 10] = R[3, 5] + R[-6, 10] = R[0, 50]$
■ 3. $R[a, b] - R[a, b] = R[a, b] + R[-a, b] = R[0, b]$

EXERCISES

1. Perform the indicated subtractions by using the definition.
  a. $R[2, 3] - R[3, 5]$ b. $R[-2, 1] - R[2, 7]$

2. Perform the operations by translation to addition first.
  a. $R[1, 2] - R[1, 2]$
  b. $R[3, 2] - R[-2, 3]$
  c. $R[-1, 2] - R[-3, 4]$

3. Show by example that subtraction on R
  a. is not commutative   b. is not associative   c. has no identity

4. Explain why subtraction can have no inverses.

5. Name the properties illustrated below.
   a. $R[a, b] - R[c, d] = R[a, b] + R[-c, d]$
   b. $R[a, b] - R[a, b] = R[0, a]$
   c. $R[a, b] - R[0, a] = R[a, b]$
   d. $(R[a, b] - R[c, d]) \in R$

SECTION D DIVISION

BEHAVIORAL OBJECTIVES

D.1 Be able to express division of rationals as a product of rationals.

D.2 Be able to define division in the set of rationals.

D.3 Be able to divide two given rational numbers.

D.4 State the properties of zero under division.

D.5 Be able to explain why we cannot divide by zero.

D.6 Be able to state the division property of $R[1, 1]$.

D.7 Be able to give an example of the division property of $R[1, 1]$.

DIVISION AND ITS PROPERTIES ON R

We have defined division on the integers as the inverse of multiplication, and we will try to carry over this definition to the rational numbers. Following the definition given for integers, we say that $R[a, b] \div R[c, d] = R[x, y]$ means that $R[a, b] = R[c, d] \times R[x, y]$. Thus, to solve a division problem we must do the following:

1. Translate a division sentence to a multiplication sentence.

2. Perform the multiplication.

3. Using the definition of equivalence, solve the sentence for $R[x, y]$.

Consider a few examples using this technique.

Example 1

$R[1, 2] \div R[3, 4] = R[x, y]$. Find $R[x, y]$; that is, find the quotient of $R[1, 2] \div R[3, 4]$.

SOLUTION   1. Translate to multiplication.
$$R[1, 2] = R[3, 4] \times R[x, y]$$
   2. Perform the multiplication.
$$R[1, 2] = R[3x, 4y]$$
   3. Now solve for $R[x, y]$.
$$[1, 2] \sim [3x, 4y]$$
$$4y = 6x$$

$$[\![4, 6]\!] \sim [\![x, y]\!]$$
$$R[\![4, 6]\!] = R[\![x, y]\!]$$
Thus, $R[\![1, 2]\!] \div R[\![3, 4]\!] = R[\![4, 6]\!]$

Example 2

$R[\![2, 3]\!] \div R[\![4, 3]\!] = R[\![x, y]\!]$. Find $R[\![x, y]\!]$.

SOLUTION $R[\![2, 3]\!] = R[\![4, 3]\!] \times R[\![x, y]\!]$
$$R[\![2, 3]\!] = R[\![4x, 3y]\!]$$
$$[\![2, 3]\!] \sim [\![4x, 3y]\!]$$
$$6y = 12x$$
$$[\![6, 12]\!] \sim [\![x, y]\!]$$
$$R[\![6, 12]\!] = R[\![x, y]\!]$$
Thus, $R[\![2, 3]\!] \div R[\![4, 3]\!] = R[\![6, 12]\!]$

Notice that the process shown above is a long and complicated one. Since we have shown that multiplication exhibits inverses in R, we can approach this operation as we did with subtraction. First, we will consider the steps involved, and then we will show that this can also be done more easily. We will use the same examples.

Example 1

■ **D.1** $R[\![1, 2]\!] \div R[\![3, 4]\!] = R[\![x, y]\!]$

SOLUTION   By translating into multiplication, $R[\![1, 2]\!] = R[\![3, 4]\!] \times R[\![x, y]\!]$. Then $R[\![4, 3]\!] \times R[\![1, 2]\!] = R[\![4, 3]\!] \times R[\![3, 4]\!] \times R[\![x, y]\!]$. By cancellation law of multiplication on R, $R[\![4, 3]\!] \times R[\![1, 2]\!] = R[\![x, y]\!]$, and by the identity property on R,

$$R[\![4, 3]\!] \times R[\![1, 2]\!] = R[\![x, y]\!]$$
$$R[\![4, 6]\!] = R[\![x, y]\!]$$

Example 2

$R[\![2, 3]\!] \div R[\![4, 3]\!] = R[\![x, y]\!]$

SOLUTION
$$R[\![2, 3]\!] = R[\![4, 3]\!] \times R[\![x, y]\!]$$
$$R[\![2, 3]\!] \times R[\![3, 4]\!] = R[\![x, y]\!] \times R[\![4, 3]\!] \times R[\![3, 4]\!]$$
$$R[\![2, 3]\!] \times R[\![3, 4]\!] = R[\![x, y]\!]$$
$$R[\![6, 12]\!] = R[\![x, y]\!]$$

■

In the examples above, we used the following steps.

1. We translated from $\div$ to $\times$ using our definition of division.
2. We multiplied both sides by the inverse, $R[\![c, d]\!]^{-1}$, of the number with which we were dividing (the divisor).
   Cancellation law allowed us to do this.
3. Since $R[\![c, d]\!]^{-1} \times R[\![c, d]\!] = R[\![1, 1]\!]$ and $R[\![1, 1]\!] \times R[\![x, y]\!] = R[\![x, y]\!]$, we produced $R[\![x, y]\!]$ on the right.
4. We multiplied on the left by $R[\![c, d]\!]^{-1}$ to produce our *quotient* or result.

Notice that the steps we followed also required that we take advantage of the properties of R under $\times$. Therefore, we did not concern ourselves with order or grouping. Notice in the first example that we associated $R[\![4, 3]\!]$ with $R[\![3, 4]\!]$ without stating the property. In the second example we commuted $R[\![4, 3]\!]$ and $R[\![x, y]\!]$ again without stating the property. This allowed us to shorten the process even further. We will supply the definition of division next and then show an even shorter method of dividing.

Definition

■ **D.2**   *For all $R[\![a, b]\!]$ and $R[\![c, d]\!] \neq 0$ in R there exists an $R[\![x, y]\!]$ in R such*
■   *that $R[\![a, b]\!] \div R[\![c, d]\!] = R[\![x, y]\!]$ if and only if $R[\![a, b]\!] = R[\![c, d]\!] \times R[\![x, y]\!]$.*

We will consider this definition in light of the previous examples and indicate a short method of performing division on R. When this process is applied to the set of common fractions in R, it is often described as "inverting and multiplying." This terminology is considered poor, because it is not precise and explains nothing about the process. However, you will see a connection between the algorithm developed and the algorithm used for division of fractions. Consider now our definition of division.

If $R[\![a, b]\!] \div R[\![c, d]\!] = R[\![x, y]\!]$, then by the definition of division, $R[\![a, b]\!] = R[\![c, d]\!] \times R[\![x, y]\!]$ since $R[\![a, b]\!] = R[\![x, y]\!] \times R[\![c, d]\!]$. Why? Then $R[\![a, b]\!] \times R[\![c, d]\!]^{-1} = R[\![x, y]\!] \times R[\![c, d]\!] \times R[\![c, d]\!]^{-1}$ by the cancellation law of multiplication on R. And using the identity property, $R[\![a, b]\!] \times R[\![c, d]\!]^{-1} = R[\![x, y]\!]$.

Notice that by combining our definition of division with the properties of multiplication, particularly the inverse property, we are able to translate a division problem directly to multiplication. In this way, the steps involved in multiplying reduce to two: Given the problem $R[\![a, b]\!] \div R[\![c, d]\!] = R[\![x, y]\!]$,

1. Change the operation to multiplication.
2. Replace $R[\![c, d]\!]$ with its inverse $R[\![c, d]\!]^{-1}$ and multiply. (Remember for $c > 0$, $R[\![c, d]\!]^{-1} = R[\![d, c]\!]$ and for $c < 0$, $R[\![c, d]\!]^{-1} = R[\![-d, -c]\!]$.)

Thus, we have the following theorem.

Theorem

■ **D.1**   *To perform the operation of division on R, translate $R[\![a, b]\!] \div R[\![c, d]\!]$*
to $R[\![a, b]\!] \times R[\![c, d]\!]^{-1}$.

Example 1

▲ **D.3** $R[\![1, 2]\!] \div R[\![3, 4]\!] = R[\![x, y]\!]$. Find $R[\![x, y]\!]$.

SOLUTION $R[\![1, 2]\!] \times R[\![4, 3]\!] = R[\![x, y]\!]$
$R[\![4, 6]\!] = R[\![x, y]\!]$

Example 2

$R[\![2, 3]\!] \div R[\![4, 3]\!] = R[\![x, y]\!]$

SOLUTION $\quad R[\![2, 3]\!] \times R[\![3, 4]\!] = R[\![x, y]\!]$

$\qquad\qquad\qquad R[\![6, 12]\!] = R[\![x, y]\!]$

Example 3

$R[\![a, b]\!] \div R[\![c, d]\!] = R[\![x, y]\!]$

SOLUTION $\quad R[\![a, b]\!] \times R[\![d, c]\!] = R[\![x, y]\!]$

$\qquad\qquad\qquad R[\![ad, bc]\!] = R[\![x, y]\!]$

Let us recall what $R[\![c, d]\!]^{-1}$ is. From the definition of rational numbers we know that $d > 0$. We must consider two cases.

Case 1 $\quad c > 0$

In this case $R[\![d, c]\!]$ is a rational number, and

$R[\![c, d]\!] \times R[\![d, c]\!] = R[\![cd, dc]\!]$

$\qquad\qquad\qquad\quad = R[\![1, 1]\!]$

So that $R[\![c, d]\!]^{-1} = R[\![d, c]\!]$.

EXAMPLE $\quad R[\![2, 3]\!]^{-1} = R[\![3, 2]\!]$

Case 2 $\quad c < 0$

Here $R[\![d, c]\!]$ is not defined, because the definition of rational number requires that the second element be positive. We note that if $c < 0$, then $-c > 0$. Thus, $R[\![-d, -c]\!]$ is a rational number, and

$R[\![c, d]\!] \times R[\![-d, -c]\!] = R[\![-cd, -dc]\!]$

$\qquad\qquad\qquad\qquad\quad = R[\![1, 1]\!]$

In this case, $R[\![c, d]\!]^{-1} = R[\![-d, -c]\!]$.

EXAMPLE $\quad R[\![-2, 3]\!]^{-1} = R[\![-3, 2]\!]$

In our definition of division, we stipulated that $c \neq 0$. However, we said nothing of $R[\![a, b]\!]$, which implies that $R[\![a, b]\!]$ can be zero. Consider now what happens when we divide zero, $R[\![0, b]\!]$, by $R[\![c, d]\!]$. Since $R[\![0, b]\!] \div R[\![c, d]\!] = R[\![0, b]\!] \times R[\![c, d]\!]^{-1}$ and $R[\![0, b]\!] \times R[\![c, d]\!]^{-1} = R[\![0, bc]\!]$, or zero, we see that the quotient is zero.

■ **D.4** *Zero Property of Division: For all* $R[\![0, b]\!]$ *in* R *and* $R[\![c, d]\!] \in R$ *with*
■ $c \neq 0, R[\![0, b]\!] \div R[\![c, d]\!] = R[\![0, 1]\!]$.

For all the sets of numbers discussed until now, the rational numbers, except zero, are the first set to be closed under division. This is shown below.

1. $R[\![a, b]\!] \div R[\![c, d]\!] = R[\![a, b]\!] \times R[\![c, d]\!]^{-1}$ for $R[\![c, d]\!] \neq 0$, that is, $c \neq 0$.

2. $R[\![c, d]\!]^{-1}$ exists because $c \neq 0$.

3. Multiplication of rationals is closed.

4. $R[\![a, b]\!] \times R[\![d, c]\!]$ exists.

■ **D.5**   The above list also shows why the divisor in division is restricted to
▲ **D.4**   be non-zero. Since $R[\![a, b]\!] \div R[\![c, d]\!] = R[\![a, b]\!] \times R[\![c, d]\!]^{-1}$, $R[\![c, d]\!]^{-1}$
must be a rational number, but if $R[\![c, d]\!] = R[\![0, d]\!]$, that is, zero, then
$R[\![0, d]\!]^{-1} = R[\![d, 0]\!]$ does not exist because $R[\![d, 0]\!]$ does not exist.

Another way of looking at the problem of division by zero is this: If
we wanted to divide $R[\![a, b]\!]$ by zero, say $R[\![a, b]\!] \div R[\![0, 1]\!] = R[\![x, y]\!]$,
then $R[\![a, b]\!] = R[\![0, 1]\!] \times R[\![x, y]\!]$, which is not true for all $R[\![a, b]\!]$ since
▲ ■ $R[\![0, 1]\!] \times R[\![x, y]\!] = R[\![0, 1]\!]$.

Notice that zero has several properties under different operations. We
list them at this point to emphasize the importance of zero in the number
system.

1. identity for addition

2. zero property of multiplication: $R[\![a, b]\!] \times R[\![0, c]\!] = R[\![0, bc]\!]$

3. zero property of subtraction: $R[\![a, b]\!] - R[\![0, c]\!] = R[\![a, b]\!]$

4. subtraction property: $R[\![a, b]\!] - R[\![a, b]\!] = R[\![0, b]\!]$

■ **D.4**   5. zero property for division: $R[\![0, b]\!] \div R[\![c, d]\!] = R[\![0, bd]\!]$

■   6. No result in R for division by zero

We have considered the commutative and associative properties for
every operation thus far. We will again state that these properties do not
hold for division. You can supply a proof of this. A counter-example is
your best bet for a proof.

We have not yet considered the identity property for division on R.
We must first indicate that since division on R is non-commutative, it
is doubtful that an identity exists. This is not a proof, simply evidence
that an identity might not exist. If an identity does exist, it must commute
with every element. Division has a right identity just as subtraction does.
You noticed that the right identity for subtraction was zero or the identity
element for addition, the inverse operation of subtraction, on R. It might
be worthwhile to consider the identity for multiplication to see if it acts
as an identity for division. Consider a few examples.

Examples

■ **D.7**   1. $R[\![1, 2]\!] \div R[\![1, 1]\!] = R[\![1, 2]\!]$, but $R[\![1, 1]\!] \div R[\![1, 2]\!] = R[\![2, 1]\!]$, the
multiplicative inverse of $R[\![1, 2]\!]$.

2. $R[\![3, 1]\!] \div R[\![1, 1]\!] = R[\![3, 1]\!]$, but $R[\![1, 1]\!] \div R[\![3, 1]\!] = R[\![1, 3]\!]$, the
multiplicative inverse of $R[\![3, 1]\!]$.

3. $R[\![a, b]\!] \div R[\![c, c]\!] = R[\![ac, bc]\!] = R[\![a, b]\!]$, but $R[\![c, c]\!] \div R[\![a, b]\!]$
■ $= R[\![cb, ca]\!]$, the multiplicative inverse of $R[\![a, b]\!]$.

You will notice that the identity for multiplication acts as a right
identity for division. In simpler terms, any number $R[\![x, y]\!]$ divided by

one, $R[\![a, a]\!]$, will produce that number, $R[\![x, y]\!]$. We will call this the division property of one and state it as follows:

D.6 ■ *Division Property of One: For any $R[\![b, c]\!]$ in R, $R[\![b, c]\!] \div R[\![1,1]\!] = R[\![b, c]\!]$.*

Let us see if we can find a rational number $R[\![x, y]\!]$ that acts as identity for division; that is, for any rational $R[\![a, b]\!]$, $R[\![x, y]\!] \div R[\![a, b]\!] = R[\![a, b]\!]$. Since $R[\![x, y]\!] \div R[\![a, b]\!] = R[\![x, y]\!] \times R[\![a, b]\!]^{-1}$, we have that $R[\![x, y]\!]$ must satisfy this equation:

$$R[\![x, y]\!] \times R[\![a, b]\!]^{-1} = R[\![a, b]\!]$$

Using the cancellation property of R,

$$R[\![x, y]\!] \times R[\![a, b]\!]^{-1} \times R[\![a, b]\!] = R[\![a, b]\!] \times R[\![a, b]\!]$$

or

$$R[\![x, y]\!] = R[\![a, b]\!] \times R[\![a, b]\!]$$
$$R[\![x, y]\!] = R[\![a^2, b^2]\!]$$

It should be obvious that there is no one rational number $R[\![x, y]\!]$ that will satisfy this last equation. If it is not obvious, try $R[\![1, 2]\!]$ and $R[\![1, 3]\!]$ for $R[\![a, b]\!]$. Now $R[\![x, y]\!] = R[\![a^2, b^2]\!]$ for $R[\![a, b]\!] = R[\![1, a]\!]$, $R[\![x, y]\!] = R[\![1, 4]\!]$, and for $R[\![a, b]\!] = R[\![1, 3]\!]$, $R[\![x, y]\!] = R[\![1, 9]\!]$. Thus, although $R[\![1, 1]\!]$ is a *right identity* for division, *there is no identity for division.*

Related to the property stated above is another property much like that of the inverse under multiplication. If we consider the property of one as an identity property, then we must consider inverse properties for division. In this case, we will find that any rational number other than zero will divide itself one time. More formally, we can state that $R[\![a, b]\!] \div R[\![a, b]\!] = R[\![b, b]\!]$. Consider a few examples of this property to assure yourself that it can be performed as stated. We will name this the division property.

Examples

D.7 1. $R[\![1, 2]\!] \div R[\![1, 2]\!] = R[\![2, 2]\!]$

2. $R[\![3, 4]\!] \div R[\![3, 4]\!] = R[\![12, 12]\!]$

■ 3. $R[\![a, b]\!] \div R[\![a, b]\!] = R[\![ab, ab]\!]$

D.6 *Division Property for Rational Numbers: For all $R[\![a, b]\!]$, $R[\![a, b]\!] \neq$ zero,*
■ $R[\![a, b]\!] \div R[\![a, b]\!] = R[\![1,1]\!]$.

EXERCISES

1. Perform the following divisions.
   a. $R[\![0, 3]\!] \div R[\![5, 2]\!]$
   c. $R[\![2, 3]\!] \div R[\![2, 3]\!]$
   e. $R[\![-5, 4]\!] \div R[\![7, 2]\!]$

   b. $R[\![0, 2]\!] \div R[\![9, 1]\!]$
   d. $R[\![5, 4]\!] \div R[\![7, 2]\!]$
   f. $R[\![5, 4]\!] \div R[\![-7, 2]\!]$

2. a. $R[\![a, b]\!] \div R[\![a, b]\!] = ?$ **b.** $R[\![0, b]\!] \div R[\![a, b]\!] = ?$

c. What is the result if a number is divided by itself?

d. What is the result if zero is divided by a non-zero number?

3. Show that the rationals do not commute with respect to division.

4. Show that the associative property does not hold for the operation of division in the set of rationals.

SECTION E   THE PROPERTIES OF ORDER ON THE SET OF RATIONAL NUMBERS

BEHAVIORAL OBJECTIVES

E.1   State the trichotomy property for rationals.

E.2   Be able to order any two rational numbers.

E.3   Be able to show that the order relation on the rationals is transitive.

E.4   Be able to show that the order relation on the rationals is not an equivalence relation.

E.5   Be able to explain the order relation on the rationals in three ways.

E.6   Be able to define the symbol $<$.

E.7   Be able to state and prove the cancellation law of addition for inequalities on the rationals.

E.8   Be able to state and prove the cancellation law of multiplication for inequalities on the rationals.

E.9   Be able to give examples of the cancellation law of addition for inequalities on the rationals.

E.10 Be able to give examples of the cancellation law of multiplication for inequalities on the rationals.

PROPERTIES OF N, W, I, **AND** R **UNDER** $+$, $-$, $\times$, **AND** $\div$

Before we continue with our development of the rational numbers, it might be instructive for you to review and compare the properties of the sets developed so far. In Figure 7.1 we offer a Venn diagram to show

Figure 7.1

Properties	Operations	Naturals	Wholes	Integers	Rationals
Closure	+	yes	yes	yes	yes
	−	no	no	yes	yes
	×	yes	yes	yes	yes
	÷	no	no	no	yes
Commutativity	+	yes	yes	yes	yes
	−	no	no	no	no
	×	yes	yes	yes	yes
	÷	no	no	no	no
Associativity	+	yes	yes	yes	yes
	−	no	no	no	no
	×	yes	yes	yes	yes
	÷	no	no	no	no
Identity	+	no	yes	yes	yes
	−	no	no	no	no
	×	yes	yes	yes	yes
	÷	no	no	no	no
Inverses	+	no	no	yes	yes
	−	no	no	no	no
	×	no	no	no	yes
	÷	no	no	no	no
Cancellation	+	yes	yes	yes	yes
	×	yes	yes	yes	yes
Distributive of × over +		yes	yes	yes	yes

Figure 7.2

the relationship between the sets N, W, I, and R. Figure 7.2 summarizes the properties of these sets. We have excluded the properties of order relations on the sets because these will be discussed in the next section.

EXERCISES

Use Figure 7.2 to answer the following questions.

1. What properties hold for subtraction?

2. What properties do not hold for subtraction?

3. What properties hold for division?

4. What properties do not hold for division?

ORDER ON R

We defined equality on R earlier in the chapter in terms of the pairs used to represent the set. With the rationals as with the integers, there is a correspondence between points on a line and the rational numbers. This means that the rational numbers have an order relation just as the points on a line do. This relation is based on the *trichotomy property*, which we will first state as an assumption for our order relation, then give some examples, and finally define an order relation in a formal manner.

■ **E.1**   *Trichotomy Property: For all* $R[\![a, b]\!]$ *and* $R[\![c, d]\!]$ *in R,* $R[\![a, b]\!]$ *relates to* $R[\![c, d]\!]$ *in one and only one of the following ways:*

1. $R[\![a, b]\!] = R[\![c, d]\!]$
2. $R[\![a, b]\!] < R[\![c, d]\!]$
■ 3. $R[\![a, b]\!] > R[\![c, d]\!]$

We have already shown that $R[\![a, b]\!]$ equals $R[\![c, d]\!]$ if and only if $[\![a, b]\!] \sim [\![c, d]\!]$. This means that we apply our definition of equivalence. Since $[\![a, b]\!] \sim [\![c, d]\!]$ if and only if $ad = bc$, we could say that $R[\![a, b]\!] = R[\![c, d]\!]$ if and only if $ad = bc$.

Consider the following examples.

Example 1

Show that $R[\![1, 2]\!] = R[\![3, 6]\!]$.

SOLUTION   We know that since $1 \times 6 = 2 \times 3$, then $[\![1, 2]\!] \sim [\![3, 6]\!]$ and $R[\![1, 2]\!] = R[\![3, 6]\!]$.

Example 2

Are the pairs $R[\![2, 3]\!]$ and $R[\![5, 6]\!]$ equal?

SOLUTION   If $2 \times 6 = 3 \times 5$, then $R[\![2, 3]\!] = R[\![5, 6]\!]$. They are obviously unequal, because $2 \times 6 = 12 \neq 15 = 3 \times 5$.

Before we define our order relations, let us turn to the number line as we did with integers and consider the set of rational fractions on the number line. We will use the more familiar a/b notation here, which uses the concept of division and is translated to mean a divided into b parts. This is easy to translate on the number line, because $\frac{1}{2}$ would mean divide the unit length into 2 equal parts such that $\frac{1}{2}$ represents one of the two parts. We would then interpret $\frac{3}{2}$ to mean $3 \times \frac{1}{2}$, or $\frac{1}{2} + \frac{1}{2} + \frac{1}{2}$. Since we will devote a chapter to fractions, we will not consider them further at this time except to illustrate order on the number line. To construct our rational number line, we first draw a line and choose a zero point and a unit of measure as shown in Figure 7.3. With this done, we can subdivide the unit 1 or $\frac{1}{1}$ into n parts to produce the following set: $S = \{\frac{1}{2}, \frac{1}{3}, \frac{1}{4}, \ldots, 1/n, \ldots\}$. This may be termed a *unit set* or the *set of unit fractions*. From

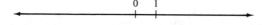

Figure 7.3

this set we can produce the set of multiples, because once we measure $\frac{1}{5}$, the set $S_{n/5} = \{\frac{1}{5}, \frac{2}{5}, \frac{3}{5}, \ldots, n/5, \ldots\}$ follows easily. Consider the number line in Figure 7.4. We have shown this number line simply to motivate

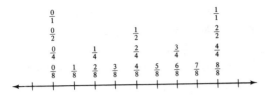

Figure 7.4

our definition of order. That is, we want to define an order relation on the set of rational numbers modeled on the intuitive notion of order as meaning "to the left of." Notice that $\frac{1}{4} < \frac{2}{4} < \frac{3}{4} < \frac{4}{4}$, and $\frac{1}{8} < \frac{2}{8} < \cdots < \frac{7}{8} < \frac{8}{8}$. Thus, we would expect that $R[\![1, 8]\!] < R[\![2, 8]\!] < \cdots < R[\![7, 8]\!] < R[\![8, 8]\!]$. Thus, we will be able to compare $R[\![a, b]\!]$ to $R[\![c, d]\!]$ by finding common second entries in the set of ordered pairs as we did earlier to show equality, and then comparing the first entries in the set of ordered pairs. Comparing the first entries would let us know which is the larger. See Figure 7.5 for a few examples on the number line again. Notice that

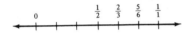

Figure 7.5

$\frac{1}{2} < \frac{2}{3}$ and $\frac{2}{3} < \frac{5}{6}$. If we translate the fractions $\frac{1}{2}$, $\frac{2}{3}$, and $\frac{5}{6}$ to elements of R, we have $R[\![1, 2]\!]$, $R[\![2, 3]\!]$, and $R[\![5, 6]\!]$.

Example 1

■ E.2   Is $R[\![1, 2]\!] < R[\![2, 3]\!]$?

SOLUTION $R[\![1, 2]\!] \times R[\![3, 3]\!] = R[\![3, 6]\!]$
$R[\![2, 3]\!] \times R[\![2, 2]\!] = R[\![4, 6]\!]$
Since $3 < 4$, we expect that
$R[\![3, 6]\!] < R[\![4, 6]\!]$, or $R[\![1, 2]\!] < R[\![2, 3]\!]$.

Example 2

Is $R[\![2, 3]\!] < R[\![5, 6]\!]$?

SOLUTION $R[\![2, 3]\!] \times R[\![6, 6]\!] = R[\![12, 18]\!]$
$R[\![5, 6]\!] \times R[\![3, 3]\!] = R[\![15, 18]\!]$
So $R[\![2, 3]\!] < R[\![5, 6]\!]$ because $R[\![12, 18]\!] < R[\![15, 18]\!]$.

Example 3

▲ E.3   From Examples 1 and 2, do you think that the "less than" relation is transitive?

SOLUTION   This can be answered by comparing $R[\![1, 2]\!]$ and $R[\![5, 6]\!]$. If the relation is transitive, then $R[\![1, 2]\!] < R[\![5, 6]\!]$.

$R[\![1, 2]\!] \times R[\![6, 6]\!] = R[\![6, 12]\!]$
$R[\![5, 6]\!] \times R[\![2, 2]\!] = R[\![10, 12]\!]$

▲ ■   Since $R[\![6, 12]\!]$ is less than $R[\![10, 12]\!]$, $R[\![1, 2]\!] < R[\![5, 6]\!]$, which gives some evidence of transitivity.

In the previous paragraph and examples, we have intuitively defined "less than." Let us consider what was involved in testing the relation.

1. We were given two elements of R, $R[\![a, b]\!]$ and $R[\![c, d]\!]$.
2. We multiplied $R[\![a, b]\!]$ by $R[\![d, d]\!]$ and we multiplied $R[\![c, d]\!]$ by $R[\![b, b]\!]$.
3. We then compared the first entries, because the second entries were the same.
4. If $ad < bc$, then we called $R[\![a, b]\!] < R[\![c, d]\!]$.

Notice that we make an attempt to justify the definition; although definitions cannot be proved, the justification is offered to clarify the definition.

Definition

■ E.6 ■   *Given $R[\![a, b]\!]$ and $R[\![c, d]\!]$ in R, $R[\![a, b]\!] < R[\![c, d]\!]$ if and only if $ad < bc$.*

Consider some examples of this definition as well as the definition of equality.

Example 1

■ E.2   How does $R[\![3, 5]\!]$ relate to $R[\![2, 8]\!]$?

SOLUTION   Since $3 \times 8 > 5 \times 2$, $R[\![3, 5]\!] > R[\![2, 8]\!]$.

Example 2

Which is larger, $R[\![1, 3]\!]$ or $R[\![-5, 8]\!]$?

SOLUTION   Since $1 \times 8 > 3 \times (-5)$, then $R[\![1, 3]\!] > R[\![-5, 8]\!]$.

Example 3

How does $R[\![2, 4]\!]$ relate to $R[\![6, 12]\!]$?

■   SOLUTION   Since $2 \times 12 = 4 \times 6$, $R[\![2, 4]\!] = R[\![6, 12]\!]$.

The definition we have developed thus far is operational in the sense that one can apply it to two given rationals and decide how they relate by performing an operation. The operation performed can be summarized as follows:

1. Given $R[\![a, b]\!]$ and $R[\![c, d]\!]$.
2. Find products ad and bc.
3. a. If $ad < bc$, then $R[\![a, b]\!] < R[\![c, d]\!]$.
   b. If $ad = bc$, then $R[\![a, b]\!] = R[\![c, d]\!]$.
   c. If $ad > bc$, then $R[\![a, b]\!] > R[\![c, d]\!]$.

When we considered a definition of "less than" on the number line, we used the idea that if $R[\![a, b]\!]$ is located to the left of $R[\![c, d]\!]$, then $R[\![a, b]\!]$ is less than $R[\![c, d]\!]$. Under these conditions there is some positive number, say $R[\![x, y]\!]$, such that $R[\![a, b]\!] + R[\![x, y]\!] = R[\![c, d]\!]$. Another way to say this is, If $R[\![a, b]\!] < R[\![c, d]\!]$, then the difference between $R[\![c, d]\!]$ and $R[\![a, b]\!]$ is positive, that is, $R[\![c, d]\!] - R[\![a, b]\!]$ is positive. Let $R[\![c, d]\!] - R[\![a, b]\!] = R[\![x, y]\!]$. Then

$$
\begin{aligned}
R[\![a, b]\!] + R[\![x, y]\!] &= R[\![a, b]\!] + R[\![c, d]\!] - R[\![a, b]\!] \\
&= R[\![a, b]\!] - R[\![a, b]\!] + R[\![c, d]\!] \\
&= R[\![0, 1]\!] + R[\![c, d]\!] \\
&= R[\![c, d]\!]
\end{aligned}
$$

■ E.5   We now have three ways to discuss order relations. The first way is by using the definition we developed first as an operational definition. The second way is to use the number line and discuss the relations in terms of "to the left" or "to the right." The third way is to show that some
■   positive rational $R[\![x, y]\!]$ exists such that $R[\![a, b]\!] + R[\![x, y]\!] = R[\![c, d]\!]$. Although these definitions are different in their approach, they all define the same concept. It would be interesting for you to try to show that these definitions are equivalent.

Definition

■ E.6   *If for $R[\![a, b]\!]$ and $R[\![c, d]\!]$ in R there exists an $R[\![x, y]\!]$ in R where $x > 0$ such that $R[\![a, b]\!] + R[\![x, y]\!] = R[\![c, d]\!]$, then we say that $R[\![a, b]\!]$ is less*
■   *than $R[\![c, d]\!]$, or $R[\![a, b]\!] < R[\![c, d]\!]$.*

Example 1

■ E.2   Find $R[\![x, y]\!]$ and explain why $R[\![2, 3]\!]$ is less than $R[\![4, 3]\!]$, given the sentence $R[\![2, 3]\!] + R[\![x, y]\!] = R[\![4, 3]\!]$.

SOLUTION
$$
\begin{aligned}
R[\![2, 3]\!] + R[\![x, y]\!] &= R[\![4, 3]\!] \\
R[\![x, y]\!] &= R[\![4, 3]\!] + R[\![-2, 3]\!]
\end{aligned}
$$
     Why?
$$
R[\![x, y]\!] = R[\![2, 3]\!]
$$
     Thus
$R[\![2, 3]\!] + R[\![2, 3]\!] = R[\![4, 3]\!]$ where $2 > 0$.
Therefore $R[\![2, 3]\!] < R[\![4, 3]\!]$ by definition.

Example 2

Using our operational definition, test $R[\![2, 5]\!]$ and $R[\![3, 7]\!]$ to see which is the larger.

▲ E.5   1. Write a sentence of the form $R[\![a, b]\!] + R[\![x, y]\!] = R[\![c, d]\!]$.
     2. Find the solution set $R[\![x, y]\!]$.
     3. Draw a diagram of this sentence on the number line.

SOLUTION   1. Since $2 \times 7 < 5 \times 3$, $R[\![2, 5]\!] < R[\![3, 7]\!]$.
2. $R[\![2, 5]\!] + R[\![x, y]\!] = R[\![3, 7]\!]$
$R[\![x, y]\!] = R[\![3, 7]\!] + R[\![-2, 5]\!]$
$R[\![x, y]\!] = R[\![1, 35]\!]$

3.

Example 3

Using the first definition, test $R[\![2, 7]\!]$ and $R[\![2, 5]\!]$ to see which is the larger. Write a sentence $R[\![a, b]\!] + R[\![x, y]\!] = R[\![c, d]\!]$, and find the solution set $R[\![x, y]\!]$.

SOLUTION   1. Since $2 \times 5 < 7 \times 2$, $R[\![2, 7]\!] < R[\![2, 5]\!]$.
2. In order to decide which rational number is larger, we must find a positive rational $R[\![x, y]\!]$ such that either $R[\![2, 7]\!] + R[\![x, y]\!] = R[\![2, 5]\!]$ or $R[\![2, 5]\!] + R[\![x, y]\!] = R[\![2, 7]\!]$. Let us solve each of these for $R(x, y)$. Solving the first, we have

$R[\![x, y]\!] = R[\![2, 5]\!] + R[\![-2, 7]\!]$
$R[\![x, y]\!] = R[\![4, 35]\!]$

Since $R[\![4, 35]\!]$ is positive, that is, $4 > 0$, we know that $R[\![2, 7]\!] < R[\![2, 5]\!]$. Suppose we had started with the second equation:

$R[\![2, 5]\!] + R[\![x, y]\!] = R[\![2, 7]\!]$
$R[\![x, y]\!] = R[\![2, 7]\!] + R[\![-2, 5]\!]$
$\qquad\quad = R[\![10, 35]\!] + R[\![-14, 35]\!]$
$\qquad\quad = R[\![-4, 35]\!]$

We note that $-4 < 0$, so that $R[\![2, 5]\!] > R[\![2, 7]\!]$.

EXERCISES

1. Use the operational definition of "less than" to determine the relationship for each pair below.
a. $R[\![3, 7]\!]$, $R[\![4, 6]\!]$
c. $R[\![5, 6]\!]$, $R[\![7, 8]\!]$
b. $R[\![4, 15]\!]$, $R[\![25, 26]\!]$
d. $R[\![-7, 9]\!]$, $R[\![-20, 24]\!]$

2. Solve the equations below for $R[\![x, y]\!]$.
a. $R[\![1, 2]\!] + R[\![x, y]\!] = R[\![2, 3]\!]$
b. $R[\![-1, 2]\!] + R[\![x, y]\!] = R[\![1, 2]\!]$
c. $R[\![3, 5]\!] + R[\![x, y]\!] = R[\![5, 4]\!]$
d. $R[\![-6, 7]\!] + R[\![x, y]\!] = R[\![-3, 7]\!]$

3. Translate each of the fractions below into the notation $R[\![a, b]\!]$, and then decide which fraction is larger.

a. $\frac{1}{2}, \frac{3}{4}$ b. $\frac{2}{3}, \frac{3}{6}$ c. $\frac{6}{7}, \frac{3}{5}$ d. $\frac{2}{3}, \frac{6}{9}$

4. Diagram each relation in Exercise 3 on a number line.

5. For Exercise 3, write fractions with a common denominator and then try to determine which is larger.

E.4 ■ 6. Test the "less than" relation for reflexive, symmetric, and transitive properties.

7. Order the sets of fractions given below.

a. $\{\frac{3}{4}, \frac{5}{6}, \frac{2}{3}, \frac{1}{2}, \frac{7}{12}\}$
b. $\{-\frac{2}{3}, -\frac{1}{2}, -\frac{5}{6}, -\frac{7}{12}, -\frac{3}{4}\}$
c. $\{\frac{2}{5}, \frac{3}{7}, \frac{4}{3}, \frac{5}{4}, \frac{6}{5}\}$
d. $\{-\frac{2}{5}, -\frac{6}{5}, -\frac{4}{3}, -\frac{5}{4}, -\frac{3}{7}\}$

8. If $R[\![x, y]\!]$ is a rational number and $x > 0$, use the definition of $<$ to show that $R[\![0, 1]\!]$ $< R[\![x, y]\!]$.

CANCELLATION LAWS FOR "LESS THAN" ON R

Before we consider the cancellation laws for "less than" on the set R, we should consider some examples to see if there is evidence for the cancellation laws. The first set of examples will relate to cancellation under addition. The cancellation law for addition would read for elements a, b, c in R that $a < c$ if and only if $a + b < c + b$.

That general statement is equivalent to two separate statements:

1. If $a < c$, then $a + b < c + b$.
2. If $a + b < c + b$, then $a < c$.

Notice that the "if and only if" statement can be read as two "if–then" statements, which we call implications. If we were to prove the "if and only if" statement, we would be required to prove both implications.

Let us now consider some examples.

Example 1

■ E.9 If $R[\![1, 3]\!] < R[\![2, 3]\!]$, is it true that $R[\![1, 3]\!] + R[\![1, 2]\!] < R[\![2, 3]\!]$ $+ R[\![1, 2]\!]$?

SOLUTION $R[\![1, 3]\!] + R[\![1, 2]\!] = R[\![5, 6]\!]$
$R[\![2, 3]\!] + R[\![1, 2]\!] = R[\![7, 6]\!]$
$R[\![5, 6]\!] < R[\![7, 6]\!]$ Why?
Therefore, $R[\![1, 3]\!] + R[\![1, 2]\!] < R[\![2, 3]\!] + R[\![1, 2]\!]$.

Example 2

If $R[\![5, 7]\!] + R[\![2, 3]\!] < R[\![8, 7]\!] + R[\![2, 3]\!]$ is $R[\![5, 7]\!] < R[\![8, 7]\!]$?

SOLUTION Since $R[\![5, 7]\!] + R[\![2, 3]\!] = R[\![15 + 14, 21]\!] = R[\![29, 21]\!]$
and $R[\![29, 21]\!] < R[\![38, 21]\!]$,
and since $R[\![38, 21]\!] = R[\![24 + 14, 21]\!] = R[\![8, 7]\!] + R[\![2, 3]\!]$,
then $R[\![5, 7]\!] + R[\![2, 3]\!] < R[\![8, 7]\!] + R[\![2, 3]\!]$.
So we see that $R[\![5, 7]\!] < R[\![8, 7]\!]$.

■

Before we define the cancellation law for "less than" under addition on R, consider one more intuitive concept. Suppose we have two boards and we considered their lengths to be $R[\![a, b]\!]$ and $R[\![c, d]\!]$ such that $R[\![a, b]\!] < R[\![c, d]\!]$.

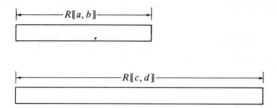

By definition, $R[\![c, d]\!]$ is larger than $R[\![a, b]\!]$ by $R[\![e, f]\!]$.

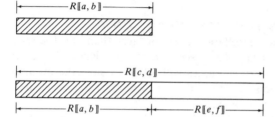

Now suppose we cut two more boards of lengths $R[\![x, y]\!]$ and join them to $R[\![a, b]\!]$ and to $R[\![c, d]\!]$.

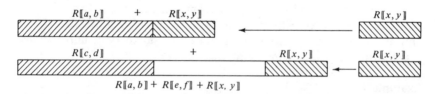

It seems reasonable to state that $R[\![x, y]\!] + R[\![a, b]\!] < R[\![x, y]\!] + R[\![c, d]\!]$ by a length of $R[\![e, f]\!]$.

Suppose, however, that we cut a piece $R[\![x, y]\!]$ from both $R[\![a, b]\!]$ and $R[\![c, d]\!]$. Consider what happens here. Each board is reduced in size by the same amount. It should follow that $R[\![x, y]\!] + R[\![a, b]\!] < R[\![x, y]\!] + R[\![c, d]\!]$. *Why do we call the operation addition in this case?* Notice $R[\![x, y]\!] \in R^-$. Notice also that the difference in length between the two boards is still $R[\![e, f]\!]$.

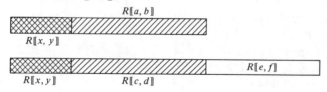

■ E.7 *Cancellation Law of "Less Than" for Addition of Rational Numbers: For $R[\![a, b]\!]$, $R[\![c, d]\!]$, $R[\![x, y]\!]$ in R, $R[\![a, b]\!] + R[\![x, y]\!] < R[\![c, d]\!] + R[\![x, y]\!]$ if and only if $R[\![a, b]\!] < R[\![c, d]\!]$.*

We will supply a partial proof for this law by showing that if $R[\![a, b]\!] + R[\![x, y]\!] < R[\![c, d]\!] + R[\![x, y]\!]$, then $R[\![a, b]\!] < R[\![c, d]\!]$.

PROOF   Given that $R[\![a, b]\!] + R[\![x, y]\!] < R[\![c, d]\!] + R[\![x, y]\!]$ and $R[\![a, b]\!]$, $R[\![c, d]\!]$, $R[\![x, y]\!]$ are in R, prove that $R[\![a, b]\!] < R[\![c, d]\!]$.

1. $R[\![a, b]\!] + R[\![x, y]\!] < R[\![c, d]\!] + R[\![x, y]\!]$ 1. given
2. $R[\![a, b]\!] + R[\![x, y]\!] + R[\![e, f]\!]$ 2. definition of $<$
   $= R[\![c, d]\!] + R[\![x, y]\!]$
3. $R[\![a, b]\!] + R[\![e, f]\!] + R[\![x, y]\!]$ 3. commutative property
   $= R[\![c, d]\!] + R[\![x, y]\!]$
4. $R[\![a, b]\!] + R[\![e, f]\!] = R[\![c, d]\!]$ 4. cancellation law
■ 5. $R[\![a, b]\!] < R[\![c, d]\!]$ 5. definition of $< R$

Notice that to execute the proof we wrote the "less than" relation and translated to the equality relation to use the properties of equality. We have pointed out this technique several times, namely, changing to a system in which the desired property holds, using the property, and then changing back. Therefore, before embarking on a proof, one must first consider that which has been proven. In this section the proofs use concepts developed for the equality relation on R. This required that we translate by using the last definition of "less than."

We have not completed the proof of the cancellation law for inequalities. This is left to you. It will help you to consider the steps used in this proof in reverse. Since every step uses a definition or theorem which was stated in "if and only if" form, the proof can be reversed.

Now let us give two examples where the cancellation law is used.

Example 1

■ E.9   If $R[\![x, y]\!] + R[\![1, 2]\!] < R[\![2, 6]\!] + R[\![1, 2]\!]$, find the solution set.

SOLUTION $R[\![x, y]\!] + R[\![1, 2]\!] < R[\![2, 6]\!] + R[\![1, 2]\!]$. $R[\![x, y]\!] < R[\![2, 6]\!]$.
Notice that we use the cancellation law here to solve an open sentence. Since $R[\![x, y]\!]$ is less than $R[\![2, 6]\!]$, the solution set is an infinite set of sets $R[\![x, y]\!]$ in R such that $R[\![x, y]\!] < R[\![2, 6]\!]$. If we graph this on the number line, we have all rationals which lie to the left of $R[\![2, 6]\!]$, or $\frac{1}{3}$.

Example 2

If $R[\![x, y]\!] + R[\![-1, 2]\!] < R[\![2, 5]\!]$, find the solution set.

SOLUTION $R[\![x, y]\!] + R[\![-1, 2]\!] < R[\![2, 5]\!]$.

$$R[x, y] + R[-1, 2] + R[1, 2] < R[2, 5] + R[1, 2]$$
$$R[x, y] < R[9, 10]$$

Notice in this example that we added $R[1, 2]$ to each side of the relation, because it is the inverse of $R[-1, 2]$. This gave us directly a statement relating $R[x, y]$ to a rational number, namely $R[9, 10]$. You should notice that $R[9, 10] + R[-1, 2] = R[2, 5]$.

■

Consider now some examples of what happens when both members of an order relation are multiplied by a rational number. You might refer to this property on I as a refresher. Remember, multiplying two ordered numbers in I by an element of I^- changes the order.

Example 1

■ E.10 $R[1, 2] < R[2, 3]$. What is the result of multiplying both members by (1) $R[1, 2]$, (2) $R[0, 2]$, and (3) $R[-1, 2]$?

SOLUTION   1. $R[1, 2] \times R[1, 2] = R[1, 4]$
$$R[2, 3] \times R[1, 2] = R[2, 6]$$
$$R[1, 4] < R[2, 6]$$
The order remains the same.
2. $R[1, 2] \times R[0, 2] = R[0, 4]$
$$R[2, 3] \times R[0, 2] = R[0, 6]$$
$$R[0, 4] = R[0, 6]$$
The order relation changes to equality.
3. $R[1, 2] \times R[-1, 2] = R[-1, 4]$
$$R[2, 3] \times R[-1, 2] = R[-2, 6]$$
$$R[-2, 6] < R[-1, 4]$$
The order relation reverses.
Why?

Example 2

$R[-1, 2] < R[1, 4]$. What is the result of multiplying both members by (1) $R[1, 2]$, (2) $R[0, 2]$, and (3) $R[-1, 2]$?

SOLUTION   1. $R[-1, 2] \times R[1, 2] = R[-1, 4]$
$$R[1, 4] \times R[1, 2] = R[1, 8]$$
$$R[-1, 4] < R[1, 8]$$
The order remains the same.
2. $R[-1, 2] \times R[0, 2] = R[0, 4]$
$$R[1, 4] \times R[0, 2] = R[0, 8]$$
$$R[0, 4] = R[0, 8]$$
The order relation changes to equality.
3. $R[-1, 2] \times R[-1, 2] = R[1, 4]$
$$R[1, 4] \times R[-1, 2] = R[-1, 8]$$
$$R[-1, 8] < R[1, 4]$$
The order relation reverses.

■

From these examples we would expect to make the following generalizations for order relations in R.

1. Multiplying both members of an order relation in R by any $R[\![a,b]\!]$ $> R[\![0,1]\!]$, that is, $a > 0$ will not affect the order relation.
2. Multiplying both members of an order relation in R by any $R[\![a,b]\!]$ $= R[\![0,1]\!]$, that is, $a = 0$ will produce an equality.
3. Multiplying both members of an order relation in R by any $R[\![a,b]\!]$ $< R[\![0,1]\!]$, that is, $a < 0$ will reverse the order relation.

We will prove the first two statements and leave the third statement to you. Since the second statement can be proved informally, we will deal with it first.

■ E.8   The zero property of multiplication tells us that the product of any rational with a zero rational produces zero. Thus, if $R[\![c,d]\!] < R[\![e,f]\!]$, then

$$R[\![c,d]\!] \times R[\![0,b]\!] = R[\![0,b]\!]$$

and

$$R[\![e,f]\!] \times R[\![0,b]\!] = R[\![0,b]\!]$$

Using the transitive property of equality we have that

$$R[\![c,d]\!] \times R[\![0,b]\!] = R[\![e,f]\!] \times R[\![0,b]\!]$$

We now wish to prove the following: Given that $R[\![a,b]\!] < R[\![c,d]\!]$ and $R[\![x,y]\!] > R[\![0,1]\!]$, prove that $R[\![a,b]\!] \times R[\![x,y]\!] < R[\![c,d]\!] \times R[\![x,y]\!]$.

1. $R[\![a,b]\!] < R[\![c,d]\!]$	1. given
2. $R[\![a,b]\!] + R[\![e,f]\!] = R[\![c,d]\!]$ where $R[\![e,f]\!]$ is positive	2. definition of $< R$
3. $R[\![x,y]\!] \times (R[\![a,b]\!] + R[\![e,f]\!])$ $= R[\![x,y]\!] \times R[\![c,d]\!]$	3. cancellation law of $=$ for $\times R$
4. $(R[\![x,y]\!] \times R[\![a,b]\!]) + (R[\![x,y]\!] \times R[\![e,f]\!])$ $= R(x,y) \times R(c,d)$	4. distribution of $\times + R$
5. $R[\![x,y]\!] \times R[\![e,f]\!]$ in R^+	5. the product of two positive rationals is positive.
6. $R[\![x,y]\!] \times R[\![a,b]\!] < R[\![x,y]\!] \times R[\![c,d]\!]$	6. definition of $< R$
7. $R[\![a,b]\!] \times R[\![x,y]\!] < R[\![c,d]\!] \times R[\![x,y]\!]$	7. Why?

We will state the cancellation law for multiplication on R below. Again the proof given above is an "if–then" proof. We have not shown that this is reversible. We will leave the completion of the proof to you. The last statement to be proved also follows this form. You must show that by multiplying by $R[\![x,y]\!] < R[\![0,1]\!]$ we produce $R[\![x,y]\!] \times R[\![e,f]\!]$ $< R[\![0,1]\!]$. This will change the direction of the order symbol.

Cancellation Law of "Less Than" for Multiplication of Rational Numbers:
For $R[\![a, b]\!]$, $R[\![c, d]\!]$, $R[\![x, y]\!]$ in R.

1. $R[\![a, b]\!] \times R[\![x, y]\!] < R[\![c, d]\!] \times R[\![x, y]\!]$ *if and only if* $R[\![a, b]\!] < R[\![c, d]\!]$
   and $R[\![x, y]\!] > R[\![0, 1]\!]$.
2. $R[\![a, b]\!] \times R[\![x, y]\!] < R[\![c, d]\!] \times R[\![x, y]\!]$ *if and only if* $R[\![c, d]\!] < R[\![a, b]\!]$
   ■   *and* $R[\![x, y]\!] < R[\![0, 1]\!]$.

EXERCISES

1. If $R[\![a, b]\!] < R[\![c, d]\!]$, which of the following statements are true? ($R[\![a, b]\!]$ and $R[\![c, d]\!]$
   not zero).
   a. $R[\![a, b]\!] \times R[\![a, b]\!] < R[\![c, d]\!] \times R[\![c, d]\!]$
   b. $R[\![a, b]\!]^{-1} < R[\![c, d]\!]^{-1}$
   c. $R[\![-a, b]\!] < R[\![-c, d]\!]$
   d. $R[\![a, b]\!] + R[\![x, y]\!] < R[\![a, b]\!] + R[\![x, y]\!]$
   e. $R[\![a, b]\!] - R[\![x, y]\!] > R[\![a, b]\!] - R[\![x, y]\!]$
   f. $R[\![-a, b]\!] > R[\![-c, d]\!]$
   g. $R[\![a, b]\!] \times R[\![c, d]\!] < R[\![a, b]\!] \times R[\![c, d]\!]$
   h. $R[\![a, b]\!] \times R[\![c, d]\!] > R[\![a, b]\!] \times R[\![c, d]\!]$

2. Find the solution set for the following statements.
   a. $R[\![x, y]\!] + R[\![2, 3]\!] < R[\![-1, 2]\!]$
   b. $R[\![3, 5]\!] - R[\![x, y]\!] < R[\![7, 10]\!]$
   c. $R[\![x, y]\!] \times R[\![1, 2]\!] < R[\![2, 3]\!]$
   d. $R[\![x, y]\!] \times R[\![-1, 2]\!] < R[\![2, 3]\!]$

3. Prove that if $R[\![a, b]\!] < R[\![c, d]\!]$, then $R[\![a, b]\!] + R[\![x, y]\!] < R[\![c, d]\!] + R[\![x, y]\!]$.

4. Prove that if $R[\![a, b]\!] < R[\![c, d]\!]$ and $x > 0$, then $R[\![a, b]\!] \times R[\![x, y]\!] < R[\![c, d]\!] \times R[\![x, y]\!]$.

5. Prove that if $R[\![a, b]\!] < R[\![c, d]\!]$ and $x < 0$, then $R[\![a, b]\!] \times R[\![x, y]\!] > R[\![c, d]\!] \times R[\![x, y]\!]$.

CHAPTER 8 RATIONAL NUMBERS AS FRACTIONS AND DECIMALS

SECTION A FRACTIONS

BEHAVIORAL OBJECTIVES

A.1 Be able to define fraction and common fraction.

A.2 Be able to illustrate common fractions using three different physical models.

A.3 Be able to identify the numerator and denominator of common fractions.

A.4 Be able to prove that two fractions are equal.

A.5 Given a fraction, be able to find other fractions equal to it.

A.6 Given two fractions, determine if they are equal.

A.7 Be able to simplify a given fraction.

COMMON FRACTIONS

In Chapter 7, we developed the set of rational numbers formally. In our everyday encounters with rational numbers, we usually deal with three forms of rational numbers, namely, fractions, decimals, and percents. The fractions with which we are familiar are sometimes called *common fractions* and can be considered a subset of the set of rational numbers.

We represent common fractions by using the notation a/b. The symbols – or / imply division and can be read a over b or a divided by b. We often name fractions in another way by reading the numerator, a, as a whole number and the denominator, b, as a unit. For example, $\frac{1}{4}$ is read "one-fourth"; $\frac{2}{3}$ is read "two-thirds"; and $\frac{5}{6}$ is read "five-sixths."

You will notice that the numerator is read as a cardinal number using the words one, two, three, etc. The denominator is read as an ordinal number, either singular or plural. Notice that the transformation from cardinal to ordinal follows a simple rule with respect to sound for all numbers except two and three; namely, add the "th" sound to the cardinal number to produce the sound of the ordinal number. This should be helpful for children who have some language difficulty. The half and third can be dealt with as a special case. A few examples follow.

four → fourth
five → fifth (notice an exception in spelling)
six → sixth
seven → seventh

In this way the numbers $\frac{1}{4}$, $\frac{1}{5}$, $\frac{1}{6}$, etc, employ a singular reading of fourth, fifth, and sixth, whereas the numbers $\frac{3}{4}$, $\frac{4}{5}$, $\frac{5}{6}$, etc., employ a plural reading of fourths, fifths, and sixths. Consider some examples: $\frac{1}{2}$ is read "one-half"; $\frac{1}{3}$ is read "one-third"; $\frac{3}{2}$ is read "three-halves"; and $\frac{2}{3}$ is read "two-thirds."

When we use this notation, the numerator is used to describe the number of units and the denominator describes the size of each unit. We can best

■ A.2   illustrate this idea by giving several physical examples. One way of physically representing fractions is by using geometric shapes or regions. Consider the diagrams in Figure 8.1. Notice that each geometric figure is

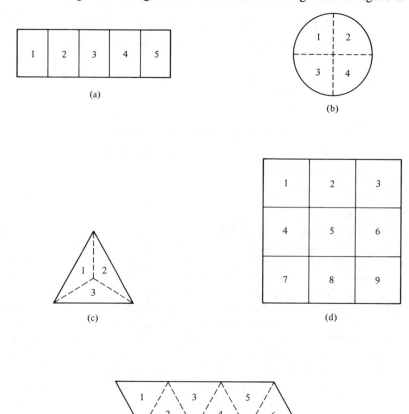

(a)

(b)

(c)

(d)

(e)

Figure 8.1

divided into congruent regions. The regions then constitute units. If we wished to discuss one of these regions, we would name it $1/b$, where b is the number of congruent regions in the whole figure. In Figure 8.1a, we name each region $\frac{1}{5}$; in Figure 8.1b, we name each region $\frac{1}{4}$; each part of the triangle (Figure 8.1c) is called $\frac{1}{3}$; each part of the square (Figure 8.1d), $\frac{1}{9}$; and each part of Figure 8.1e (the parallelogram), $\frac{1}{6}$. If we wish to represent a number of regions in this manner, we follow the convention of shading the regions we wish to illustrate. Consider Figure 8.2; count the

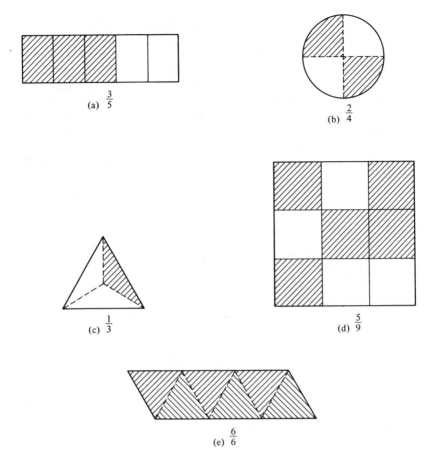

Figure 8.2

number of regions and then count the number of shaded regions. The number of shaded regions will represent the numerator of the fraction being illustrated and the number of congruent regions will represent the denominator.

You can summarize the results of your work by considering the table in Figure 8.3; it refers directly to Figure 8.2.

Figure	Shaded Regions	Congruent Regions	Fraction
Rectangle	3	5	$\frac{3}{5}$
Circle	2	4	$\frac{2}{4}$
Triangle	1	3	$\frac{1}{3}$
Square	5	9	$\frac{5}{9}$
Parallelogram	6	6	$\frac{6}{6}$

Figure 8.3

The representation illustrated in Figure 8.2 is a bit unwieldly because of the requirement of congruent figures, but it is one way of illustrating the fraction concept. Another illustration which is often misunderstood relates a fraction to a fractional part of a set of objects. For example, three books in a set of ten can be represented as $\frac{3}{10}$, or five people from a class of twenty-five can be represented as $\frac{5}{25}$, or $\frac{1}{5}$. This is a slightly more difficult situation, because we are dealing with objects in a set rather than congruent figures. Actually, we consider the set of books as a whole or the class of people as a whole, in which case each element in the set is again much like a congruent region in a geometric figure (Figure 8.4).

■ A.2

■

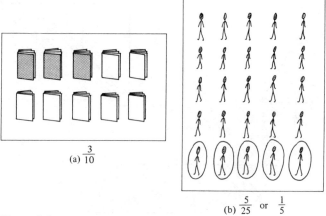

(a) $\frac{3}{10}$

(b) $\frac{5}{25}$ or $\frac{1}{5}$

Figure 8.4

■ A.2 One of the most useful models for representing numerical concepts geometrically is that of the number line. We introduced this model earlier, but we will illustrate it once more in Figure 8.5.

■

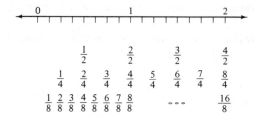

Figure 8.5

Notice that throughout this section we have discussed positive fractions only. We stated earlier that we would discuss common fractions; actually common fractions are non-negative fractions that are less than 1 and can be defined as follows.

Definition

A.1 ■ *Fractions, $\frac{a}{b}$, are rational numbers $R[\![a, b]\!]$, where a is in I and b is in N.*

The use of the terms *numerator* and *denominator* in common fractions relates to the above physical models as follows:

■ **A.3** Numerator: The first number, *a* in the pair *a/b* or $R[\![a, b]\!]$ reports the number of "congruent" units.

    Denominator: The second number, *b*, in the pair *a/b* or $R[\![a, b]\!]$

■ reports the size of each "congruent" unit.

Definition

◀ **A.1** ■ *Common fractions are fractions, $\frac{a}{b}$, such that $0 \le R[\![a, b]\!] < 1$.*

EXERCISES

1. Name the numerator and denominator of each fraction below.
   a. $\frac{2}{6}$ b. $\frac{21}{13}$ c. $\frac{5}{9}$ d. $\frac{7}{2}$

2. Represent the fractions below as regions in a rectangle, a triangle, and a circle.
   a. $\frac{2}{3}$ b. $\frac{4}{9}$ c. $\frac{5}{6}$ d. $\frac{5}{3}$

3. Represent the fractions shown below on a number line.
   a. $\frac{2}{5}$ b. $\frac{5}{2}$ c. $\frac{12}{16}$ d. $\frac{6}{8}$

4. Locate on a number line each set of fractions below.
   a. $\{\frac{2}{3}, \frac{6}{9}, \frac{18}{27}\}$ b. $\{\frac{1}{2}, \frac{2}{4}, \frac{4}{8}\}$

5. What did the fractions in Exercise 4 have in common? Make a geometric diagram for the fractions in Exercise 4, and compare the results.

6. Write a fraction for each description given below.
   a. six of nine boys on the baseball team
   b. two cows from a herd of twenty
   c. three classes from a school with twenty-seven classes
   d. a flat tire on an automobile

7. Multiply the numerator and the denominator of the given fractions by 2. Find the given fractions and the ones produced on the number line and compare them.
   a. $\frac{1}{2}$ b. $\frac{2}{3}$ c. $\frac{3}{4}$ d. $\frac{2}{6}$

8. Divide the numerator and denominator of each fraction by 3 and compare the results.
   a. $\frac{3}{9}$ b. $\frac{3}{6}$ c. $\frac{9}{6}$ d. $\frac{6}{12}$

9. What conclusions can you draw from Exercises 7 and 8?

10. Name the points indicated on the number line below by fraction and rational number.

EQUALITY OF FRACTIONS

In Chapter 7 we showed that each number $R[\![a, b]\!]$ represents a set of ordered pairs or an equivalence class. A fraction *a/b* has many names and can be represented in an infinite number of ways. For example, the fraction $\frac{1}{2}$ can be written as

$$\frac{1}{2}, \frac{2}{4}, \frac{3}{6}, \frac{4}{8}, \ldots, n/2n$$

Instead of calling these representations equivalent fractions, we will name them equal fractions, because they each represent the rational number $R[\![1, 2]\!]$. If we apply the definition for equivalence of rationals, we find that it will produce the desired result here.

Claim

■ A.4   Two fractions a/b and c/d are equal if and only if $ad = bc$.

PROOF OF CLAIM

"*Only if.*" Since $a/b = R[\![a, b]\!]$ and $c/d = R[\![c, d]\!]$, if $a/b = c/d$, then $R[\![a, b]\!] = R[\![c, d]\!]$. In this case $ad = bc$.
"*If.*" If $ad = bc$, then $R[\![a, c]\!] = R[\![b, d]\!]$, and by definition of common ■   fractions, $a/c = b/c$.

The above proof actually uses the definition of fractions to translate the claim into rational number notation. Then the work is done using the properties of rational numbers.

Examples

■ A.6   1. $\frac{1}{2} = \frac{4}{8}$, since $1 \times 8 = 2 \times 4$.
   ■   2. $\frac{6}{8} = \frac{9}{12}$, since $6 \times 12 = 8 \times 9$.
         3. $6 \times 12 = 24 \times 3$, then $\frac{6}{24} = \frac{3}{12}$.
         4. $9 \times 8 = 18 \times 4$, then $\frac{9}{18} = \frac{4}{8}$.

If two fractions are equal, then we can write both fractions in simplest form by dividing both numerator and denominator by the G.C.D. of the numerator and denominator. This is a second way of comparing fractions.

■ A.6   **Example**
▲ A.4   Show that $\frac{4}{8}$ and $\frac{9}{18}$ are equal.

SOLUTION   The G.C.D. of 4 and 8 is 4. Therefore, $\frac{4}{8}$ simplifies to $\frac{1}{2}$. The
           G.C.D. of 9 and 18 is 9. Therefore, $\frac{9}{18}$ simplifies to $\frac{1}{2}$. Since
▲ ■ $\frac{4}{8} = \frac{1}{2}$ and $\frac{9}{18} = \frac{1}{2}$, then $\frac{4}{8} = \frac{9}{18}$.

We can also produce sets of equal fractions simply by multiplying the numerator and denominator of a given fraction by a natural number, just as we produced the set of ordered pairs that form the rational number.

Example

■ A.5   Given the fraction $\frac{1}{3}$, produce a set of fractions by multiplying numerator and denominator by each natural number. Check to see if the set produced contains equal fractions.

$$\frac{1 \times 1}{3 \times 1} = \frac{1}{3}$$
$$\frac{1 \times 2}{3 \times 2} = \frac{2}{6}$$
$$\frac{1 \times 3}{3 \times 3} = \frac{3}{9}$$
$$\frac{1 \times 4}{3 \times 4} = \frac{4}{12}$$
$$\vdots$$
$$\frac{1 \times n}{3 \times n} = \frac{n}{3n}$$

The set produced in this manner is

■ $\{\frac{1}{3}, \frac{2}{6}, \frac{3}{9}, \ldots, \frac{n}{3n}\}$

Each fraction in the set is equal to the other fractions in the set, since $(1 \times n) \times (3n) = (3 \times n) \times n$. Notice that $n/3n = \frac{3}{9}$, since $n \times 9 = 3n \times 3$.

We have just indicated two properties of fractions which we now state:

■ **A.6** 1. The fraction a/b is equal to the fraction $\frac{a/n}{b/n}$ where n is in N.
■   2. The fraction a/b is equal to an/bn, where n is in N.

We indicated earlier that a fraction can be written in simplest form by dividing the numerator and denominator by their G.C.D. We should point out that if the G.C.D. is 1, then the fraction is already in simplest form. This should be obvious, because dividing both numerator and denominator of the fraction a/b by 1 will produce a/b. Consider a few examples involving writing fractions in simplest form.

Example 1
■ **A.7** Simplify $\frac{8}{12}$.

SOLUTION Since the G.C.D. of 8 and 12 is 4, $\frac{8 \div 4}{12 \div 4} = \frac{2}{3}$.

Example 2
Simplify $\frac{34}{51}$.

SOLUTION The G.C.D. of 34 and 51 is 17, so $\frac{34 \div 17}{51 \div 17} = \frac{2}{3}$.

Example 3
Simplify $\frac{56}{144}$.

■ SOLUTION Since 8 is the G.C.D. of 56 and 144, $\frac{56 \div 8}{144 \div 8} = \frac{7}{18}$.

EXERCISES

1. Which of these pairs of fractions are equal by definition?
  a. $\frac{7}{8}$ and $\frac{63}{72}$ b. $\frac{6}{14}$ and $\frac{18}{52}$
  c. $\frac{10}{8}$ and $\frac{50}{40}$ d. $\frac{9}{12}$ and $\frac{27}{48}$

2. From the fractions below, make a set of equal fractions.
  a. $\frac{5}{6}$ b. $\frac{3}{4}$ c. $2a/b$ d. a/b

3. Simplify the fractions below.
  a. $\frac{49}{56}$ b. $\frac{25}{40}$ c. $\frac{42}{24}$ d. $\frac{8}{8}$

4. Find the values of n which satisfy the sentences below.

   a. $\frac{3}{4} = n/64$ b. $n/16 = \frac{1}{2}$ c. $\frac{3}{5} = 9/n$ d. $3n = \frac{24}{16}$

5. Use the definition of fraction to prove that $a/b = an/bn$ for n in N.

SECTION B   ADDITION AND SUBTRACTION

BEHAVIORAL OBJECTIVES

B.1 Be able to illustrate addition of fractions using physical models.

B.2 Be able to show how to add two fractions.

B.3 Be able to explain why the rule for subtracting fractions works.

B.4 Be able to subtract fractions.

B.5 Be able to distinguish between common and improper fractions.

B.6 Given two fractions, be able to find their least common denominator.

B.7 Be able to write improper fractions as mixed numbers.

B.8 Be able to write mixed numbers as improper fractions.

ADDITION OF FRACTIONS

We can illustrate addition of rational numbers by using the geometric ideas developed earlier. Let us consider the examples below before we define the operation.

Example 1

■ **B.1**   Find the sum of $\frac{3}{4}$ and $\frac{2}{4}$.

SOLUTION   We can model this sum using the number line.

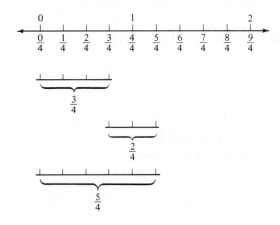

Example 2

Find the sum of $\frac{2}{8}$ and $\frac{3}{8}$.

SOLUTION   We can model this sum using congruent rectangles.

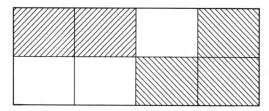

$\frac{2}{8}$ is represented by the region cross-hatched to the left and $\frac{3}{8}$ is represented by the region cross-hatched to the right. The number of shaded regions is 5. Therefore, $\frac{2}{8} + \frac{3}{8} = \frac{5}{8}$.

You will notice that the examples given had one thing in common. In each problem the denominators of the numbers to be added were the same. You will notice that the sum was found by adding the numerators and writing that sum above the common denominator; that is,

$$\frac{2}{8} + \frac{3}{8} = \frac{2+3}{8}, \quad \text{or} \quad \frac{5}{8}$$

This geometric argument is very suitable as an explanation of addition of fractions in the elementary grades, but we have defined a common fraction as a rational number, and since we know how to add rational numbers, we should know how to add fractions; that is, adding fractions is really adding rational numbers. So we will state the rule for addition of fractions as a theorem and leave the proof as an exercise.

Theorem

B.2   *The sum of two common fractions $\dfrac{a}{b}$ and $\dfrac{c}{b}$ is $\dfrac{a+c}{b}$*

Example 1

Find the sum of $\frac{2}{3}$ and $\frac{5}{3}$.

SOLUTION $\dfrac{2}{3} + \dfrac{5}{3} = \dfrac{2+5}{3}$, or $\dfrac{7}{3}$

Example 2

Find the sum of $\frac{3}{12}$ and $\frac{4}{12}$

SOLUTION $\dfrac{3}{12} + \dfrac{4}{12} = \dfrac{3+4}{12}$, or $\dfrac{7}{12}$.

Since not all fractions have the same denominators, we must be able to add fractions with unlike denominators. Again, we can use the fact that common fractions are rational numbers to prove the following theorem on addition of fractions.

Theorem

■ **B.2**   *The sum of two common fractions $\dfrac{a}{b}$ and $\dfrac{c}{d}$ is $\dfrac{ad+bc}{bd}$ that is,*

$$\frac{a}{b} + \frac{c}{d} = \frac{ad+bc}{bd}$$

Example

Find the sum of $\frac{3}{4}$ and $\frac{5}{8}$ using the theorem.

SOLUTION   1. $\dfrac{3}{4} + \dfrac{5}{8} = \dfrac{24+20}{32}$, or $\dfrac{44}{32}$

2. The L.C.M. of 4 and 8 is 8, so that $\frac{3}{4} = \frac{6}{8}$. Thus, $\frac{6}{8} + \frac{5}{8} = \frac{11}{8}$.

Notice that the second answer is in a simpler form and that the first answer can be simplified to $\frac{11}{8}$.

■

SUBTRACTION OF FRACTIONS

Since the common fractions are defined as rational numbers, the operation of subtraction of fractions is really subtraction of rational numbers. Again, we leave the proof of the following theorem to you and simply give several examples. You should also be able to apply to subtraction the model that we used to illustrate addition (see pages 308–309).

Theorem

■ **B.4**   *The difference of two fractions $\dfrac{a}{b} - \dfrac{c}{d}$ is $\dfrac{ad-bc}{bd}$;   that is,*

$$\frac{a}{b} - \frac{c}{d} = \frac{ad-bc}{bd}$$

Example 1

Find the difference $\frac{3}{4} - \frac{1}{4}$.

SOLUTION $\dfrac{3}{4} - \dfrac{1}{4} = \dfrac{3-1}{4}$, or $\dfrac{2}{4}$

Example 2

Find the difference $\frac{5}{8} - \frac{1}{2}$.

SOLUTION $\dfrac{5}{8} - \dfrac{1}{2} = \dfrac{(5 \times 2) - (8 \times 1)}{8 \times 2}$

$$= \frac{10-8}{16}$$

$$= \frac{2}{16}$$

$$= \frac{1}{8}$$

Example 3

Find the difference $3\frac{3}{4} - \frac{1}{2}$

SOLUTION $\quad 3 + \frac{3}{4} - \frac{1}{2} = 3 + (\frac{3}{4} - \frac{2}{4}).$
$$= 3 + \frac{1}{4}$$

■

■ **B.3** At this point we should state that these theorems about fractions are
no more than theorems about rational numbers, which we discussed in
■ Chapter 7.

EXERCISES

1. Find the following differences.

a. $\frac{3}{4} - \frac{1}{4}$ b. $\frac{5}{6} - \frac{4}{6}$ c. $\frac{3}{5} - \frac{2}{5}$ d. $\frac{6}{1} - \frac{2}{1}$

2. Find the following differences.

a. $\frac{3}{4} - \frac{2}{3}$ b. $\frac{5}{8} - \frac{1}{4}$ c. $\frac{12}{5} - \frac{6}{15}$ d. $\frac{3}{8} - \frac{5}{24}$

3. Describe a physical model for subtraction of fractions.

4. Use the definition of fractions to prove the following.

a. $\dfrac{a}{b} + \dfrac{c}{b} = \dfrac{a+c}{b}$ b. $\dfrac{a}{b} + \dfrac{c}{d} = \dfrac{ad+bc}{b}$

c. $\dfrac{a}{b} - \dfrac{c}{b} = \dfrac{a-c}{b}$ d. $\dfrac{a}{b} - \dfrac{c}{d} = \dfrac{ad-bc}{bd}$

5. If $a < c,$

a. how do the rational numbers $R[\![a, b]\!]$ and $R[\![c, b]\!]$ compare?
b. would $a/b - c/b$ be a positive or negative rational number?
c. is the set of common fractions closed under subtraction?

6. Based on Exercise 5, how would you define a negative fraction?

COMMON NOTATION USED WITH FRACTIONS

In some of the problems illustrated, we have used fractions of the form
$\frac{20}{3}$ or $\frac{5}{4}$. Notice that these fractions have a common characteristic, namely,
the numerator is greater than the denominator. This type of fraction is
often called an *improper fraction.* This choice of name is probably a poor
one since there is nothing improper about such a fraction; however, the
term is used in many texts and so we include it here.

■ **B.5** Every improper fraction can be written in a different form, namely, as the
sum of two rational numbers, one of which is an integer and the second
a common fraction. A common fraction is a fraction which is greater
■ than or equal to 0 but less than 1, that is, $\frac{0}{5}, \frac{2}{3}, \frac{3}{20}$, and so forth.

Example 1

■ **B.7** $\frac{20}{3} = R[\![20, 3]\!]$
$= R[\![18, 3]\!] + R[\![2, 3]\!]$
$= R[\![6, 1]\!] + R[\![2, 3]\!]$
$= 6 + \frac{2}{3}$

Example 2

$$\frac{8}{5} = R[\![8, 5]\!]$$
$$= R[\![5, 5]\!] + R[\![3, 5]\!]$$
$$= R[\![1, 1]\!] + R[\![3, 5]\!]$$
$$= 1 + \frac{3}{5}$$

■

By convention, we write the numbers $6 + \frac{2}{3}$ and $1 + \frac{3}{5}$ without the addition sign, that is, as $6\frac{2}{3}$ and $1\frac{3}{5}$, respectively. Thus, $\frac{20}{3} = 6\frac{2}{3}$ and $\frac{8}{5} = 1\frac{3}{5}$. A fraction written in this new form is called a *mixed number*, since the form is a mixture of integers and common fractions.

Let us now translate mixed number notation to fraction notation. Note that $6\frac{2}{3}$ means $6 + \frac{2}{3}$, not $6 \times \frac{2}{3}$!

Example 1

■ **B.8**   Translate $3\frac{1}{4}$ to fraction form.

SOLUTION   Since 3 can be written as $\frac{3}{1}$, $3\frac{1}{4} = \frac{3}{1} + \frac{1}{4}$.
But $\frac{3}{1} + \frac{1}{4} = \frac{12}{4} + \frac{1}{4} = \frac{13}{4}$, so $3\frac{1}{4} = \frac{13}{4}$.

Example 2

Write $6\frac{2}{3}$ as a fraction.

SOLUTION $6 + \frac{2}{3} = \frac{6}{1} + \frac{2}{3}$, and since the L.C.M. of 3 and 1 is 3, we rewrite $\frac{6}{1}$ as $\frac{18}{3}$. Now $\frac{18}{3} + \frac{2}{3} = \frac{20}{3}$.

Example 3

Translate $2\frac{3}{4}$ to fraction form.

SOLUTION $2 + \frac{3}{4} = \frac{2}{1} + \frac{3}{4}$. The L.C.M. of 1 and 4 is 4, so $\frac{2}{1}$ is rewritten as $\frac{8}{4}$, and $\frac{8}{4} + \frac{3}{4} = \frac{11}{4}$.

■

Notice that we used the concept of the L.C.M. in this section in conjunction with the denominators of fractions as a means of finding a common denominator to be used in the addition operation. When this is done, we call this the *least common denominator*. We could, of course, call this the least common multiple of denominators, but this is so wordy it has been shortened to the abbreviation L.C.D.

Definition

The least common denominator, L.C.D., of two fractions is the least common multiple, L.C.M., of the two denominators.

Now that we have discussed the translation of mixed numbers to fractions and the translation of improper fractions to mixed numbers, we should be able to add mixed numbers as well as mixed numbers and fractions. Consider some examples.

Example 1

■ B.2   Find the sum of $5\frac{1}{2}$ and $3\frac{3}{4}$.

SOLUTION   1. We can translate and use our definition as shown below.

$$5\frac{1}{2} = \frac{5}{1} + \frac{1}{2} = \frac{10}{2} + \frac{1}{2} = \frac{11}{2}$$

and

$$3\frac{3}{4} = \frac{3}{1} + \frac{3}{4} = \frac{12}{4} + \frac{3}{4} = \frac{15}{4}$$

. B.6 ▲   then we add $\frac{11}{2} + \frac{15}{4}$. The L.C.D. is 4, and $\frac{11}{2}$ can be written $\frac{22}{4}$. Therefore, $\frac{22}{4} + \frac{15}{4} = \frac{37}{4}$. When 37 is divided by 4 we get 9, with a remainder of 1. Therefore, $\frac{37}{4} = 9\frac{1}{4}$.

Notice that since this solution was very long and involved many steps, we will offer an alternate which is more direct.

2. Since we are adding two mixed fractions, we can use our properties to rewrite as follows:

$$5\frac{1}{2} + 3\frac{3}{4} = (\frac{5}{1} + \frac{1}{2}) + (\frac{3}{1} + \frac{3}{4})$$

B.6 ▲   By commutative and associative properties of R, this becomes $(\frac{5}{1} + \frac{3}{1}) + (\frac{1}{2} + \frac{3}{4})$. Finding the L.C.D. for 2 and 4, we have $(\frac{5}{1} + \frac{3}{1}) + (\frac{2}{4} + \frac{3}{4})$. Then adding we have $\frac{8}{1} + \frac{5}{4}$, and translating we get $\frac{8}{1} + 1\frac{1}{4}$. Adding again we have $(\frac{8}{1} + \frac{1}{1}) + \frac{1}{4} = \frac{9}{1} + \frac{1}{4}$, or $9\frac{1}{4}$.

Example 2

Find the sum of $5\frac{3}{4}$ and $\frac{2}{3}$.

SOLUTION   We will shorten the solution considerably at this point.

$$5\frac{3}{4} + \frac{2}{3} = \frac{5}{1} + \frac{3}{4} + \frac{2}{3}$$

B.6 ▲   The L.C.M. of 4 and 3 is 12; therefore,

$$\frac{5}{1} + (\frac{9}{12} + \frac{8}{12}) = \frac{5}{1} + \frac{17}{12}$$

Translating, $\frac{5}{1} + 1\frac{5}{12}$ gives $6\frac{5}{12}$.

Example 3

Find the sum of $2\frac{1}{3}$, $\frac{5}{6}$, and $3\frac{1}{4}$.

SOLUTION $2\frac{1}{3} + \frac{5}{6} + 3\frac{1}{4} = (2 + 3) + (\frac{1}{3} + \frac{5}{6} + \frac{1}{4})$

B.6 ▲   The L.C.D. of 3, 6, and 4 is 12. Therefore, the problem be-
■   comes $(2 + 3) + (\frac{4}{12} + \frac{10}{12} + \frac{3}{12})$. Then, $5 + \frac{17}{12} = 5 + 1\frac{5}{12}$, or $6\frac{5}{12}$.

As cumbersome and perhaps illogical as mixed number notation is, the fact that it is commonly used both in elementary school texts and in practical situations justifies our including it here.

EXERCISES

1. Find the following sums.

a. $\frac{3}{8} + \frac{2}{8}$

b. $\frac{4}{7} + \frac{1}{7}$

c. $\frac{6}{9} + \frac{3}{9}$

d. $\frac{8}{5} + \frac{5}{5}$

2. Translate the following to mixed fractions.

a. $\frac{25}{12}$

b. $\frac{13}{4}$

c. $\frac{18}{3}$

d. $\frac{7}{2}$

3. Find the L.C.M. for the given number pairs.

a. 3 and 5

b. 3 and 12

c. 4 and 6

d. 15 and 10

4. Translate the following mixed numbers to improper fractions.

a. $2\frac{3}{4}$

b. $5\frac{5}{6}$

c. $8\frac{3}{12}$

d. $4\frac{2}{3}$

5. Find the L.C.D. for the pairs of fractions given below.

a. $\frac{2}{3}$ and $\frac{3}{4}$

b. $\frac{1}{12}$ and $\frac{2}{3}$

c. $\frac{3}{15}$ and $\frac{1}{2}$

d. $\frac{5}{8}$ and $\frac{5}{6}$

6. Translate the following whole numbers to fractions.

a. 5

b. 3

c. 8

d. 9

7. Find the following sums.

a. $\frac{2}{5} + \frac{1}{4}$

b. $\frac{1}{6} + \frac{1}{3}$

c. $\frac{2}{8} + \frac{1}{6}$

d. $\frac{3}{5} + \frac{4}{5}$

8. Find the sums given, and translate to mixed number form.

a. $\frac{3}{4} + \frac{5}{6}$

b. $\frac{7}{8} + \frac{3}{4}$

c. $\frac{7}{9} + \frac{9}{4}$

d. $\frac{8}{15} + \frac{7}{10}$

9. Find the sums given, and write your answers as mixed numbers.

a. $3\frac{1}{2} + 4\frac{3}{4}$

b. $5\frac{3}{5} + 6\frac{1}{8}$

c. $4\frac{5}{8} + 2\frac{5}{6}$

d. $1\frac{1}{2} + 3\frac{1}{5}$

SECTION C FUNDAMENTAL PROPERTIES OF ADDITION AND SUBTRACTION

BEHAVIORAL OBJECTIVES

C.1 Be able to state and illustrate the fundamental properties of addition and subtraction of fractions

C.2 Given an example of a property of addition and subtraction of fractions, be able to identify the property by name.

PROPERTIES OF + and − FOR FRACTIONS

Since a fraction is a rational number, addition of fractions has all the properties of addition of rational numbers. Thus, the following properties will hold for the set of fractions. These properties can be intuitively justified by examples as shown in this chapter.

■ **C.1** Closure for $+$ and $-$

Commutativity for $+$

Associativity for $+$

Identity for $+$, $(0/a)$

Zero property for $-$

Right identity for $-$, $(0/a)$

■ Additive inverse for $+$

Using the definition of fractions and the properties of rational numbers, you should be able to prove that each of the above properties holds for the set of fractions, that is, $\{a/b \mid a$ is an integer and b is a natural number$\}$.

EXERCISES

■ **C.2** 1. Name the property illustrated below.

a. $\frac{2}{3} + \frac{3}{4} = \frac{3}{4} + \frac{2}{3}$

b. $\frac{2}{3} - \frac{0}{3} = \frac{2}{3}$

c. $(\frac{2}{3} + \frac{3}{4}) + \frac{1}{2} = \frac{2}{3} + (\frac{3}{4} + \frac{1}{2})$

■ d. $\frac{0}{2} + \frac{1}{2} = $, and $\frac{1}{2} + \frac{0}{2} = \frac{1}{2}$

2. Is subtraction closed for positive fractions? Why or why not?

3. Is addition closed for positive fractions? Why or why not?

4. Illustrate the commutative, associative, and identity properties for fractions.

5. Find the additive inverse of the numbers below.

a. $\frac{3}{4}$ b. $\frac{2}{3}$ c. $-\frac{3}{5}$ d. $\frac{0}{2}$

6. Find rational fractions which complete the statements below.

a. $\frac{2}{3} + \square = 0$ b. $\frac{2}{3} + \frac{3}{4} = \square$

c. $\frac{1}{2} - \square = \frac{1}{2}$ d. $\square - \frac{1}{2} = 0$

7. Show that the set of fractions can be described by $\{x/y \mid x$ is an integer and y is a natural number$\}$. That is, show that there is a 1–1 correspondence between the set of fractions and the above set that holds under the operations of addition and multiplication.

SECTION D   MULTIPLICATION AND DIVISION

BEHAVIORAL OBJECTIVES

D.1 Be able to show why the rule for multiplying fractions works.

D.2 State a rule for multiplying and dividing fractions.

D.3 Be able to multiply fractions.

D.4 Be able to use the distributive property to multiply mixed numbers.

D.5 Be able to show why the rule for division of fractions works.

D.6 Be able to illustrate multiplication and division of fractions using a physical model.

D.7 Be able to divide fractions employing the definition.

D.8 Be able to divide fractions by using the common denominator method.

MULTIPLICATION OF FRACTIONS

The operation of multiplication of fractions is based on multiplication of rational numbers, because we have defined a fraction as a rational number. Thus, it is very easy to see how we multiply fractions. As usual we will leave the proof of the general statement of multiplication of fractions to you, but this time we will hint at the proof by way of an example.

■ **D.1** ▲ **D.3**   **Example**

Since $\frac{2}{3} = R\,[\![2, 3]\!]$ and $\frac{7}{11} = R\,[\![7, 11]\!]$,

$$\frac{2}{3} \times \frac{7}{11} = R\,[\![2, 3]\!] \times R\,[\![7, 11]\!]$$
$$= R\,[\![2 \times 7, 3 \times 11]\!]$$
$$= \frac{2 \times 7}{3 \times 11}$$

▲ ■ That is, $\frac{2}{3} \times \frac{7}{11} = \frac{14}{33}$.

Before we state the theorem for multiplication, consider several examples of multiplication which utilize a model.

■ **D.6** ▲ **D.3**   **Example 1**

Find $\frac{1}{4} \times \frac{1}{2}$.

SOLUTION   First this can be read as follows: what is $\frac{1}{4}$ of $\frac{1}{2}$. Consider $\frac{1}{2}$ as a half of a unit.

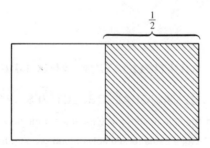

To find $\frac{1}{4}$ of this we draw horizontal lines to separate the unit into four equal parts. If we cross hatch 1 of the 4 equal parts, the double hatched area represents the product. The result, 1 out of 8 blocks of equal area, or $\frac{1}{8}$, is cross-hatched.

$$\tfrac{1}{4} \times \tfrac{1}{2} = \tfrac{1}{8}$$

Example 2

Find $\tfrac{2}{3} \times \tfrac{3}{4}$

SOLUTION   Using the same convention, we represent $\tfrac{3}{4}$ as shown below:

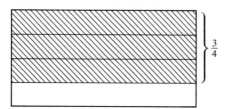

Now, divide the unit vertically into thirds and hatch 2 of the thirds as shown.

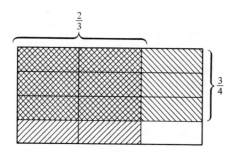

The result, 6 out of 12 blocks of equal area, or $\tfrac{1}{2}$, is cross-hatched.

▲ ■ $\tfrac{2}{3} \times \tfrac{3}{4} = \tfrac{6}{12}$, or $\tfrac{1}{2}$

Theorem

D.2 ■ $For \dfrac{a}{b} \ and \ \dfrac{c}{d} \ fractions, \dfrac{a}{b} \times \dfrac{c}{d} = \dfrac{ac}{bd}.$

Example 1

■ D.3 $\dfrac{1}{1} \times \dfrac{3}{4} = \dfrac{1 \times 3}{1 \times 4} = \dfrac{3}{4}.$

Example 2

$$3 \times \frac{3}{4} = \frac{3}{1} \times \frac{3}{4}$$

$$= \frac{3 \times 3}{4}$$

∎

$$= \frac{9}{4}$$

It is a common procedure to write fractions in lowest terms, meaning that the numerator and denominator are relatively prime. We will illustrate how this may be done by an example and ask you to justify the general case as an exercise. (See A.7 earlier in this chapter.)

Example

$$1. \frac{3}{4} \times \frac{2}{3} = \frac{3 \times 2}{4 \times 3}$$

$$= \frac{6}{12}$$

Dividing numerator and denominator by 6, we produce

$$= \frac{1}{2}$$

$$2. \frac{3}{4} \times \frac{2}{3} = \frac{3 \times 2}{4 \times 3}$$

$$= \frac{2 \times 3}{4 \times 3}$$

$$= \frac{2}{4} \times \frac{3}{3}$$

Dividing numerator and denominator of the first fraction by 2 and the second fraction by 3, we produce

$$= \frac{1}{2} \times \frac{1}{1}$$

Then we have

$$\frac{3}{4} \times \frac{2}{3} = \frac{1}{2}$$

$$3. \frac{3}{4} \times \frac{2}{3} = \frac{3 \times 2}{4 \times 3}$$

$$= \frac{3 \times 2}{(3 \times 2) \times 2}$$

Dividing numerator and denominator by the common factors 3×2, we produce

$$= \frac{1 \times 1}{1 \times 1 \times 2}$$

Then

$$\frac{3}{4} \times \frac{2}{3} = \frac{1}{2}$$

In 1 we multiplied and implicitly used the fact that $R[\![6, 12]\!] = R[\![1, 2]\!]$; in 2 we rewrote the product, simplified, and then performed the multiplication. In 3 we divided out the common factors in the numerator and denominator of the implied product. You can use any of these methods.

EXERCISES

1. Find the products below

   a. $\frac{5}{6} \times \frac{1}{2}$ b. $\frac{7}{8} \times \frac{4}{5}$ c. $\frac{3}{5} \times \frac{10}{12}$

2. Find the products below by dividing out common factors first.

   a. $\frac{3}{4} \times \frac{8}{12}$ b. $\frac{5}{8} \times \frac{16}{25}$ c. $\frac{5}{3} \times \frac{12}{20}$

3. Find the products below in two ways, and compare your results. Which method is easier? Why?

   a. $\frac{8}{12} \times \frac{6}{4}$ b. $\frac{3}{3} \times \frac{5}{5}$ c. $\frac{2}{5} \times \frac{15}{7}$ d. $\frac{1}{4} \times \frac{7}{12}$

4. Find the products below in any manner.

   a. $2\frac{1}{2} \times 3\frac{3}{4}$ b. $\frac{5}{6} \times \frac{2}{23}$ c. $\frac{5}{7} \times \frac{21}{12}$

5. a. Give reasons why $an/b \times c/dn = ac/bd$.

   b. Give reasons why $an/bn \times c/d = ac/bd$.

   (*Hint*: Use the definition of fractions and properties of rational numbers.)

MULTIPLYING MIXED NUMBERS

The concept of multiplying mixed numbers is based on the properties of the rational numbers. All we need to do is remember that the notation $3\frac{2}{5}$ means $3 + \frac{2}{5}$ and then use the distributive properties. For example, $3\frac{1}{2} \times 2\frac{2}{3}$, since $3\frac{1}{2}$ is $3 + \frac{1}{2}$, then $3\frac{1}{2} \times 2\frac{2}{3} = (3 + \frac{1}{2}) \times (2 + \frac{2}{3})$. If we consider this in terms of a distributive property then it reads $a \times (b + c)$, where $(3 + \frac{1}{2})$ is a, 2 is b, and $\frac{2}{3}$ is c. Then $(3 + \frac{1}{2}) \times (2 + \frac{2}{3}) = [(3 + \frac{1}{2}) \times 2] + [(3 + \frac{1}{2}) \times \frac{2}{3}]$. Now apply the distributive property to $(3 + \frac{1}{2}) \times 2$ and $(3 + \frac{1}{2}) \times \frac{2}{3}$. Thus,

$$[(3 + \tfrac{1}{2}) \times 2] + [(3 + \tfrac{1}{2}) \times \tfrac{2}{3}] = [(3 \times 2) + (\tfrac{1}{2} \times 2)] + [(3 \times \tfrac{2}{3}) + (\tfrac{1}{2} \times \tfrac{2}{3})]$$
$$= [6 + 1] + [2 + \tfrac{1}{3}]$$
$$= 9\tfrac{1}{3}$$

Of course, most of the work involved in multiplying mixed numbers is eliminated if we first write the mixed numbers as improper fractions and then multiply. For example,

$$3\tfrac{1}{2} \times 2\tfrac{2}{3} = \tfrac{7}{2} \times \tfrac{8}{3}$$
$$= \tfrac{56}{6}$$
$$= 9\tfrac{2}{6}$$
$$= 9\tfrac{1}{3}$$

Down with mixed numbers!

EXERCISES

1. Write the following as mixed numbers.

   a. $\frac{23}{4}$ b. $\frac{12}{8}$ c. $\frac{31}{12}$ d. $\frac{17}{6}$

2. Write the following as fractions.

   a. $2\frac{3}{5}$ b. $3\frac{5}{8}$ c. $7\frac{2}{15}$ d. $3\frac{1}{7}$

3. Find the products below by using the distributive property.

   a. $2\frac{2}{3} \times 4\frac{5}{6}$ b. $3\frac{3}{8} \times 2\frac{4}{5}$ c. $4\frac{7}{8} \times 3\frac{4}{9}$

4. Find the products from Exercise 3 by rewriting the mixed numbers as improper fractions.

DIVISION OF FRACTIONS

Recall how we defined division of rational numbers:

$$R[\![a, b]\!] \div R[\![c, d]\!] = R[\![a, b]\!] \times R[\![c, d]\!]^{-1}$$

■ **D.5** Since fractions are rational numbers, it is reasonable to expect that division of fractions will work like division of rationals. Suppose that $a/b = R[\![a, b]\!]$ and $c/d = R[\![c, d]\!]$. Then,

$$\begin{aligned} a/b \div c/d &= R[\![a, b]\!] \div R[\![c, d]\!] \\ &= R[\![a, b]\!] \times R[\![c, d]\!]^{-1} \\ &= a/b \times (c/d)^{-1} \end{aligned}$$

We need to know what $(c/d)^{-1}$ is. Thus, we need to know what fraction $R[\![c, d]\!]^{-1}$ is, namely, $R[\![d, c]\!]$ if c is a positive integer or $R[\![-d, -c]\!]$ if c is a negative integer. That is,

$$R[\![1, 2]\!]^{-1} = R[\![2, 1]\!]$$

and

$$R[\![-1, 2]\!]^{-1} = R[\![-2, 1]\!]$$

Thus, $(c/d)^{-1}$ is d/c if c is a positive integer and $-d/-c$ if c is a negative integer. For example, $(\frac{1}{2})^{-1} = \frac{2}{1}$, whereas $(\frac{-1}{2})^{-1} = \frac{-2}{1}$. Now let us look at some examples involving division of fractions.

Examples

■ **D.7**

1. $\dfrac{2}{3} \div \dfrac{1}{5} = \dfrac{2}{3} \times \dfrac{5}{1}$

            $= \dfrac{10}{3}$

2. $\dfrac{-9}{7} \div \dfrac{-2}{3} = \dfrac{-9}{7} \times \dfrac{-3}{2}$

              $= \dfrac{-9 \times -3}{7 \times 2}$

              $= \dfrac{27}{14}$

We state the rule for dividing two fractions as a theorem.

Theorem

D.2 ■ *The quotient of two fractions* $\dfrac{a}{b} \div \dfrac{c}{d} = \dfrac{a}{b} \times \left(\dfrac{c}{d}\right)^{-1}$

The theorem gives us a means for operating, and can be described as follows:

1. Given $a/b \div c/d$.
2. Translate to $a/b \times d/c$ or $a/b \times -d/-c$.
3. Multiply.

We can now look at a model which gives a physical justification for the division rule. As we have said before, this model usually provides the "proof," or reason, for the rule given in the elementary schools. First, consider the question of $\frac{1}{1} \div \frac{1}{2}$. That is, how many $\frac{1}{2}$'s are in 1?

Example 1

■ **D.6**

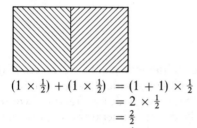

$$(1 \times \tfrac{1}{2}) + (1 \times \tfrac{1}{2}) = (1 + 1) \times \tfrac{1}{2}$$
$$= 2 \times \tfrac{1}{2}$$
$$= \tfrac{2}{2}$$
$$= 1$$

That is, there are two $\frac{1}{2}$'s in 1, or $\frac{1}{1} \div \frac{1}{2} = 2$.

Example 2

$\frac{3}{4} \div \frac{1}{4}$. Construct $\frac{3}{4}$.

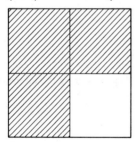

■ How many $\frac{1}{4}$'s in $\frac{3}{4}$? Notice that there are three $\frac{1}{4}$'s in $\frac{3}{4}$.

Example 1

Another method that can be used to divide fractions is to rewrite each of the fractions with a common denominator. This approach to division of fractions is often clearer to children than simply explaining the definition of division. Consider these examples.

◄ **D.8** Find the quotient $\frac{3}{5} \div \frac{1}{4}$.

SOLUTION   Since the L.C.D. of the denominator is 20, we translate the problem to $\frac{12}{20} \div \frac{5}{20} = \frac{12}{5}$, because $\frac{12}{5} \times \frac{5}{20} = \frac{12}{20}$.

Example 2

Find the quotient $1\frac{1}{2} \div \frac{1}{3}$, or $\frac{3}{2} \div \frac{1}{3}$.

SOLUTION   Again, the L.C.D. of the denominator is 6. We can translate the problem to $\frac{9}{6} \div \frac{2}{6}$, $\frac{9}{6} \div \frac{2}{6} = \frac{9}{2}$.

For the examples above, you can check the results by applying the definition of division. Why is it that the quotient of two fractions a/b and c/d can be found in this manner? Consider this short proof.

■ D.5   Since $a/b \div c/d$ can be represented with like denominators as $ad/bd \div cb/db$, and since by definition

$$\frac{ad}{bd} \div \frac{cb}{db} = \frac{ad}{bd} \times \frac{db}{cb} = \frac{ad \times db}{bd \times cb}$$

which, in turn, can be written

$$\frac{ad \times db}{cb \times bd} = \frac{ad}{cb}$$

by the commutative property and the definition of equality, we can see that the quotient is ad/cb, which is the same quotient produced by our definition.

Notice that we are simply asserting that if we divide two fractions whose denominators are alike, we need only divide numerators, because the units described by the denominators are alike in each fraction.

EXERCISES

1. Find the multiplicative inverses of the fractions given.

  a. $\frac{3}{5}$ b. $\frac{10}{7}$ c. $2\frac{1}{2}$ d. $\frac{4}{7}$

2. Find the products below.

  a. $\frac{2}{3} \times \frac{3}{2}$ b. $\frac{4}{5} \times \frac{5}{4}$
  c. $\frac{7}{3} \times \frac{3}{7}$ d. $2\frac{1}{3} \times \frac{3}{7}$

3. Find the quotients given by multiplying numerator and denominator by the inverse of the denominator.

  a. $\frac{2}{3} \div \frac{5}{6}$ b. $\frac{3}{4} \div 2\frac{1}{2}$ c. $\frac{5}{7} \div \frac{7}{5}$
  d. $3\frac{1}{4} \div 3\frac{1}{4}$ e. $\frac{0}{6} \div \frac{2}{3}$

4. Find the quotient by finding common denominators first.

  a. $2\frac{1}{4} \div \frac{3}{4}$ b. $\frac{3}{5} \div \frac{2}{5}$ c. $\frac{4}{7} \div 1\frac{1}{4}$
  d. $\frac{5}{8} \div \frac{5}{6}$ e. $\frac{0}{4} \div \frac{1}{2}$ f. $\frac{3}{4} \div \frac{1}{3}$

5. Make a geometric diagram to illustrate the following quotients.

  a. $\frac{7}{4} \div \frac{3}{2}$ b. $\frac{5}{8} \div \frac{3}{8}$
  c. $\frac{3}{2} \div \frac{3}{4}$ d. $\frac{3}{5} \div \frac{3}{5}$

6. Explain the relationship between Exercises a and b, and c and d.

  a. $\frac{3}{4} \div \frac{2}{5}$ b. $\frac{2}{5} \div \frac{3}{4}$
  c. $\frac{5}{8} \div \frac{3}{4}$ d. $\frac{3}{4} \div \frac{5}{8}$

7. Why can we not divide by the fraction $\frac{0}{2}$?

8. Explain why the quotient of two fractions with the same denominator can be found by dividing the numerators.

SECTION E FUNDAMENTAL PROPERTIES OF MULTIPLICATION AND DIVISION

BEHAVIORAL OBJECTIVES

E.1 Be able to summarize the fundamental properties of multiplication and division of fractions.

E.2 Be able to illustrate the fundamental properties of multiplication and division of fractions.

E.3 Given an example of a fundamental property of multiplication and division of fractions, be able to identify the property by name.

PROPERTIES OF MULTIPLICATION AND DIVISION OF FRACTIONS

To summarize the properties of the set of fractions F, we need only note that the operations of multiplication and division on F are really the operations on the set of rational numbers. Since fractions were defined as rational numbers, the set F has all the properties of R under multiplication and division. We summarize the properties of F below.

■ **E.1** Closure property for $\times$ and $\div$
Commutative property for $\times$
Associative property for $\times$
Identity for $\times$
Inverse for $\times$
Zero property for $\times$
Zero property for $\div$
Right identity for $\div$
Any number divides itself once
■ Distributive property of $\times$ over $+$

EXERCISES

■ **E.3** **1.** Name the property illustrated below.

　　a. $\frac{3}{4} \div \frac{3}{4} = \frac{1}{1}$　　b. $\frac{2}{3} \times \frac{0}{3} = \frac{0}{9}$　　c. $\frac{2}{3} \times \frac{3}{5} = \frac{3}{5} \times \frac{2}{3}$

■　　d. $\frac{3}{4} \times \frac{2}{2} = \frac{3}{4}$　　e. $\frac{4}{5} \div \frac{1}{1} = \frac{4}{5}$

■ **E.2** **2.** Illustrate the properties below.

　　a. Associative $\times$ F　　　　　　b. Inverse $\times$ F
　　c. Commutative $\times$ F　　　　　　d. Zero $\div$ F

3. Explain why the set F is a commutative group under $\times$.

4. Does the set of fractions form a field?

5. Using the definition of fractions and the corresponding property of the rationals, prove that F has the properties listed on page 323.

SECTION F **FRACTIONS AND DECIMALS**

BEHAVIORAL OBJECTIVES

F.1 Be able to correspond fractions, exponential notation, and decimal fractions.

F.2 Be able to name the digit positions in base ten.

F.3 Be able to translate decimal fractions to fractions.

F.4 Be able to give an example of a fraction that has a terminating decimal expression.

F.5 Be able to give an example of a fraction that has a non-terminating decimal expression.

F.6 Be able to explain why some fractions have terminating decimal expressions and some do not.

F.7 Given an expression in repeating decimal form, be able to write the fraction in the form a/b.

DECIMAL FRACTIONS

Decimal fractions are simply fractions written in different form. Most fractions are written as a/b; however, decimal fractions are fractions in which b is a power of 10 and b is implied rather than written. Instead of writing $\frac{32}{100}$, we write .32, and instead of $\frac{62}{1000}$, we write .062.

When we discussed numeration systems, we stressed the fact that place value employs powers of 10 as multipliers. When we consider fractions whose denominators are powers of 10, we are simply extending the concept further. If we consider the unit's place as a center and the ten's place as one place to the left of center, then the tenths' ($\frac{1}{10}$) place is simply one place to the right of center. The fact that decimal points are used to set off fractions from whole numbers often causes confusion. It might have been better if a bar were used to mark the unit's place. Consider Figure 8.6, which relates powers of 10 to our decimal system.

As we mentioned earlier, the places are symmetrical with respect to the unit's place. Also, a numeral in a particular place represents a multiple of a power of 10. The power of 10 is positive if it is located to the left of the unit's place and negative if it is located to the right of the unit's place. A negative power of 10 can be translated as the inverse of the same positive

F.2

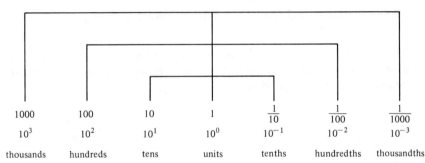

1000	100	10	1	$\frac{1}{10}$	$\frac{1}{100}$	$\frac{1}{1000}$
10^3	10^2	10^1	10^0	10^{-1}	10^{-2}	10^{-3}

■ thousands hundreds tens units tenths hundredths thousandths

Figure 8.6

power. In this manner, we can translate negative powers of 10 to unit fractions whose denominators are powers of 10. Consider some examples.

Examples

■ F.1   1. $10^{-2} = \dfrac{1}{10^2} = \dfrac{1}{100} = .01$

2. $10^{-3} = \dfrac{1}{10^3} = \dfrac{1}{1000} = .001$

■   3. $10^{-1} = \dfrac{1}{10} = .1$

Now if we combine these ideas, we produce the following statements:

1. Places in our decimal system are symmetric with respect to the unit's place.
2. Positive powers of 10 represent whole numbers and are located to the left of the unit's place.
3. Negative powers of 10 represent fractions and are located to the right of the unit's place.
4. A negative power of 10 represents a fraction whose denominator is a positive power of 10, such as $10^{-3} = \dfrac{1}{10^3}$.
5. We can represent a negative power of 10 as a fraction or as a decimal fraction.
6. The number of places to the right of the decimal indicates the implied denominator.
7. A decimal fraction is a multiple of a negative power of 10.

We can consider several examples of decimal fractions and translate them to common fractions by using the list given above.

Examples

◀ F.3   1. .307 can be written $\dfrac{307}{10^3}$ or $\dfrac{307}{1000}$. This also means $\dfrac{3}{10} + \dfrac{0}{100} + \dfrac{7}{1000}$.

2. 2.85 means $2\frac{85}{100}$ and can be expanded to read $2 + \frac{8}{10} + \frac{5}{100}$, or $2 + \frac{80}{100} + \frac{5}{100}$.

We use decimal notation because it is simpler to operate with, since decimal notation allows us to perform operations in a similar manner to the operations for whole numbers.

EXERCISES

1. Write the following fractions as decimals.
   a. $\frac{23}{100}$ b. $\frac{321}{1000}$ c. $3\frac{3}{100}$ d. $25\frac{34}{100}$

2. Write the following decimals as fractions.
   a. 2.305   b. 3.42   c. 23.04   d. .3283

3. Write the following powers of 10 as fractions.
   a. 10^{-3} b. 10^{-2} c. 10^{-1}

4. Write the following decimals in expanded form.
   a. 32.458   b. 253.6042   c. 23.003   d. 132.032

REPEATING AND TERMINATING DECIMAL FRACTIONS

When a fraction is translated to decimal form, the operation performed is division, because a/b can be interpreted as $a \div b$. In some cases this division process terminates, that is, all the digits after some non-zero digit are zeros. In other cases the division process continues with non-zero digits. Consider some examples.

Examples

■ F.4 ■   1. $\frac{1}{5} = 1 \div 5$, or $5\overline{)1.0}$ (quotient .2)

■ F.5 ■   2. $\frac{1}{3} = 1 \div 3$, or $3\overline{)1.00000}$ (quotient .3333...)

$$\begin{array}{r} 9 \\ \hline 10 \\ 9 \\ \hline 10 \end{array}$$

Notice in the first example that the decimal representation is .2 and terminates at the 10^{-1} place. In the second example, however, the decimal representation repeats. We represent decimals such as .333... as $.\overline{3}$. The bar written over the numeral or numerals indicates what part of the number repeats. For example, $1/27 = .037037037...$ can be written $.\overline{037}$.

Complete the table in Figure 8.7. Can you see a relationship between the fractions which form repeating decimals and the factors of their denominators? Can you see a relationship between the fractions which form terminating decimals and the factors of their denominators? After completing the table, see if you can answer the questions above.

Fraction	Denominator	Factors of the Denominator	Decimal Fraction
$\frac{3}{8}$	8	$2 \times 2 \times 2$	.375
$\frac{3}{14}$	14	2×7	.2142857
$\frac{5}{16}$	16	$2 \times 2 \times 2 \times 2$	
$\frac{5}{6}$	6	2×3	
$\frac{3}{5}$	5		.6
$\frac{7}{15}$	15		.4$\overline{6}$
$\frac{7}{20}$	20	$2 \times 2 \times 5$	
$\frac{8}{30}$	30	$3 \times 2 \times 5$	
$\frac{3}{25}$	25		.12
$\frac{5}{18}$	18		.2$\overline{7}$

Figure 8.7

■ **F.6**   You might consider what causes one fraction to terminate and another to repeat. Suppose that a/b is a terminating fraction. Then $a/b = .x_1x_2x_3 \ldots x_n000$, where the x_i are integers, but

$$.x_1x_2x_3 \ldots x_n = \frac{x_1x_2x_3 \ldots x_n}{10^n}$$

since there are n places to the right of the decimal point. Thus

$$a/b = \frac{x_1x_2x_3 \ldots x_n}{10^n}$$

That is, the fraction can be written with a denominator that is a power of 10. Since $10 = 2 \times 5$,

$$\frac{a}{b} = \frac{x_1x_2x_3 \ldots x_n}{2^n5^n}$$

From this last expression we can see that if a fraction has a decimal expression that terminates, then the denominator is a multiple of 2 or a multiple of 5 *only*. In this manner we can see that the decimal expression for fractions like $\frac{3}{8}, \frac{5}{16}, \frac{3}{5}, \frac{7}{20}$, and $\frac{3}{25}$ will terminate. Fractions like $\frac{3}{14}, \frac{5}{6}, \frac{7}{15}$, $\frac{8}{30}$, and $\frac{5}{18}$ will repeat, since their denominators contain factors other than
■ 2 or 5.

We have shown how one finds the decimal representation of a fraction in the form a/b. Next let us consider how we show the fractional representation of a repeating decimal in the form $.a_1a_2a_3 \ldots a_n, \ldots$. Consider the examples below. In these examples, we use a bar over the digits of the decimal fraction that repeats. That is, $.\overline{257}$ means $.257257257 \ldots$, and $.\overline{33}$ means $.3333 \ldots$.

Example 1

◀ **F.7**   Write the decimal fraction $.\overline{3}$ as a fraction.

SOLUTION   If $a/b = .\overline{3}$, then $10 \times a/b = 3.\overline{3}$.
Now $10 \times a/b - a/b = 3.\overline{3} - .\overline{3} = 3.00000$.
Therefore, $9 \times a/b = 3$, and $a/b = \frac{3}{9}$, or $\frac{1}{3}$.

Example 2

Write the decimal fraction $.\overline{037}$ as a fraction.

SOLUTION   If $a/b = .\overline{037}$ and $1000 \times a/b = 37.\overline{037}$, then $1000 \times a/b - a/b$
$= 37.\overline{037} - .\overline{037}$. Therefore, $999 \times a/b = 37$, and $a/b = \frac{37}{999}$
or $\frac{1}{27}$.

You might check this process by dividing 27 into 1 to see the pattern of
digits that results.

EXERCISES

1. Write the following repeating decimal fractions in the form a/b.
  a. 11...   b. $.\overline{35}$ c. $.\overline{9}$ d. $.\overline{123}$

2. Which of the following fractions should repeat? Why?
  a. $\frac{7}{15}$ b. $\frac{4}{21}$ c. $\frac{8}{30}$ d. $\frac{5}{8}$ e. $\frac{6}{33}$

3. Write the following decimal fractions in the form a/b.
  a. .1234   b. $.27\overline{15}$ c. .625

SECTION G OPERATIONS WITH DECIMALS

BEHAVIORAL OBJECTIVES

G.1   Given a pair of decimal fractions, be able to find the sum
using expanded notation and the distributive property.

G.2   Given a pair of decimal fractions, be able to find the
difference using expanded notation and the distributive
property.

G.3   Be able to multiply fractions using expanded notation and
distributive properties.

G.4   Be able to state and use the algorithm for addition and
subtraction of decimal fractions.

G.5   Be able to state and use the algorithm for multiplication of
decimal fractions.

G.6   Be able to multiply decimal fractions by powers of 10 by
shifting the decimal place.

G.7   Be able to explain why multiplication of decimal fractions by
powers of 10 by shifting the decimal place works.

G.8   Be able to divide decimal fractions using the division algorithm.

G.9   Be able to explain how the division algorithm works.

G.10 Be able to explain whether the division algorithm will produce
a repeating or non-repeating decimal.

ADDITION AND SUBTRACTION OF DECIMAL FRACTIONS

If we consider the fact that the sum of two fractions a/b and c/d is defined in terms of the fractions rewritten with like denominators, you should immediately see why we can easily combine decimal fractions. If we wish to add .3 and .4, we know that each fraction has an implied denominator of 10 even though it is not expressed. The fractions above represented in decimal form can be written as $\frac{3}{10}$ and $\frac{4}{10}$; therefore, the sum is automatically $\frac{7}{10}$, or .7. Suppose we have the decimal fractions .24 and .356. The first decimal fraction means $\frac{24}{100}$, or $\frac{240}{1000}$, and the second decimal fraction means $\frac{356}{1000}$. Therefore, we can add them by definition to produce $(240 + 356)/1000$, or $\frac{596}{1000}$, which is written .596.

Since subtraction operates much like addition with respect to employing like denominators, we subtract decimal fractions in a like manner. Suppose we wish to subtract .23 from .35. Again, .23 means $\frac{23}{100}$ and .35 means $\frac{35}{100}$; therefore, the difference is found by writing $(35 - 23)/100$, or $\frac{12}{100}$, which is .12. We may also consider the difference $.5 - .034$, which translates to $\frac{5}{10} - \frac{34}{1000}$ or $\frac{500}{1000} - \frac{34}{1000}$, and the difference is given as $(500 - 34)/1000$, or $\frac{466}{1000}$, which is .466.

Consider each of the examples above in several forms. We can always translate before we add or subtract; however, since the notation immediately allows us to have like denominators, we can add or subtract directly.

■ G.1　**Example 1**
▲ G.4　　Find the sum of .3 and .4.

■　　SOLUTION　1. Translating, $\frac{3}{10} + \frac{4}{10} = \frac{7}{10}$, or .7.
　　　　　　　2. Distributing, $(3 \times 10^{-1}) + (4 \times 10^{-1}) = (3 + 4) \times 10^{-1}$,
　　　　　　　　or 7×10^{-1}, which is .7.
　　　　　　　3. Using the decimal notation,

▲
$$\begin{array}{r} .3 \\ +.4 \\ \hline .7 \end{array}$$

◄ G.1　**Example 2**
▲ G.4　　Find the sum of .24 and .356.

SOLUTIONS　1. Translating, $.24 = \dfrac{24}{100}$

$$.356 = \frac{356}{1000}$$

$$\frac{24}{100} + \frac{356}{1000} = \frac{240}{1000} + \frac{356}{1000}$$

$$\frac{(240 + 356)}{1000} = \frac{596}{1000}, \text{ or } .596$$

Notice that .24 and .240 are names for the same number.

2. Distributing, $.24 = (2 \times 10^{-1}) + (4 \times 10^{-2})$ and
$$.356 = (3 \times 10^{-1}) + (5 \times 10^{-2})$$
$$+ (6 \times 10^{-3}).$$

Therefore, we have by commutative, associative, and distributive properties in the set F,

$$.24 + .356 = [(2 \times 10^{-1}) + (3 \times 10^{-1})] + [(4 \times 10^{-2})$$
$$+ (5 \times 10^{-2})] + (6 \times 10^{-3})$$
$$= [(2 + 3) \times 10^{-1}] + [(4 + 5) \times 10^{-2}]$$
$$+ (6 \times 10^{-3})$$
$$= (5 \times 10^{-1}) + (9 \times 10^{-2}) + (6 \times 10^{-3})$$
$$= .596$$

3. Using the decimal notation,

$$\begin{array}{r} .24 \\ + .356 \\ \hline .596 \end{array}$$

■ **G.2 Example 3**
▲ **G.4**   Find the difference $.35 - .23$.

SOLUTION 1. Translating, $\frac{35}{100} - \frac{23}{100} = \frac{12}{100}$.
2. Distributing,

$$.35 = (3 \times 10^{-1}) + (5 \times 10^{-2})$$
$$.23 = (2 \times 10^{-1}) + (3 \times 10^{-2})$$
$$= [(3 \times 10^{-1}) + (5 \times 10^{-2})]$$
$$- [(2 \times 10^{-1}) + (3 \times 10^{-2})]$$
$$= [(3 \times 10^{-1}) - (2 \times 10^{-1})]$$
$$+ [(5 \times 10^{-2}) - (3 \times 10^{-2})]$$
$$= [(3 - 2) \times 10^{-1}] + [(5 - 3) \times 10^{-2}]$$
$$= (1 \times 10^{-1}) + (2 \times 10^{-2})$$
$$= .12$$

3. Using the decimal notation without the powers of 10,

$$\begin{array}{r} .35 \\ - .23 \\ \hline .12 \end{array}$$

■ **G.2 Example 4**
▲ **G.4**   Find the difference $.5 - .034$.

SOLUTION 1. Translating,
$$.5 \quad = \tfrac{5}{10} = \tfrac{50}{100} = \tfrac{500}{1000}$$
$$.034 = \tfrac{34}{1000}$$

Then

$$\frac{500}{1000} - \frac{34}{1000} = \frac{466}{1000}, \text{ or } .466$$

2. Distributing,

$$.5 = 5 \times 10^{-1}$$
$$.034 = (3 \times 10^{-2}) + (4 \times 10^{-3})$$
$$.5 - .034 = (5 \times 10^{-1}) - [(3 \times 10^{-2}) + (4 \times 10^{-3})]$$
$$= (4 \times 10^{-1}) + (10 \times 10^{-2})$$
$$- [(3 \times 10^{-2}) + (4 \times 10^{-3})]$$

Since $1 \times 10^{-1} = 10 \times 10^{-2}$,

$$= (4 \times 10^{-1}) + [(10 - 3) \times 10^{-2}]$$
$$- (4 \times 10^{-3})$$
$$= (4 \times 10^{-1}) + (7 \times 10^{-2}) - (4 \times 10^{-3})$$
$$= (4 \times 10^{-1}) + (6 \times 10^{-2})$$
$$+ [(10 \times 10^{-3}) - (4 \times 10^{-3})]$$
$$= (4 \times 10^{-1}) + (6 \times 10^{-2})$$
$$+ [(10 - 4) \times 10^{-3}]$$
$$= (4 \times 10^{-1}) + (6 \times 10^{-2}) + (6 \times 10^{-3})$$
$$= .466$$

3. Using the decimal notation,

$$\begin{array}{r} .5 \\ - .034 \\ \hline .466 \end{array}$$

You will notice that we offered three types of solutions to the problems above. The first involves translating and is used only to clarify the meaning of decimal fraction and operations on decimal fractions. There would be no point in having decimal fractions if we had to translate to another notation system to operate.

The second method shown was an illustration of the meaning of the operations on decimal fractions. We could state this solution in general as follows:

Find the sum of the number $.abc \ldots n$ and $.a_i b_i c_i \ldots n_i$, where they are written in decimal form and where $a, b, c, \ldots, n$ and $a_i, b_i, c_i, \ldots, n_i$ represent whole numbers. Since $.a = a \times 10^{-1}$ and $.a_i = a_i \times 10^{-1}$,

$$.abc \ldots n = a \times 10^{-1} + b \times 10^{-2} + c \times 10^{-3} + \cdots + n \times 10^{-n}$$

and

$$.a_i b_i c_i \ldots n_i = a_i \times 10^{-1} + b_i \times 10^{-2} + c_i \times 10^{-3} + \cdots + n_i \times 10^{-n}$$

the sum is written

$$[(a + a_i) \times 10^{-1}] + [(b + b_i) \times 10^{-2}] + \cdots + [(n + n_i) \times 10^{-n}]$$

A similar result to the one shown above can be developed for subtraction. To summarize its meaning, we might simply state an algorithm for the operation of addition and subtraction for decimal fractions. This algorithm will allow us to use the notation as we did in the third solution to the examples above.

Algorithm

■ G.4   *To add or subtract two decimal fractions, write them with the decimal*
■   *points over each other, and then simply add or subtract the digits.*

Note: This procedure assures that we add fractions with like denominators, that is, tenths to tenths, hundreths to hundredths, etc.

EXERCISES

1. Perform the operations below by translating and adding.
a. .23 + .03   b. .34 + .076
c. 2.507 + .78   d. .567 + .007

2. Perform the operations below by expanding and using the distributive property.
a. 1.25 + .06   b. .056 + .72
c. .246 + .135   d. .006 + .207

3. Perform the operations below by using the decimal notation and the algorithm.
a. .234 + .305   b. 1.2 + 3.46
c. 2.67 + .08   d. .207 + .098

4. Perform the operations below by translating.
a. 2.34 + .04   b. .54 − .23
c. .1204 − .0006   d. 305 − .206

5. Perform the operations below by expanding and using the distributive property.
a. .35 − .14   b. .36 − .29
c. 2.05 − 1.27   d. .059 − .041

6. Perform the operations below by definition.
a. 2.34 − 1.56   b. 3.78 − .986
c. .057 − .009   d. .204 − .034

MULTIPLICATION OF DECIMAL FRACTIONS

We will consider the multiplication operation as we did in the last section by finding the result of our operation on fractions in the form a/b. If you recall multiplication of fractions, you will remember that the product of a/b and c/d is the fraction $(a \times c)/(b \times d)$. Consider how this will work if we wish to multiply $.2 \times .3$. We know that $.2 = \frac{2}{10}$ and that $.3 = \frac{3}{10}$. Thus, $.2 \times .3 = \frac{2}{10} \times \frac{3}{10} = \frac{6}{100}$, or .06. In order to develop an algorithm for multiplication of decimal fractions, let us observe that the product is formed by multiplying the two numbers as though they were whole numbers ($2 \times 3 = 6$), and the placement of the decimal is determined by the product of the implied denominators, namely, ($10 \times 10 = 10^2$); since the product of implied denominators is 10^2, we need two places for our answer.

Often students get confused at this point and write .60 rather than .06. Since .60 is $\frac{60}{100}$ or $\frac{6}{10}$ rather than $\frac{6}{100}$, we can see that there is an error in this line of reasoning.

Consider now some examples of the multiplication operation. Again, we will consider solutions by (1) translation, (2) properties, and (3) using the decimal notation without the powers of 10.

■ **G.3** **Example 1**
▲ **G.5**   Find the product of .25 and .05.

SOLUTION   1. Translating,

$$\frac{25}{100} \times \frac{5}{100} = \frac{125}{10^4}$$

Therefore,

$$.25 \times .05 = .0125$$

Notice that since the product was a *three*-digit product, a zero is inserted to the left of the three-digit product and to the right of the decimal point.

2. Using properties,

$$.25 = (2 \times 10^{-1}) + (5 \times 10^{-2})$$
$$.05 = 5 \times 10^{-2}$$

Therefore,

$$
\begin{aligned}
(.25) \times (.05) &= [(2 \times 10^{-1}) + (5 \times 10^{-2})] \\
&\quad \times (5 \times 10^{-2}) \\
&= [(2 \times 10^{-1}) \times (5 \times 10^{-2})] \\
&\quad + [(5 \times 10^{-2}) \times (5 \times 10^{-2})] \\
&= [(2 \times 5) \times (10^{-1} \times 10^{-2})] \\
&\quad + [(5 \times 5) \times (10^{-2} \times 10^{-2})] \\
&= [(10 \times 10^{-3}) + (25 \times 10^{-4})] \\
&= (1 \times 10^{-2}) + [(20 + 5) \times 10^{-4}] \\
&= (1 \times 10^{-2}) + [(20 \times 10^{-4}) \\
&\quad + (5 \times 10^{-4})] \\
&= (1 \times 10^{-2}) + [2 \times (10 \times 10^{-4}) \\
&\quad + (5 \times 10^{-4})] \\
&= (1 \times 10^{-2}) + (2 \times 10^{-3}) + (5 \times 10^{-4}) \\
&= .01 + .002 + .0005 \\
&= .0125
\end{aligned}
$$

3. Using the decimal notation,

$$
\begin{array}{r}
.25 \\
\times\ .05 \\
\hline
.0125
\end{array}
$$

■ **G.3** **Example 2**
▲ **G.5**   Find the product of 2.3 and 3.02.

    SOLUTION   1. Translating,

$$2\tfrac{3}{10} \times 3\tfrac{2}{100}$$

    Rewriting,

$$\frac{23}{10} \times \frac{302}{100} = \frac{6946}{10^3}$$

    Therefore,

$$2.3 \times 3.02 = 6.946$$

    2. Using properties,

$$2.3 = 2 + (3 \times 10^{-1})$$
$$3.02 = 3 + (2 \times 10^{-1})$$

    Therefore,

$$
\begin{aligned}
(2.3) \times (3.02) &= [2 + (3 \times 10^{-1})] \times [3 + (2 \times 10^{-2})] \\
&= \{[2 + (3 \times 10^{-1})] \times 3\} \\
&\quad + \{[2 + (3 \times 10^{-1})] \times (2 \times 10^{-2})\} \\
&= \{(2 \times 3) + (3 \times 10^{-1}) \times 3\} \\
&\quad + \{2 \times (2 \times 10^{-2}) + (3 \times 10^{-1}) \\
&\quad \times (2 \times 10^{-2})\} \\
&= \{6 + (9 \times 10^{-1})\} + \{(4 \times 10^{-2}) \\
&\quad + (3 \times 2) \times (10^{-1} \times 10^{-2})\} \\
&= \{6 + (9 \times 10^{-1})\} + \{(4 \times 10^{-2}) \\
&\quad + (6 \times 10^{-3})\} \\
&= 6 + .9 + .04 + .006 \\
&= 6.946
\end{aligned}
$$

    3. Using the decimal notation,

$$
\begin{array}{r}
2.3 \\
\times\ 3.02 \\
\hline
46 \\
690 \\
\hline
6.946
\end{array}
$$

    You will again notice that methods 1 and 2 are much more cumbersome than the method 3. However, the first two methods do explain the process to some extent, whereas the last method is simply an algorithm to be used. We will now state the algorithm for multiplying decimal fractions.

Algorithm
■ **G.5**   *Multiplication of decimal fractions is performed by finding the product of the factors in question as though they were whole numbers. The decimal*

point is placed in position by finding the sum of decimal places for each
■ *factor and then giving the product that number of decimal places.*

EXERCISES

1. Find the products below by translating to fractions.
    a. .23 × .06   b. 2.35 × .035
    c. .35 × .25   d. 23.6 × 25.5

2. How many places should be in the products below?
    a. .005 × .0302   b. .5 × .05
    c. 245.2 × 345   d. 25.08 × 23.001

3. Find the products below.
    a. .23 × .002   b. .29 × .3
    c. 3.23 × .024   d. .78 × .09
4. Find the products by using the algorithm.
    a. 2.34 × 3.67   b. .084 × .052
    c. .307 × .67   d. 3.502 × 2.003

5. Find the products below.
    a. 2.4 × 10   b. 3.24 × 100
    c. 6.52 × 1000   d. 23.245 × 10^3

6. What rule does Exercise 5 suggest?

7. Try to find the products below by using the rule you found from Exercise 6.
    a. 32.25 × 10^2 b. 102.234 × 10^3
    c. 35.2 × 10   d. 35.2 × 100

MULTIPLICATION BY POWERS OF 10

In the last exercise set, we asked that you develop a rule for multiplication by 10 and powers of 10. Let us consider several examples below and then consider a rule.

Examples

1. The product 10 × 2.7 is 27.0.
2. The product 100 × 2.7 is 270.0, or 10^2 × 2.7 = 270.0.
3. The product 1000 × 2.7 is 2700.0, or 10^3 × 2.7 = 2700.0.

Notice in each example above that the power of 10 related directly to the final product. When 2.7 was multiplied by the first power of 10, the product was 27.0; by the second power, the product was 270.0; and in the third case, the product was 2700.0. If we look at the products in relation to powers of 10, we can immediately see the rule involved.

Factor	Power of 10	Product
2.7	1	27.0
2.7	2	270.0
2.7	3	2700.0

Before stating a rule, we might ask what effect a negative power of 10 has on a product. Consider a few more examples.

Examples

1. Find the product 2.7×10^{-1}. (Remember that $10^{-1} = \frac{1}{10}$.)

$$2.7 \times 10^{-1} = .27$$

2. Find the product 2.7×10^{-2}. ($10^{-2} = \frac{1}{100}$, or .01.)

$$2.7 \times 10^{-2} = .027$$

3. Find the product 2.7×10^{-3}.

$$2.7 \times .001 = .0027$$

Now we can extend the table above to look like this.

Factor	Power of 10	Product
2.7	3	2700.0
2.7	2	270.0
2.7	1	27.0
2.7	0*	2.7
2.7	−1	0.27
2.7	−2	0.027
2.7	−3	0.0027

*(Remember, $10^0 = 1$)

■ **G.7**   From the examples above, you should be intuitively convinced of the following conclusions. In fact, you should be capable of demonstrating or proving these facts.

1. If we multiply any rational number by a positive power of 10, we move the decimal point to the left n places, where n represents the power of 10.
2. If we multiply a rational number by 10^0, we produce that rational number in the product, because 10^0 is an identity for the rational under multiplication.
3. If we multiply any rational number by a negative power of 10, we move The decimal point to the left n places, where n represents the power of 10.
4. Multiplying by 10 to a negative power has the same effect as dividing by 10 to the same positive power.

■

EXERCISES

■ **G.6**   1. Find the products below without performing the operation.
   a. 2.879×100 b. 293.09×1000
   c. $289.35 \times .001$ d. $2.3 \times .0001$

2. Translate the following to powers of 10.
  a. 10000   b. .01   c. .00001   d. 100000

3. Translate the following to decimal numbers.
  a. 10^3 b. 10^{-2} c. 10^6 d. 10^{-4}

■ **G.6** **4.** Find the products below without performing the operation.
  a. 23.67×10^{-2} b. 3.24×10^3
■ c. 45.98×10^4 d. 1.002×10^{-3}

DIVISION OF DECIMAL FRACTIONS

Before we consider division of decimal fractions, it might help to consider a division problem which involves whole number division. We know, for example, that $\frac{75}{25}$ is a number in the form a/b which we call a rational number. We also know that there is an operation implied by the / or the –, namely, division. Therefore, $\frac{75}{25}$ can be written

$$25\overline{\smash{\big)}75} \quad \begin{array}{r} 3 \\ \underline{75} \end{array}$$

We usually name $\frac{75}{25}$, 3. There are many such examples. Consider now the rational number $\frac{1000}{8}$. If we operate below, we find that $\frac{1000}{8}$ can be written

$$8\overline{\smash{\big)}1000} \quad \begin{array}{r} 125 \\ \underline{8} \\ 20 \\ \underline{16} \\ 40 \\ \underline{40} \end{array}$$

These examples cause very little problem for most adults and most children by the fifth grade. However, ask someone to divide 1 by 8 and you may find that they have difficulty. You may hear someone say that it cannot be done. We know that $1 \div 8$ can be written $\frac{1}{8}$. But consider the problem of actually dividing. We can rewrite $\frac{1}{8}$ as $\frac{1000}{8} \times \frac{1}{1000}$. This simply requires that we multiply $\frac{1}{8}$ by $\frac{1000}{1000}$, which is the identity for multiplication. Now, since

$$8\overline{\smash{\big)}1000}^{\,125} \quad \text{and} \quad \frac{125}{1000} = .125$$

we can see that

$$\frac{1000}{8} \times \frac{1}{1000} = 125 \times \frac{1}{1000}$$
$$= \frac{125}{1000}$$
$$= .125$$

■ **G.8** To shorten this process, we use the following technique:

$$8\,\overline{\smash{\big)}\,1}$$

translates to

$$
\begin{array}{r}
.125 \\
8\,\overline{\smash{\big)}\,1.0000} \\
\underline{8} \\
20 \\
\underline{16} \\
40 \\
\underline{40}
\end{array}
$$

■

Next let us consider the problem of dividing .1 by .08. Again, we can write this as .1/.08. We can also multiply the numerator and denominator by powers of 10 as long as we multiply both by the same power. Consider what happens to the problem in this situation.

■ **G.8**   If we multiply .1/.08 by $\frac{100}{100}$, we produce $\frac{10}{8}$, since $100 \times .1 = 10$ and
▲ **G.9** $100 \times .08 = 8$. Following our previous short method for division, we write this as $8\,\overline{\smash{\big)}\,10.00}$. Notice that the original problem, $.08\,\overline{\smash{\big)}\,.1}$, translates to $8\,\overline{\smash{\big)}\,10}$, or $8\,\overline{\smash{\big)}\,10.00}$.

We usually indicate this by using a mark to indicate that we have multiplied the divisor and dividend or denominator and numerator by 100, as shown below.

$$.08.\,\overline{\smash{\big)}\,.10.00}$$

Then we divide.

$$
\begin{array}{r}
1.25 \\
.08.\,\overline{\smash{\big)}\,.10.00} \\
\underline{8} \\
20 \\
\underline{16} \\
40 \\
\underline{40}
\end{array}
$$

■

We can summarize all this by stating that if a decimal appears in the divisor, we multiply the divisor and dividend by the power of 10 necessary to produce a whole number in the divisor. Then we divide as shown in the first example by adding as many zeros to the dividend as we need to
▲ complete the division.

In an earlier section, we discussed repeating decimals. We may produce a repeating decimal by dividing. For example, consider the fraction $\frac{3}{7}$. If we wish to write the decimal equivalent of $\frac{3}{7}$, we must divide as shown below.

$$\begin{array}{r} .428571 \\ 7\overline{\smash{\big)}3.000000} \\ \underline{28} \\ 20 \\ \underline{14} \\ 60 \\ \underline{56} \\ 40 \\ \underline{35} \\ 50 \\ \underline{49} \\ 10 \\ \underline{7} \\ 3 \end{array}$$

G.10   In the problem above there are six possible remainders which can result if the decimal is to repeat rather than end. Obviously, a remainder of 0 means that the decimal terminates at some point. The other remainders are 1, 2, 3, 4, 5, and 6, since a remainder of 7 is equivalent to a 0 remainder. Now, as soon as one of these six remainders repeats itself, then a whole sequence begins again. Notice that we stopped dividing when the 3 appeared, since the next remainder will be 2 and the sequence will begin again. Thus, $\frac{3}{7}$ can be written as .428571428571 . . . .

EXERCISES

1. Find the following quotients.
   a. $23\overline{\smash{\big)}25.3}$ b. $2.7\overline{\smash{\big)}5.94}$
   c. $12.5\overline{\smash{\big)}134.25}$ d. $13.6\overline{\smash{\big)}149.872}$

2. Without computing the quotients. find those examples which will have the same quotients, and explain why.
   a. $2.5\overline{\smash{\big)}2800}$ b. $25\overline{\smash{\big)}280}$ c. $25\overline{\smash{\big)}28}$
   d. $25\overline{\smash{\big)}280}$ e. $.025\overline{\smash{\big)}2800}$ f. $2.5\overline{\smash{\big)}2.8}$
   g. $.025\overline{\smash{\big)}.028}$ h. $25\overline{\smash{\big)}28000}$

3. Find the quotients below. and explain how you know that some of the quotients are repeating decimals.
   a. $14\overline{\smash{\big)}2.5}$ b. $.06\overline{\smash{\big)}.032}$ c. $.25\overline{\smash{\big)}.326}$
   d. $.02\overline{\smash{\big)}2.54}$ e. $.05\overline{\smash{\big)}16.8}$

CAPSULE V **MATHEMATICS LABORATORY**

From the results of learning theory reported by researchers such as Piaget, there is an indication that children can learn mathematics better through exploration of environment rather than from lecture techniques. As mentioned earlier, the work of Bruner, Piaget, Davis, and others leads us to believe that a discovery approach to learning might best facilitate learning of mathematics in the elementary school. As a result, many people have become interested in what is termed the *laboratory approach*. Such an approach concerns itself as much with how one learns concepts as it does with what concepts are learned.

We define a laboratory approach as a way of thinking rather than a place to work. This approach can be used in a classroom or in a corner of a classroom or in a special room in the school. It may utilize a great deal of specialized equipment, or it may require that children construct their own equipment. The main characteristics of the laboratory approach are:

1. Materials are provided by the school and/or invented by children along with the teacher.
2. The activities are open ended and the atmosphere is child centered.
3. Reference materials are available in the form of books, films, tapes, filmstrips, as well as the concrete materials.
4. Flexibility prevails, and no set schedule of activities is required.

This listing of characteristics is certainly not complete, but it covers the main concerns of people such as Lola May, Marguirite Kluttz, Lore Rasmussen, and others (see Suggested Readings). The research results of the validity of the laboratory approach are very mixed. Most of the studies cited in our bibliography have indicated no significant differences in learning through this method. We feel that with this method children will be more excited about mathematics, and ultimately this will lead to greater learning. Many people take issue with this point of view. Fitzgerald reported that the Gagne-Ausubel position indicates that the more efficient teacher-textbook approach to learning mathematics is better than a laboratory approach. This point of view is also directed toward the use of the discovery approach in general. The main issue is that of time and end product. It may be that one can produce more learning of mathematics by conventional textbook teaching than one can through laboratory and/or discovery approaches to mathematics. On the other hand, our position is that the enjoyment of learning is a goal which must be considered. Children in laboratory situations seem to be more interested, and if the excitement can be continued, the result will be a love for learning. It would take a long-term research project to uncover the result which we suggest, and none is available at this time. We would hope that some research group would consider an in-depth study of the laboratory approach over several years which measures both cognitive and affective results. For a recent report of research about laboratory settings in mathematics see Vance and Kieren's article in *The Arithmetic Teacher*.

SUGGESTED READINGS

Bartnik, Lawrence P. *Designing the Mathematics Classroom.* Washington, D.C.: National Council of Teachers of Mathematics, 1965.

Clarkson, David M. "A Mathematics Laboratory for Prospective Teachers," *The Arithmetic Teacher*, January, 1970, pp. 75–78.

Davidson, Patricia S. "An Annotated Bibliography of Suggested Manipulative Devices," *The Arithmetic Teacher*, October, 1968.

————— and Fair, Arlene. "A Mathematics Laboratory—Dream to Reality," *The Arithmetic Teacher*, February, 1970, pp. 105–110.

Fitzgerald, William M. "A Mathematics Laboratory for Prospective Elementary School Teachers," *The Arithmetic Teacher*, October, 1958.

Friauf, Frances C. and Hobbs, Peggy B. *Games for Learning Mathematics*, Project Mid-Tenn, 1970.

Hamilton, E. W. "Manipulative Devices," *The Arithmetic Teacher*, October, 1966 pp. 461–465.

Kidd, K. P., Myers, S. S., and Cilley, D. M. *The Laboratory Approach to Mathematics.* Chicago: Science Research Associates, 1970.

Kluttz, Marguirite. "The Mathematics Laboratory—A Meaningful Approach to Mathematics Instruction," *The Mathematics Teacher*, March, 1963.

May, Lola J. "Individualizing Instruction in a Learning Laboratory Setting," *The Arithmetic Teacher*, November, 1968, pp. 609–612.

—————. "Learning Laboratories in Elementary Schools in Winnetka," *The Arithmetic Teacher*, October, 1968, pp. 501–505.

—————. "Mathematics in the Elementary School," *Illinois Journal of Education*, January, 1967, pp. 43–45.

Rasmussen, Lore. "Creating a Mathematics Laboratory Environment in the Elementary School," *The School District of Philadelphia Journal*, 1968.

————— and Rasmussen, Don. "The Miquan Mathematics Program," *The Arithmetic Teacher* April, 1962, pp. 180–187.

Reys, R. E. "Considerations for Teachers Using Manipulative Materials," *The Arithmetic Teacher*, December, 1971, pp. 551–558.

Schaefer, Anne and Mauther, Albert H. "Problem Solving with Enthusiasm—The Mathematics Laboratory," *The Arithmetic Teacher*, January, 1970, pp. 7–14.

Swart, William L. "A Laboratory Plan for Teaching Measurement in Grades 1–8," *The Arithmetic Teacher*, January, 1967, pp. 652–653.

Vance, J. H. and Kieren, T. E. "Laboratory Settings in Mathematics: What Does Research Say to the Teacher?" *The Arithmetic Teacher*, December, 1971, pp. 585–589.

Williams, Catherine M. "Portable Mathematics Laboratory for In-Service Teacher Education," *Mathematics in Elementary Education*, edited by Nicholas J. Vigilante. London: Macmillan, 1969, pp. 300–304.

CHAPTER 9 **REAL NUMBERS**

SECTION A **FUNDAMENTAL PROPERTIES OF OPERATIONS ON THE REALS**

BEHAVIORAL OBJECTIVES

A.1 Be able to describe the correspondence between the points on a line and the real numbers.

A.2 Be able to define a real and an irrational number.

A.3 Be able to show that irrational numbers exist.

A.4 Be able to approximate irrational numbers by rational numbers.

A.5 Be able to multiply two reals raised to the same power.

A.6 Be able to multiply equal numbers raised to powers.

A.7 Be able to raise real numbers to a power.

A.8 Be able to operate with and explain the meaning of rational powers of real numbers.

A.9 Be able to state and give examples of the field properties for the real numbers under addition and multiplication.

INTRODUCTION TO REAL NUMBERS

■ **A.1**   The basic assumption we are going to be working under is that there is a 1–1 correspondence between the set of real numbers and the points on a line. This line is called the *real number line*, or simply the number line. Up to now, we have considered sets of numbers (rationals, integers, whole, natural) that correspond to points that are proper subsets of the real number line.

If we were to remove all points that correspond to rational numbers from the number line, we would find that an infinite number of points would still remain. This set is called the *irrational points*. The numbers that correspond to this set are called the *irrational numbers*. We will show in this chapter that the set of irrational numbers is not empty. That is, we will show that there are really numbers corresponding to points on
■ the number line that are not rational.

Definitions
■ **A.2**   *A real number is a number that corresponds to a point on the number line.*

■   *An irrational number is a real number that is not rational.*

Now, defining an irrational number does not guarantee the existence of irrational numbers any more than defining a pink-eared elephant guarantees the existence of such an animal. The reason that we introduce irrational numbers is to produce the set of real numbers. It turns out that every rational number is a real number (i.e., the rationals are a subset of the reals), but not every real number is a rational number. However, the union of the rational numbers and irrational numbers is the set of real numbers.

Let us now show that there are really irrational numbers. To do this we must show that there is a point on the number line that does not correspond to some rational number. This will show that the set of irrational numbers is not an empty set.

First, let us describe the point that we wish to consider. Place a right triangle whose legs have lengths 1 and 2 on the number line with the right angle at the point corresponding to 1. See Figure 9.1. Now, using

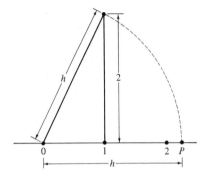

Figure 9.1

the hypotenuse of this right triangle as a radius, draw a circle that intersects the number line at a point P. According to our basic assumption, there is a real number that corresponds to P. We will show that this number is not rational. The length h of the hypotenuse can be determined using the theorem of Pythagoras; that is, the square of the length of the hypothenuse of a right triangle equals the sum of the squares of the lengths of the legs (see Chapter 4):

$$h^2 = 1^2 + 2^2$$

or

$$h^2 = 1 + 4$$
$$h^2 = 5$$

So we say that $h = \sqrt{5}$, that is, $h \cdot h = 5$. We use the symbol $\sqrt{a}$ to indicate that number which when squared is a. Thus, $\sqrt{a} \cdot \sqrt{a} = a$.

Before showing that h is not a rational number, we need to prove a lemma. A *lemma* is a statement that is needed to prove some other theorem, in this case to prove that h is irrational.

Lemma
    If a^2 is a multiple of 5, then a is a multiple of 5.

PROOF   If a^2 is a multiple of 5, then $a \times a = 5 \times n$ for n some natural number. The fundamental theorem of arithmetic (Chapter 4) tells us that a has a unique factorization of primes $a = p_1 \times p_2 \times \cdots \times p_k$. Thus, $(p_1 \times p_2 \times \cdots \times p_k) \times (p_1 \times p_2 \times \cdots \times p_k) = 5 \times n$. This tells us that $p_1 \times p_2 \times \cdots \times p_k \times p_1 \times p_2 \times \cdots \times p_k$ is a prime factorization of $5 \times n$, which means that some combination of the p's must multiply to give 5 and the others must multiply to give n. Since 5 is prime, the only combination of p's that multiply to 5 must consist of 5 itself. Thus, one of the prime factors of a is 5. This means that a is a multiple of 5.

Before proceeding with the proof that $\sqrt{5}$ or h is not a rational number, let us describe how the proof will go.

We want to show that there is no rational number (fraction) corresponding to the point p in Figure 11.2. To do this we will assume that there *is* a rational number corresponding to p. This rational number can be represented by the basic pair $R[\![a, b]\!]$, or a/b. The proof that follows will show that *there is not a rational in lowest terms* that corresponds to p. If there is no rational in lowest terms that does the job, then since every rational can be written as a basic pair, there is no rational number corresponding to p.

Now we are ready to show that $\sqrt{5}$, or h, is not a rational number.

Theorem
5 is irrational.

PROOF   Suppose that $\sqrt{5}$ is rational. Then $\sqrt{5} = R[\![a, b]\!]$, where a and b is the basic pair; that is, a and b have no common divisor. Then $\sqrt{5} = a/b$ and $5 = a^2/b^2$, so $5b^2 = a^2$. Thus, a^2 is a multiple of 5, and by our lemma a must be a multiple of 5. That is, $a = 5k$ for some integer k. If so, then $a^2 = 5 \cdot 5k^2$, and since

$$5b^2 = a^2$$
$$5b^2 = 5 \cdot 5k^2$$

or

$$b^2 = 5k^2$$

Therefore, b^2 is a multiple of 5. Again, using our lemma, b must be a multiple of 5. Thus, we have shown that both a and b are multiples of 5. But this contradicts the statement that a and b have no common divisor. Thus, if $\sqrt{5}$ is rational, we reach a contradiction. Hence, $\sqrt{5}$ cannot be rational.

■ A.3   The above theorem shows that the set of irrational numbers is not empty. There are in fact an infinite number of irrationals. We list here a few irrationals with which you may already be familiar: $\sqrt{2}, \sqrt{3}, \sqrt{5}, \sqrt{7}, \sqrt{8}$ are irrational. Notice that each of these are square roots of natural numbers which are not squares. We also list numbers such as $\pi, \sqrt{\pi}, 4\pi$, and π^2
■ as examples of irrational numbers.

EXERCISES

1. Construct the point on the number line that corresponds to the following numbers.
   a. $\sqrt{2}$ b. $\sqrt{10}$ c. $\sqrt{6}$ d. $\sqrt{3}$

2. Prove that $\sqrt{2}$ is not rational.

3. Show that there are an infinite number of rational numbers between 2 and 3.

4. Does the set of irrational numbers contain an additive identity?

5. Is the set of irrational numbers closed under multiplication?

APPROXIMATING IRRATIONAL NUMBERS

Real numbers have been defined as a set of numbers which correspond to points on a number line, and irrationals have been defined as the non-rational numbers in the set of reals. We now consider a method for approximating an irrational with rationals using roots of rational numbers.

In the previous section, we showed that $\sqrt{5}$ was not a rational number. We now will try to find a rational number which is very close to an irrational number. That is, we will try to approximate an irrational by a rational.

Example

■ A.4   Approximate $\sqrt{2}$ by a rational number.

SOLUTION   It is obvious that $1 < \sqrt{2} < 2$, since $1^2 = 1$, $(\sqrt{2})^2 = 2$, and $2^2 = 4$; that is, $1^2 < (\sqrt{2})^2 < 2^2$. How does $\sqrt{2}$ compare with 1.5? $(\sqrt{2})^2 = 2$, while $(1.5)^2 = 2.25$. Thus $\sqrt{2} < 1.5$.

Notice that the interval from 1 to 2 is 1. Since $(1.5)^2$ is larger than 2, we try the interval between 1.4 and 1.5. $(1.4)^2$ is 1.96, which tells us that $\sqrt{2}$ is greater than 1.4, thus $1.4 < \sqrt{2} < 1.5$.

The square of 1.41 and 1.42 are 1.9881 and 2.0164, respectively. Now $\sqrt{2}$ is greater than 1.41 and less than 1.42. The interval is .01. Therefore, $1.41 < \sqrt{2} < 1.42$.

Our next approximation should fall in an interval of .001. Consider this interval as follows: Since 1.414^2 is 1.999396 and 1.415^2 is 2.002225, we have come even closer to the value we seek. We can state this as follows:

$1.414 < \sqrt{2} < 1.415$.

Now consider the sequence used to approximate the $\sqrt{2}$.

First approximation	$1 < \sqrt{2} < 2$	since $1^2 < (\sqrt{2})^2 < 2^2$
Second approximation	$1.4 < \sqrt{2} < 1.5$	since $(1.4)^2 < (\sqrt{2})^2 < (1.5)^2$
Third approximation	$1.41 < \sqrt{2} < 1.42$	since $(1.41)^2 < (\sqrt{2})^2 < (1.42)^2$
Fourth approximation	$1.414 < \sqrt{2} < 1.415$	since $(1.414)^2 < (\sqrt{2})^2 < (1.415)^2$

In using the above method, we can approximate an irrational number as closely as we wish. Numbers such as $\sqrt{2}, \sqrt{3}$, and $\sqrt{7}$ can all be approximated in this way. Since each positive number on the real line has an inverse which is negative, we can also find a series of intervals to approximate $-\sqrt{2}, -\sqrt{3}, -\sqrt{7}$, etc.

Many times we wish to work with rationals and irrationals without finding their decimal approximations. We will assume (the proof is beyond the scope of this book) that the real numbers are a field under $+$ and $\times$. This assumption is not too unreasonable, because a subset of the reals, namely the rationals, forms a field under $+$ and $\times$ and permits us to operate with the real numbers easily.

The roots of some rational numbers are irrational, while the roots of other rationals are rational. Consider the rational roots below.

$$\sqrt{1/4} = \tfrac{1}{2}, \quad \text{because} \quad (\tfrac{1}{2})^2 = \tfrac{1}{4}$$
$$\sqrt{4/9} = \tfrac{2}{3}, \quad \text{because} \quad (\tfrac{2}{3})^2 = \tfrac{4}{9}$$
$$\sqrt{25} = 5, \quad \text{because} \quad (5)^2 = 25$$

We would like to be able to add and multiply such numbers; in the next section we will state some simple rules for operating with roots.

EXERCISES

1. Approximate the following roots to two decimal places.
   a. $\sqrt{3}$ b. $\sqrt{2/2}$ c. $\sqrt{5}$

2. Which of the following numbers are rational? Why?
   a. $\sqrt{36}$ b. $\sqrt{10}$ c. $\sqrt{5/25}$ d. $\sqrt{2/8}$

3. Find the sums of the following numbers.
   a. $\sqrt{4} + \sqrt{9}$ b. $\sqrt{25} + \sqrt{36}$ c. $\sqrt{1/4} + \sqrt{9/36}$

4. Find the products of the following pairs.
   a. $\sqrt{4} \times \sqrt{4}$ b. $\sqrt{36} \times \sqrt{25}$ c. $\sqrt{1/4} \times \sqrt{4}$

5. Compare your answers from Exercise 4 to the products of the following rational numbers. Explain the relationship.
   a. 4×4 b. 36×25 c. $\tfrac{1}{4} \times 4$

OPERATIONS ON REAL NUMBERS

We have already discussed the operations on the rational subset of the reals. The next question is, How do we operate on the irrational numbers? To discuss this, we will consider only a subset of the irrationals, namely, the roots of rational numbers, which are not necessarily rational.

We will first consider how to operate with powers of rational numbers. To add 2^2 and 3^2, we must first square the numbers and then find the sums.

■ **A.5** $2^2 = 4$ and $3^2 = 9$; therefore $2^2 + 3^2 = 13$. When we multiply 2^2 and 3^2 we see that $2^2 \times 3^2 = 4 \times 9 = 36 = (6)^2 = (2 \times 3)^2$. The fact that $2^2 \times 3^2 = (2 \times 3)^2$ is no coincidence. It is only an example of the general rule:

■ $a^n \times b^n = (a \times b)^n$ *for a and b* $\in R$ *and* $n \in W$.

A.6 Consider next what happens when we multiply equal numbers that are raised to powers. What is the product of 5^2 and 5^3? We know from previous work that $5^2 \times 5^3 = 25 \times 125 = 5^5$. We now state this property as a rule.

$a^n \times a^m = (a)^{n+m}$ *for* $a \in R$, *n, and* $m \in W$.

The rule stated above simply says that to multiply equal numbers that are raised to powers, we raise the number to the sum of the exponents.

■ Consider some examples of each rule.

Example 1

A.5 Find the product of 5^2 and 3^2.

SOLUTION $5^2 \times 3^2 = 15^2$, or 225.

Example 2

Find the product of $(\frac{1}{4})^3 \times (2)^3$.

SOLUTION $(\frac{1}{4})^3 \times (2)^2 = (\frac{1}{2})^3$, or $\frac{1}{8}$. Notice that $(\frac{1}{4})^3 = \frac{1}{64}$ and $2^3 = 8$,
■   and that $\frac{1}{64} \times 8 = \frac{1}{8}$.

Example 3

A.6 Find the product of 3^2 and 3^1.

SOLUTION $3^2 \times 3^1 = 3^3$, or 27.

Example 4

Find the product of $(\frac{1}{3})^2$ and $(\frac{1}{3})^3$.

■ SOLUTION $(\frac{1}{3})^2 \times (\frac{1}{3})^3 = (\frac{1}{3})^5$, or $\frac{1}{243}$.

A.7 We have already written some rules for multiplying numbers raised to a power. Consider one more rule here. Since $(2^3)^2$ means $2^3 \times 2^3$, or 2^6, and since $(2^2)^3$ means $2^2 \times 2^2 \times 2^2$, or 2^6, you can see that raising a number of one power to another power may involve multiplication. We will simply define this operation as follows:

$(a^n)^m = a^{nm}$, $a \in R$, *n, and* $m \in W$.

It is reasonable for n a positive integer that a^n means the number composed of n factors of a. If we raise that number, a^n, to a power m, that is, $(a^n)^m$, we are naming the number with m factors, a^n. Thus, for n and m in I, $(a^n)^m$ means m factors of a^n. Consider how this looks.

$$(a^n)^m = \underbrace{a^n \times a^n \times \cdots \times a^n}_{m \text{ factors}}$$

and

$$a^n = \underbrace{a \times a \times a \times \cdots \times a}_{n \text{ factors}}$$

Therefore, $(a^n)^m$ is a number composed of $n \times m$ factors of a; that is,

$$(a^n)^m = \underbrace{\underbrace{a \times \cdots \times a}_{n} \times \cdots \times \underbrace{a \times \cdots \times a}_{n}}_{m}$$

$$\blacksquare \qquad = a^{mn}$$

Since this rule holds for exponents from the set W, we might expect that it would also be true for fractions or rational exponents of a number. The question then is, What does a fractional exponent mean? Consider what happens when we raise $\sqrt{2}$ to the second power, or $(\sqrt{2})^2$. We know that the result is 2.

■ A.8   Let us now try to determine to what power 2 should be raised to be equivalent to $\sqrt{2}$. That is, we want to find a p such that $2^p = \sqrt{2}$. Since $(\sqrt{2})^2 = 2$, $(2^p)^2 = 2$. Using the rule above, we have $2^{2p} = 2$. If $p = \frac{1}{2}$, then $2^{2 \times 1/2} = 2^1 = 2$. Now this means that $2^{1/2}$ is another name for $\sqrt{2}$. Consider some similar examples.

Example 1
Write $\sqrt{20}$ in terms of exponents.

SOLUTION   Let $\sqrt{20} = 20^p$. Then $(\sqrt{20})^2 = (20^p)^2$, or $20 = 20^{2p}$. Let $p = \frac{1}{2}$; then $\sqrt{20} = 20^{1/2}$.

Example 2
Write $\sqrt[3]{8}$ as 8 to a power.

SOLUTION   Let $\sqrt[3]{8} = 8^p$. Since $(\sqrt[3]{8})^3 = 8$, $(8^p)^3 = 8$, or $8^{3p} = 8$. Thus, $p = \frac{1}{3}$, because $8^{3 \times 1/3} = 8^1$.

Example 3
Write $\sqrt[4]{16}$ as 16 to a power.

SOLUTION $\sqrt[4]{16} = 16^p$. Since $(\sqrt[4]{16})^4 = 16$, $(16^p)^4 = 16$, but $(16^p)^4 = 16^{4p}$, so if $p = \frac{1}{4}$, $16^{4 \times 1/4} = 16$, or $\sqrt[4]{16} = 16^{1/4}$.

In general, $\sqrt[n]{a} = a^{1/n}$.
If we combine fractional powers with powers which are integers, we produce an interesting result. Consider the following examples.

Example 1
■ A.8   Write $\sqrt[3]{8}$ as 2 to a power.

SOLUTION   Since $8 = 2^3$, $\sqrt[3]{8} = \sqrt[3]{(2^3)} = (2^3)^{1/3}$. Now $(2^3)^{1/3} = 2^{3/3}$, or simply 2^1.

Example 2
Write $\sqrt[4]{16}$ as 2 to a power.

SOLUTION   Since 16 is 2^4, $\sqrt[4]{16} = \sqrt[4]{(2^4)} = (2^4)^{1/4}$. Thus, $\sqrt[4]{16} = (2^4)^{1/4}$ $= 2^{4/4}$, or 2^1.

We can consider the meaning of fractional powers in general, namely, any number $a^{m/n}$. Since we know that the fraction m/n can be considered the product of m with $1/n$, we can call the numerator the power and the denominator the root. In this way, $a^{3/4}$ would mean the fourth root of a^3, or the third power of $a^{1/4}$. If we use the rules which we have developed for powers, we are in a position to translate fractional powers to roots and roots to fractional powers, and multiply numbers whose bases are the same and whose powers are fractions as well as whole numbers. We should also note here that fractional powers comply with the properties of fractions. Thus, $y^{3/4} = y^{6/8} = y^{12/16}$, and $y^{3/4} \times y^{5/6} = y^{3/4 + 5/6}$, or $y^{19/12}$.

Example 1
A.8   Write $3^{3/4}$ as a root.

SOLUTION $3^{3/4} = \sqrt[4]{3^3}$, or $\sqrt[4]{27}$.

Example 2
Write $3^{4/3}$ as a root.

SOLUTION $3^{4/3} = \sqrt[3]{3^4}$, or $\sqrt[3]{81}$.

Example 3
Find the product of $3^{3/4}$ and $3^{4/3}$.

SOLUTION $3^{3/4} \times 3^{4/3} = 3^{(3/4)+(4/3)} = 3^{25/12} = 3^{2+1/12}$, which is $3^2 \times 3^{1/12}$, or $9\sqrt[12]{3}$. Thus, $3^{3/4} \times 3^{4/3} = 9\sqrt[12]{3}$.

EXERCISES

1. Find the products below by writing your answers in the form b^n.
a. $3^2 \times 3^5$ b. $2^3 \times 2^2$ c. $3^2 \times 12^2$
d. $4^3 \times (\frac{1}{2})^3$ e. $a^x \times a^y$ f. $a^x \times b^x$

2. Write the following quotients in simplest form, using b^n to report your result.
a. $3^4/3^2$ b. $6^3/6^5$ c. $3^4/2^4$
d. $8^6/4^6$ e. a^b/a^x f. a^b/x^b

3. Find the products and quotients indicated below.
a. $(2x^3y)(4x^2)$ b. $(9a^2b^3)(8ab^3)$
c. $(5a)^2(5a)^3$ d. $(3^3a^4b^2)/(3^2a^2b^4)$
e. $(6^3xy^3)/(6^5y^2z)$ f. $(5^2ab)^3/(5a^2b)^3$

4. Translate the following roots to power notation.
a. $\sqrt{2}$ b. $\sqrt[3]{a^2}$ c. $\sqrt[4]{a^2b}$ d. $\sqrt{a^3b^2}$

5. Translate the following terms from power notation to radical form.
a. $6^{1/2}a^{1/2}$ b. $4^{1/2}a^{1/2}b^{1/2}$ c. $12^{2/3}a^{2/3}$
d. $3^{3/4}a^{3/4}b^{3/4}$ e. $6^{1/4}a^{1/2}$ f. $a^{2/3}b^{3/4}$

PROPERTIES OF REAL NUMBERS

Since we are looking at the real numbers in an intuitive fashion, it is not possible for us to develop the properties of this set of numbers deductively. Therefore, we will postulate and illustrate the properties of this set without offering rigorous proofs. We can concisely state all the properties of the real numbers by postulating that the set of real numbers forms a field under the operations of addition and multiplication. Remember that subtraction and division can be defined in terms of addition and multiplication. We will now state the properties of the field of real numbers.

■ **A.9** *The Real Numbers Are Closed Under Addition and Multiplication. For*
■ *a, b ∈ real, (a + b) is real and (a × b) is real.*

Let us now decide which operations on the reals are commutative. We find real numbers to be commutative under both multiplication and addition. However, a few examples should easily convince us that subtraction is not commutative and neither is division. Remember that some subsets of the reals are not commutative under subtraction or division.

■ **A.9** *The Real Numbers Are Commutative Under the Operations of Multiplica-*
■ *tion and Addition. For all a and b ∈ real, a + b = b + a and ab = ba.*

Examples

1. $2 + \sqrt{2} = \sqrt{2} + 2$
2. $\sqrt{2} \times \sqrt{8} = \sqrt{8} \times \sqrt{2}$

The associative property is one which we have seen as a property of addition and multiplication on N, W, I, and R. We would then expect that since it was not a property of subtraction and division on these sets, it could not be a property of subtraction and division for the set of real numbers. You should be able to supply counter-examples to show that subtraction and division of real numbers are not associative. We will again simply state this property for addition and multiplication.

■ **A.9** *The Real Numbers Are Associative Under the Operations of Multiplica-*
tion and Addition. For all a, b, c ∈ real, (a + b) + c = a + (b + c) and (ab)c
= a(bc).

Examples

1. $(\sqrt{2}+\sqrt{2}) + 2\sqrt{2} = \sqrt{2} + (\sqrt{2} + 2\sqrt{2})$
2. $(\sqrt{3} \times \sqrt{3}) \times \sqrt{9} = \sqrt{3} \times (\sqrt{3} \times \sqrt{9})$

The real numbers have identities for addition and multiplication, which are, respectively, 0 and 1.

Zero Is the Identity Element in the Reals Under Addition: For all a ∈ real,
a + 0 = a and 0 + a = a.

1 Is the Identity Element in the Reals Under Multiplication: For all $a \in$
real, $a \times 1 = a$ *and* $1 \times a = a.$

We should note

The Zero Property of Multiplication: For all $a \in real,$ $a \times 0 = 0$ *and*
■ $0 \times a = 0.$

In earlier chapters we discussed the concept of an inverse. The first
inverse to appear was the additive inverse in the set *I*. Later in the set of
rationals we introduced the multiplicative inverse. Our definition of inverse required that the set have an identity under the operation being
discussed. This means that we can consider inverses for addition and multiplication only, because the operations of subtraction and division have
no identities. Again, we will state without proof that both multiplication
and addition have inverses.

■ **A.9** *The Set of Real Numbers Has Inverses for each Element Under Addition
and for each Non-Zero Element Under Multiplication. For all* $a \in real,$ *the
additive inverse of a is* $-a$ *such that* $a + -a = 0.$ *For all* $a \in real,$ *the multiplicative inverse of a is* $1/a,$ *that is,* $a \cdot 1/a = 1, a \neq 0.$

The last set of properties to be discussed under the real numbers is the
distributive properties. These properties always refer to two operations
and in a sense connect the two operations.

*The Distributive Properties of Multiplication over Addition and Multipli-
cation over Subtraction of Real Numbers: For all a, b, c* $\in real,$ $a \times (b + c)$
$= ab + ac$ *and* $(b + c) \times a = ba + ca.$ *For all a, b, c real,* $a \times (b - c)$
■ $= ab - ac$ *and* $(b - c) \times a = ba - ca.$

Since the set of real numbers forms a commutative group under addition
and multiplication and has a distributive property of multiplication over
addition, the real numbers form a field under these operations. This
structure is one of the most important structures in elementary mathematics. A foundation in this area should be of great help to those who
instruct the young.

EXERCISES

1. Give an example to show that the set of reals is not commutative under subtraction.
(*Hint:* Look at whole numbers, a subset of the reals.)

2. Give an example to show that the set of reals is not commutative under division.

3. Show by example that subtraction in the set of reals is not associative.

4. Show by example that division in the set of reals is not associative.

5. Give an example of the distributive property of multiplication over addition.

6. Give an example of the distributive property of multiplication over subtraction.

SECTION B **SOLUTIONS AND SOLUTION SETS**

BEHAVIORAL OBJECTIVES

B.1 Be able to decide whether a statement is or is not a mathematical sentence.

B.2 Be able to decide when a sentence is true.

B.3 Be able to define the solution set of an open sentence.

B.4 Be able to graph solution sets of open sentences.

B.5 Be able to find the solution set of an open sentence involving equality by using the field properties of the real numbers.

B.6 Be able to find and graph the solution sets of an open sentence involving inequalities by using the field properties and cancellation laws for inequalities.

B.7 Be able to find the solution sets of open sentences involving "and" and "or."

B.8 Be able to describe a given set using open sentences in several ways.

B.9 Be able to use graphs to find the union and intersection of given sets.

MATHEMATICAL SENTENCES AND SOLUTION SETS

A mathematical sentence is simply a sentence involving mathematical objects and relations. Sometimes the sentence may be true, for example, "2 is an integer" or "$2 = 6 - 4$." Sometimes the sentence may be false, such as "$4 = 2$" or "$\sqrt{2}$ is a rational number." Mathematical sentences like those above that have a truth value, that is, true or false, are often called "statements."

■ **B.2**   Some mathematical sentences are not statements, for instance, "x is a rational number." In this sentence, x is a place holder for a number. If we replace x by $1/2$, then the sentence is true. But if x is replaced by $\sqrt{2}$, the sentence is false. This type of sentence, where the truth or falsity of the sentence depends upon the object that replaces the place holder, is called an *open sentence*. For example, the sentence "$\pi + x = \pi$" is an open sentence. If x is replaced by 2, the sentence is false. If x is replaced by 0, the sentence is true. Let us look at some more examples of mathematical sentences.

Example 1

▲ **B.1**   Is "$2 + 3$" a mathematical sentence?

SOLUTION   Since "$2 + 3$" is not even a sentence, it is not a mathematical sentence. On the other hand, $2 + 3 = 5$ is a true mathematical sentence, while $2 + 3 = \pi$ is a false mathematical sentence.

Example 2

Is "$3 \times 2 + 5 < 10$" a mathematical sentence?

SOLUTION Since $3 \times 2 + 5 = 11$, the statement $3 \times 2 + 5 < 10$ means that $11 < 10$, which is false. Therefore, this is a mathematical sentence, but it is false.

Example 3

Is "$3 \times 2 + 5 \neq 10$" a true mathematical sentence?

SOLUTION Since $3 \times 2 + 5 = 11$ and $11 \neq 10$, $3 \times 2 + 5 \neq 10$ is a true mathematical sentence.

Example 4

Is "$2p + 3 < 4$" a mathematical sentence?

SOLUTION Since $2p + 3 < 4$ is a false sentence for $p = 1$ but a true sentence for $p = 0$, "$2p + 3 < 4$" is an open sentence.

When we replace the variable in an open sentence by a number such that the open sentence becomes true, we say that the number is a *solution* of the open sentence.

Definition

B.3 *The set of all solutions to a given open sentence is called the solution set, or truth set, of the sentence.*

Consider some examples of open sentences and their truth sets. Notice that a particular universe is defined for the sentences in question. The universe is extremely important, because solutions to some sentences are found in a universe which may be beyond the student in the elementary school or secondary school. When this happens we say that the solution set is empty. The universe is N for all the examples below.

Example 1

Find the solution set for the sentence $\Box + 6 = 12$.

SOLUTION Since we know that $6 + 6 = 12$, the solution set $S = \{6\}$. Notice that 6 is in N and makes the sentence true.

Example 2

Find the solution set for the sentence $\Box + 6 < 12$.

SOLUTION Again, since we know that $6 + 6 = 12$, it is the case that any member of N which is less than 6 will make the sentence true. Therefore, the solution set $S = \{5, 4, 3, 2, 1\}$. Notice that 0 is not in N. We could simply write the set $S = \{x \mid x \in N \text{ and } x < 6\}$. Notice that *and* means x has both properties described.

Example 3

Find the solution set for the sentence $\Box + 6 \geq 12$

SOLUTION   The solution set in this case is infinite and includes 6, since the sentence involves the relation $\geq$. Thus $S = \{6, 7, 8, 9, \ldots\}$, or $S = \{x | x \in N$ and $x \geq 6\}$.

You will notice in the preceding examples that the solution sets were not empty and that there were three kinds of sets used as solutions, namely, a set containing one element, a set which was finite with more than one element, and an infinite set. This indicates that solution sets can contain a finite or an infinite number of elements. Next let us consider some sentences whose solution set is empty or infinite. Notice that we are using p to indicate the variable, or place holder, in these sentences. In the early grades, $\square$ or frames are used as place holders in open sentences. In the examples below $U = N$.

Example 1
Find the solution for $p - 6 \geq 4$.

SOLUTION   Since we know that $10 - 6 \geq 4$, $S = \{10, 11, 12, 13, \ldots\}$.

Example 2
Find the solution for $6p = 4$.

SOLUTION   Since $6 \times \frac{2}{3} = 4$, $S = \{\frac{2}{3}\}$. If the universe is N, then $S = \varnothing$, since $\frac{2}{3} \notin N$.

Example 3
Find the solution to the sentence $p + 6 \leq p + 7$.

SOLUTION   Since $p + 6$ is always less than $p + 7$, the solution set is the set of all numbers in the universe.

So far we have considered these sentences on an intuitive level only. We did not apply prior definitions, proofs, or properties to the sentences to solve them. As sentences become more complicated, it is necessary to apply principles to them before a solution set can be found. Before we consider this manner of solving sentences, we will look at some diagrams of sentences and consider ways of solving sentences informally. One way of solving sentences is to use graphs or diagrams. We usually employ the number line for this purpose. Consider such solutions below.

Example 1
■ B.4   Find the solution set on the number line for the sentence $x + 5 = 10$. Let $U = $ reals.

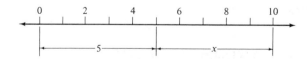

SOLUTION   Notice that the number of units in which we describe x is 5. We state this solution in several ways. We can write $S = \{5\}$ or $S = \{x \mid x = 5\}$, or we state it on a number line by placing a *closed* point at 5, as shown below.

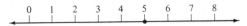

Example 2

Find the solution set on the number line for the sentence $x + 3 < 7$. Let $U = N$.

SOLUTION   Notice that x is 4 units in length. This means that the solution is $S = \{x \mid x \in N \text{ and } x < 4\}$, or $S = \{3, 2, 1\}$. On the number line this appears as

Notice that the points at 1, 2, and 3 are closed. The spaces between 1 and 2 and between 2 and 3 are left to indicate that the solution is in the set N only.

EXERCISES

1. Find and tabulate solution sets for the following sentences. Let $U = W$.
   a. $x + 3 = 6$ b. $x - 2 = 5$ c. $x \cdot 6 = 24$
   d. $2 \cdot x = 12$ e. $x/2 = 3$

2. Find the solution sets and use set builder notation for the following. Let $U = W$.
   a. $2 \cdot x < 10$ b. $x + 5 \geq 12$ c. $x - 6 \leq 5$
   d. $x + 1 < 0$ e. $2x - 2 < 6$

3. Find solution sets for the sentences below by using the number line. Let $U = N$.
   a. $2x = 6$ b. $2x < 6$ c. $x + 5 < 6$ d. $x - 3 > 6$

4. If $U = $ reals, find the solution set of the following.
   a. $\sqrt{3}x = \sqrt{27}$ b. $x - 3\sqrt{5} = \sqrt{5}$ c. $3x - 5 = 13$

FINDING SOLUTION SETS TO OPEN SENTENCES OF EQUALITY

In the previous section we discussed open sentences and their solution sets. Solution sets were tabulated, reported by use of set builder notation, or shown on the number line. We did not indicate how the solution sets were found except by intuition. The problem with finding solutions by intuition is that often our intuition fails us. In this section we will consider some of the properties that allow us to find a solution set to open sentences.

We will first illustrate the use of additive and multiplicative inverses to find solutions of open sentences. In the examples below, $U = I$.

Example 1

■ B.5   Find the solution set for the sentence $p + 3 = 5$.

SOLUTION $p + 3 = 5$
$$p + 3 + (-3) = 5 + (-3)$$
$$p + 0 = 5 + (-3)$$
$$p = 2$$
The solution set is $\{2\}$.

Example 2

Find the solution set for the sentence $p + 5 = 3$.

SOLUTION $p + 5 = 3$
$$p + 5 + (-5) = 3 + (-5)$$
$$p + 0 = 3 + (-5)$$
$$p = -2$$
■   The solution set is $\{-2\}$.

It should be obvious that in the second example if the universe did not contain the negative integers, then the solution set would have been empty. That is, if the universe for Example 2 were N, then $S = \emptyset$, since $-2 \notin N$.

We will illustrate the use of multiplicative inverses to find solutions of open sentences. In the examples that follow, $U = $ real numbers.

Example 1

■ B.5   Find the solution set for the sentence $3p = -9$.

SOLUTION   Since the sentence is such that 3 is a factor of the left member, we multiply both sides by $\frac{1}{3}$ to get p alone.
$$3p = -9$$
$$\tfrac{1}{3}3p = \tfrac{1}{3}(-9)$$
$$1p = \tfrac{-9}{3}$$
$$p = -3$$
Thus, $S = \{-3\}$.

Example 2

Find the solution set for the open sentence $3p = 5$.

SOLUTION   Again we multiply by $\frac{1}{3}$, the multiplicative inverse of 3, in order to get p.

$$3p = 5$$
$$\tfrac{1}{3} \times 3p = \tfrac{1}{3} \times 5$$
$$1p = \tfrac{5}{3}$$
$$p = \tfrac{5}{3}$$

■ Thus, $S = \{\tfrac{5}{3}\}$.

In the second example the solution set is a subset of the rationals, whereas in the first example the solution is an integer. It should be obvious that if we restrict the universe, the solution set of an open sentence may vary. Consider now what happens if our universe is too small to contain the solution set. This often happens in school, when children are told that they cannot subtract 6 from 4 or that 3 will not divide 4. This is one good reason for using the universe and learning that a solution may not exist in our universe but that other universes may contain the solution we seek. Consider the next two examples.

Example 1

■ **B.5**   Find the solution for $p + 5 = 2$.

SOLUTION $p + 5 = 2$
$$(p + 5) - 5 = 2 - 5$$
$$p + (5 - 5) = 2 - 5$$
$$p + 0 = 2 - 5$$
$$p = 2 - 5$$
$$p = -3$$

1. If the universe is I, then the solution set is $\{-3\}$.
2. If the universe is N, then the solution set is $\emptyset$.
3. What is the solution set if the universe is W? R?

Notice that our solution set is empty in one universe but in others there are solutions.

Example 2

Find the solution set for $3x = 10$ when the universe is W and when the universe is R.

SOLUTION $3x = 10$
$$\tfrac{1}{3} \times 3x = \tfrac{1}{3} \times 10$$
$$\tfrac{3x}{3} = \tfrac{10}{3}$$
$$x = \tfrac{10}{3}$$

■ If the universe is W, $S = \emptyset$, because $\tfrac{10}{3} \notin W$. If the universe is R, the set is $S = \{\tfrac{10}{3}\}$.

EXERCISES

1. Find solution sets for the open sentences below when $U = N$.

a. $2x = 8$ b. $3x = 8$ c. $n + 5 = 7$

d. $x + 5 = 5$ e. $x - 2 = 6$ f. $x/3 = 2$

g. $x + 6 = x + 2$ h. $x + 4 = 2$

2. In what universe will the following sentences have solutions? Why?

a. $2z = 9$ b. $3q = -2$ c. $2s = 12$

d. $x - 5 = 2$ e. $x + 5 = 2$ f. $x + 12 = x + 12$

g. $2 - x = 6$ h. $2x = 1$

3. Find the solution set for the sentences below.

a. $\{x \mid x \in N \text{ and } 2x = 6\}$ b. $\{x \mid x \in I \text{ and } x + 5 = 1\}$

c. $\{x \mid x \in W \text{ and } 3x = 7\}$ d. $\{x \mid x \in Ra \text{ and } x - 5 = 1/2\}$

e. $\{x \mid x \in W \text{ and } x - 5 = x - 5\}$ f. $\{x \mid x \in Ra \text{ and } 4a = 2/3\}$

4. If $U = $ reals, find the solution set of the following open sentences.

a. $\sqrt{2p} = \sqrt{8}$ b. $\sqrt{2p} = 8$ c. $p - \sqrt{2} = \sqrt{8}$

d. $p - \sqrt{2} = 8\sqrt{2}$

FINDING SOLUTION SETS FOR OPEN SENTENCES WITH MIXED OPERATIONS

More often than not, open sentences are such that the operations within one or both of the members are mixed or multiple operations. For example, the sentence $3x - 6 = 3$ is a sentence in which multiplication by 6 and then subtraction of 3 is found in the left member. Since finding solution sets involves two principles at this point, namely, the cancellation laws and the use of inverses, we need to decide on an efficient means of finding solutions. Consider the following example of multiplication and subtraction in a sentence and the possible means of solution.

Example

■ **B.5**   Find the solution set for $\{p \mid p$ is real and $3p - 6 = 3\}$.

SOLUTIONS   There are two ways of solving this problem.

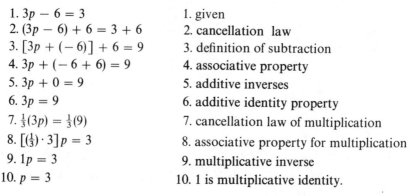

1. $3p - 6 = 3$ 1. given
2. $(3p - 6) + 6 = 3 + 6$ 2. cancellation law
3. $[3p + (-6)] + 6 = 9$ 3. definition of subtraction
4. $3p + (-6 + 6) = 9$ 4. associative property
5. $3p + 0 = 9$ 5. additive inverses
6. $3p = 9$ 6. additive identity property
7. $\frac{1}{3}(3p) = \frac{1}{3}(9)$ 7. cancellation law of multiplication
8. $[(\frac{1}{3}) \cdot 3]p = 3$ 8. associative property for multiplication
9. $1p = 3$ 9. multiplicative inverse
10. $p = 3$ 10. 1 is multiplicative identity.

We get $S = \{3\}$. As a check, since p and 3 are names for the same number,

$$3p - 6 = 3 \leftrightarrow 3(3) - 6 = 3$$
$$9 - 6 = 3$$
$$3 = 3$$

In our second method,

1. $3p - 6 = 3$	1. given
2. $\frac{1}{3}(3p - 6) = (\frac{1}{3}) \cdot 3$	2. cancellation law of multiplication
3. $[(\frac{1}{3}) \cdot 3p] - [(\frac{1}{3}) \cdot 6] = 1$	3. distribution of multiplication over subtraction
4. $1p - 2 = 1$	4. multiplicative inverse
5. $p - 2 = 1$	5. multiplicative identity
6. $(p - 2) + 2 = 1 + 2$	6. cancellation law of addition
7. $[p + (-2)] + 2 = 3$	7. definition of subtraction
8. $p + (-2 + 2) = 3$	8. associative property of addition
9. $p + 0 = 3$	9. inverses under addition
10. $p = 3$	10. 0 is additive identity.

■ And so $S = \{3\}$.

Notice in the first solution that we used the cancellation law of addition first and the additive inverse to clear the left-hand member of the number being subtracted. Then we used the cancellation law for multiplication and the multiplicative inverse to clear the left-hand member of the number multiplied times the variable. This first method is often easier than the second. In the second method, we reversed the order of the procedure; however, we were forced to use the distributive property since the multiplicative inverse was multiplied times the left member, which contained two terms; namely, $3p$ and (-6). This is usually where a student makes an error. There is a tendency to multiply $\frac{1}{3}$ times $3p$ only. For this reason, we will show only one solution to the examples which follow. It is often instructive for students to consider at least two solutions, as well as a check, when first dealing with these concepts.

Example 1 (multiplication and addition)

■ B.5   For U the reals, solve $3p + 5 = 12$.

SOLUTION

1. $3p + 5 = 12$	1. given
2. $(3p + 5) + (-5) = 12 + (-5)$	2. cancellation law
3. $3p + [5 + (-5)] = 7$	3. associative property
4. $3p + 0 = 7$	4. adding inverses
5. $3p = 7$	5. identity property of 0
6. $(\frac{1}{3})(3p) = (\frac{1}{3})7$	6. cancellation property of multiplication

7. $(\frac{1}{3}) \times (3)\, p = \frac{7}{3}$

       7. associative property of multiplication

8. $1p = \frac{7}{3}$

       8. multiplicative inverses

9. $p = \frac{7}{3}$

       9. 1 is multiplicative identity.

As a check, since p and $\frac{7}{3}$ are names for the same number,

$$3p + 5 = 12$$
$$3(\tfrac{7}{3}) + 5 = 12$$
$$\tfrac{21}{3} + 5 = 12$$
$$7 + 5 = 12$$
$$12 = 12$$

Note that if the universe for the place holders of the open sentence were W, then $S = \varnothing$, since $\frac{7}{3} \notin W$.

Example 2 (multiplication and subtraction)

Find the solution set for $16 - 5p = 9$. Notice in this sentence that $5p$ is subtracted from a constant. Let $U = $ reals.

SOLUTIONS   We will consider the solution in two ways. First,

1. $16 - 5p = 9$

       1. given

2. $(16 - 5p) + 5p = 9 + 5p$

       2. cancellation law

3. $[(16 + (-5p)] + 5p = 9 + 5p$

       3. definition of subtraction

4. $16 + (-5p + 5p) = 9 + 5p$

       4. associative property of addition

5. $16 + 0 = 9 + 5p$

       5. additive inverse of addition

6. $16 = 9 + 5p$

       6. additive identity of addition

7. $-9 + 16 = -9 + (9 + 5p)$

       7. cancellation law

8. $7 = (-9 + 9) + 5p$

       8. associative property of addition

9. $7 = 0 + 5p$

       9. adding inverses

10. $7 = 5p$

       10. identity property of 0

11. $(\frac{1}{5})(7) = (\frac{1}{5})(5p)$

       11. cancellation property of multiplication

12. $\frac{7}{5} = (\frac{1}{5})(5)\, p$

       12. associative property for multiplication

13. $\frac{7}{5} = 1p$

       13. multiplying inverses

14. $\frac{7}{5} = p$

       14. 1 is multiplicative identity.

Thus, $S = \{\frac{7}{5}\}$. As a check, Since $\frac{7}{5}$ is a name for p,

$$16 - 5p = 9$$
$$16 - 5(\tfrac{7}{5}) = 9$$
$$16 - 7 = 9$$
$$9 = 9$$

In our second solution, we consider removing 16 from the left-hand member by using cancellation laws and additive inverse. Essentially, this is opposite to the solution shown above.

1. $16 - 5p = 9$	1. given
2. $-16 + (16 - 5p) = -16 + 9$	2. cancellation law for addition
3. $(-16 + 16) + (-5p) = -7$	3. associative property
4. $0 + (-5p) = -7$	4. additive inverse
5. $-5p = -7$	5. identity property of 0
6. $(\frac{-1}{5}) \times (-5p) = (\frac{-1}{5}) \times (-7)$	6. cancellation law for multiplication
7. $1p = \frac{7}{5}$	7. multiplicative inverse
8. $p = \frac{7}{5}$	8. identity property of multiplication

■ We get $S = \{\frac{7}{5}\}$.

The two examples above indicate that there are alternate methods of solution for open sentences. In each of the examples we indicated the step-by-step procedures used. We will eventually shorten this process by combining steps. In the examples below, we will combine the use of cancellation laws, inverses, and identities to produce the solution set. If you do not find the examples clear, reread the previous examples and reconstruct the steps involved.

Example 1 (division and addition)

■ **B.5** $\{p \mid p \text{ real and } \dfrac{p}{5} + 2 = 3\}$. Find the solution set for p.

SOLUTION

1. $\dfrac{p}{5} = 1$	1. adding -2 to both members: $2 + -2 = 0$ and $\dfrac{p}{5} + 0$ is $\dfrac{p}{5}$
2. $5\left(\dfrac{p}{5}\right) = 5 \cdot 1$	2. cancellation law of multiplication
3. $[5 \cdot (\frac{1}{5})]p = 5$	3. associative property of multiplication
4. $p = 5$	4. inverse of multiplication and identity of multiplication.

Thus, $S = \{5\}$. To check,

$$\frac{p}{5} + 2 = 3$$
$$\frac{5}{5} + 2 = 3$$
$$1 + 2 = 3$$

Example 2 (multiplication and addition)

Find the solution set for $\dfrac{(3p)}{4} + 5 = 12$, $U = $ reals.

SOLUTION $\dfrac{(3p)}{4} + 5 = 12$

$$\dfrac{(3p)}{4} = 7$$

$$4 \times \dfrac{(3p)}{4} = 4 \times 7$$

$$3p = 28$$

$$\dfrac{1}{3}(3p) = \dfrac{1}{3} \times 28$$

$$p = \dfrac{28}{3}$$

$$S = \left\{ \dfrac{28}{3} \right\}$$

To check,

$$\begin{aligned}
\tfrac{3}{4}p + 5 &= 12 \\
\tfrac{3}{4} \times \tfrac{28}{3} + 5 &= \tfrac{28}{4} + 5 \\
&= 7 + 5 \\
&= 12
\end{aligned}$$

∎

Notice in Example 2 that we indicated that the operations of multiplication and addition are employed in the sentence. We could easily have considered this sentence an example of multiplication, division, and addition, since $(3p)/4 = (\tfrac{3}{4})p$.

Before we summarize this section, we should consider sentences that contain variables in more than one member. Such sentences are handled in the same manner as those shown thus far. The only deviation from the process we have used is that we bring variables together in one member and constants together in the other member. After we have accomplished this, we use the distributive property to combine the variables. To refresh your memory, a few examples of the distributive property are given below. Combine the following by employing the distributive properties of multiplication over addition or multiplication over subtraction.

Example 1

$2p + 4p$

SOLUTION $(2 + 4)p = 6p.$

Example 2

$5p - 2p$

SOLUTION $(5 - 2)p = 3p.$

Example 3

$\tfrac{3}{4}p - \tfrac{2}{3}p$

SOLUTION $\left(\dfrac{3}{4}\right) - \left(\dfrac{2}{3}\right)p = \left(\dfrac{9-8}{12}\right)p,$ which equals $\dfrac{1}{12}$ $(p),$ or $\dfrac{p}{12}$

Example 4
   $-3p + 2p$

SOLUTION $(-3+2)p = -1p,$ or $-p.$

The examples above should be of some help in understanding the examples which follow below.

Example 1
   $5p + 8 = 2p - 1.$ Let U = reals.

■ **B.5**

SOLUTION $5p + 8 = 2p - 1.$

1. $(5p + 8) + [-2p + (-8)]$ 1. cancellation law of addition
   $= [2p + (-1)] + [-2p + (-8)]$
2. $[5p + (-2p)] + [8 + (-8)]$ 2. commutative and additive
   $= [2p + (-2p)] + [-1 + (-8)]$ properties of addition
3. $5p + (-2p) = -1 + (-8)$ 3. inverse of addition, identity
        of addition
4. $3p = -9$ 4. definition of addition,
        distributive property of
        multiplication over addition
5. $\dfrac{3p}{3} = \dfrac{-9}{3}$ 5. cancellation law of addition
6. $p = -3$ 6. definition of division

To check,
$$5p + 8 = 2p - 1$$
$$5(-3) + 8 = 2(-3) - 1$$
$$-15 + 8 = -6 - 1$$
$$-7 = -7$$

Example 2
$\{p\,|\,p \in \text{reals}\ \frac{3}{4}p - 6 = \frac{2}{3}p + 4\}.$ Solve for $p.$

SOLUTION $\frac{3}{4}p - 6 = \frac{2}{3}p + 4$
$[\frac{3}{4}p + (-6)] + [6 + (-\frac{2}{3}p)] = (\frac{2}{3}p + 4) + [6 + (-\frac{2}{3}p)]$
$[\frac{3}{4}p + (-\frac{2}{3}p)] + (-6 + 6) = [\frac{2}{3}p + (-\frac{2}{3}p)] + (4 + 6)$
$[\frac{3}{4}p + (-\frac{2}{3}p)] = (4 + 6)$
$[\frac{3}{4} + (-\frac{2}{3})]\,p = 10$
$\frac{1}{12}p = 10$
$12(\frac{1}{12}p) = 12(10)$
$p = 120$
$S = \{120\}$

As a check,

$$\tfrac{3}{4}p - 6 = \tfrac{2}{3}p + 4$$
$$\tfrac{3}{4}(120) - 6 = \tfrac{2}{3}(120) + 4$$
$$90 - 6 = 80 + 4$$
$$84 = 84$$

■

In summary, we suggest the following thoughts about finding the solution set of a sentence which is an equality relation.

1. Consider the operations performed on the variables.
2. Use the cancellation law, which allows you to add, subtract, multiply, or divide both members of the sentence by like quantities.
3. Use the inverse operation to undo the operation observed.
4. It is usually easier to perform addition and subtraction before multiplication or division.
5. Some sentences have no solution because of the universe defined. In these cases we say that the solution set is empty.
6. Usually solution sets for equality relations are empty or contain a single element; however, some sentences are true for all replacements of the variables.

EXERCISES

1. Find solutions to the following in W, I, and R where they exist.
   a. $p + 5 = 12$ b. $p + 7 = 4$
   c. $2p - 5 = 15$ d. $3p/4 + 6 = -6$
   e. $3p/5 + 7 = 10$ f. $p/3 + 6 = -9$

2. Find solution sets in W, I, and R where they exist.
   a. $2p + 5 = 2p$ b. $3p - 2 = 3p$
   c. $p + 7 = 2 - p$ d. $p + 4 = 4 + p$

3. Find solution sets for the following.
   a. $\{p \mid p \in W \text{ and } p + 5 = 12\}$
   b. $\{p \mid p \in W \text{ and } 2p + 6 = 12\}$
   c. $\{p \mid p \in W \text{ and } 3p + 5 = 7\}$
   d. $\{p \mid p \in W \text{ and } 2p + 6 = p - 6\}$

4. Find the solution sets for the following.
   a. $\{p \mid p \in I \text{ and } 3p - 5 = 4p + 2\}$
   b. $\{p \mid p \in I \text{ and } \tfrac{2}{3}p - 6 = 4\}$
   c. $\{p \mid p \in I \text{ and } 3p - 7 = 5p - 12\}$
   d. $\{p \mid p \in I \text{ and } \tfrac{3}{4}p + 5 = 12\}$

5. Find solution sets for the following.
   a. $\{p \mid p \in \text{reals and } \tfrac{5}{6}p + 2 = 7\}$
   b. $\{p \mid p \in \text{reals and } \tfrac{2}{3}p + 3 = \tfrac{3}{5}p + 2\}$
   c. $\{p \mid p \in \text{reals and } 3p - 5 = p/2 + 7\}$
   d. $\{p \mid p \in \text{reals and } 4p + 6 = 15\}$

SOLUTION SETS FOR SENTENCES
WHICH ARE NOT EQUALITY RELATIONS

All the sentences discussed in the previous sections expressed equality. Sometimes, however, we are interested in a sentence which expresses inequality. A simple example is $x < 3$. This sentence could have a solution set $S = \{1, 2\}$ if $U = N$; $S = \{0, 1, 2\}$ if $U = W$; $S = \{\ldots -2, -1, 0, 1, 2\}$ if $U = I$; $S = \{p \mid p \in R$ and $p < 3\}$ if $U = R$; or $S = \{p \mid p \in$ reals and $x < 3\}$ if $U =$ reals. Obviously, this solution set is directly related to the universe defined. Notice that as the universe gets larger, the set increases from a finite to an infinite number of elements. The importance of defining that universe should not be taken lightly, because it is apparent that the same statement has many meanings in different universes. This is true with any language: Statements often have different interpretations in different contexts.

Before we discuss the process of finding solution sets for open sentences involving inequalities, we will consider ways of describing such solution sets. Consider the statement "p is less than 3," that is, $p < 3$. If the solution set is finite, we can describe this by tabulation or by words or by set builder notation. Or, we can describe this set by graphing. Consider the examples below in which the uses of set builder notation and graphs are compared.

Examples

1. $\{p \mid p \in N$ and $p < 3\}$

2. $\{p \mid p \in W$ and $p < 3\}$

3. $\{p \mid p \in I$ and $p < 3\}$

4. $\{p \mid p \in R$ and $p < 3\}$

5. $\{p \mid p \in$ reals and $p < 3\}$

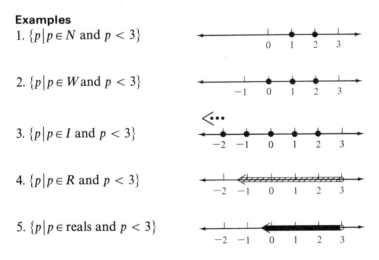

Notice that the number line indicates a universe of real numbers in that it is a complete unbroken line; however, we place only the numbers in the universe under consideration on the line. In the set I, we used an arrow and three dots on the left, written above the line, to indicate that the set was infinite. Once we reached the set R, we indicated all rationals in the set that are less than 3 by placing an open circle on the 3 and a hatched ray going to the left. The use of an open circle indicates that there is no element 3 in the solution set but that everything to the left is in the solution set.

Since we are dealing with inequalities, we must use our cancellation laws for inequalities in the solution process. The solution process for inequalities parallels that of solving open sentences which are equalities. Actually, the solution set for an equality partitions the universe due to the trichotomy property. By this we mean that if there is a set whose solution is $p = 3$ in a certain universe, then there is a set whose solution is $p < 3$ and one whose solution is $p > 3$. We will initially consider solutions from this point of view. Consider the following example.

Example

$\{p \mid p \in W \text{ and } p + 3 = 5\}$

SOLUTION $S = \{2\}$.

Now consider the partition produced by this equality. Since either $p = 2$, $p < 2$, or $p > 2$, we can write at least two more statements which relate to the original example. These are shown below.

Example 1

■ B.6 $\{p \mid p \in W \text{ and } p + 3 < 5\}$

SOLUTIONS

1. $p + 3 < 5$	1. given
2. $p + (3 - 3) < 5 - 3$	2. cancellation
3. $p + 0 < 2$	3. inverse operations
4. $p < 2$	4. identity $+ W$

Thus, $S = \{0, 1\}$, or $S = \{p \mid p \in W \text{ and } p < 2\}$, or, graphically,

To check, notice that $p + 3 < 5$ is true for replacements of p by 0 or 1, since $3 < 5$ and $4 < 5$.

Example 2

$\{p \mid p \in W \text{ and } p + 3 > 5\}$

SOLUTIONS $p + 3 > 5$

$p + (3 - 3) > 5 - 3$

$p > 2$

Thus,

$S = \{3, 4, 5, \ldots\}$, or $S = \{p \mid p \in W \text{ and } p > 2\}$, or, graphically,

From what you have already learned about inequalities, you should realize that we could have considered the relations $p \neq 2$, $p \leq 2$, $p \geq 2$, and so on. The changes in solution sets that these relations will produce may not be obvious at this point; therefore, we will summarize them below. We will only consider the combinations that must be mentioned, because the others follow easily from these.

Example 1

■ B.7 $\{p \mid p \in W \text{ and } p + 3 \neq 5\}$

SOLUTIONS $S = \{0, 1, 3, 4, 5, \ldots\}$, or $S = \{p \mid p \in W \text{ and } p \neq 2\}$, or, graphically,

Example 2

$\{p \mid p \in W \text{ and } p + 3 \leq 5\}$

SOLUTIONS $S = \{0, 1, 2\}$, or $S = \{p \mid p \in W \text{ and } p \leq 2\}$, or, graphically,

Notice that since the relation $\leq$ means less than or equal to, the solution set is the union of two sets $\{p \mid p < 2\}$ and $\{p \mid p = 2\}$. This simply translates the word "or" to the operation union.

Example 3

$\{p \mid p \in W \text{ and } p + 3 \geq 5\}$ translates to $\{p \mid p \in W \text{ and } p + 3 > 5\}$ $\cup \{p \mid p \in W \text{ and } p + 3 = 5\}$.

SOLUTIONS $S = \{2, 3, 4, \ldots\}$, or $S = \{p \mid p \in W \text{ and } p > 2\}$
$\cup \{p \mid p \in W \text{ and } p = 2\}$
$= \{p \mid p \in W \text{ and } p \geq 2\}$, or, graphically,

■

Notice that we have translated the word "or" to mean union. This is a very specialized use of the word, and it indicates that "or" is an *inclusive* rather than *exclusive* "or." By inclusive we mean that "or" implies everything in a set A or B or both A and B. This, of course, implies the definition of union. We also use the word "and" to mean everything that is in both A and B. Thus "and" implies intersection. Sometimes this usage confuses students; therefore, you should get into the habit of translating "and" to mean "intersects with" and "or" to mean "union with." Before we continue into more complicated sentences, it might be constructive for you to consider several translations of sentences which involve such words as "and" and "or" used in this specialized way.

Example 1
■ **B.7**   Translate $p \in W$ and $p < 5$.

SOLUTION $W \cap \{p \mid p < 5\} = \{0, 1, 2, \ldots\} \cap \{p \mid p < 5\}$
$= \{0, 1, 2, 3, 4\}$

Graphically, if we intersect the graph of W and the graph of $\{p \mid p < 5\}$, we will have the graph of $\{p \mid p \in W$ and $p < 5\}$.

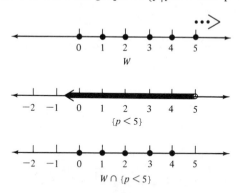

▲ B.9 ▲

Example 2
Translate $\{p \in W\}$ and $[\{p < 5\}$ or $\{p = 5\}]$.

SOLUTION $W \cap [\{p < 5\} \cup \{5\}] = \{0, 1, 2, 3, 4, 5\}$, or, graphically,

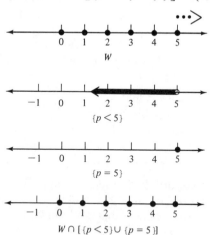

▲ B.9
▲ ■

In the previous examples, we expanded the sentences to clarify their meaning. It is not really necessary to show $p \leqq 5$ as the union of $p < 5$ with $p = 5$, nor is it necessary to show $p \in W$ and $p < 5$ as an intersection. It is true, however, that these words and phrases mean exactly that. Later, we will look at other compound statements which will require that we union or intersect solution sets to produce a final solution set. For now we will interpret $p \in U$, where U is N, W, I, etc., as a condition on a simple sentence. We will also consider a sentence like $p + 3 \leq 5$ as a simple sentence and deal with the relations $\leq$, $\geq$, and $\neq$ as though they were simple relations as opposed to compound relations. It is important to remember that these relations translate respectively as "less than *or* equal to," "greater than *or* equal to," and "not equal to," which means "less than or greater than." We will consider some examples at this point which are generally similar to the examples in the previous section.

■ B.6   **Example 1**
▲ B.7 $\{p \mid p \in R \text{ and } 3p - 6 < 3\}$

SOLUTIONS

1. $3p - 6 < 3$ 1. given
2. $[3p + (-6)] + 6 < 3 + 6$ 2. cancellation law of addition in R for inequalities
3. $3p < 9$ 3. associative property of addition in R, additive inverses in R, identity property of addition in R
4. $\frac{1}{3}(3p) < \frac{1}{3}(9)$ 4. cancellation law of multiplication in R for inequality
5. $p < 3$ 5. associative property of multiplication in R, multiplicative inverses of multiplication in R, identity property of multiplication in R.

▲ ■   Thus, $S = \{p \mid p \in R \text{ and } p < 3\}$, or, graphically,

To check, notice that in checking we find that replacements of p in $3p - 6 < 3$ by a number greater than or equal to 3 makes the sentence false. Anything less than 3 makes the sentence true.

■ B.6   **Example 2**
▲ B.7 $\{p \mid p \in R \text{ and } 16 - 5p \geq 9\}$

SOLUTION

1. $16 - 5p \geq 9$ 1. given
2. $-16 + (16 - 5p) \geq -16 + 9$ 2. cancellation law of addition in R for inequalities
3. $-5p \geq -7$ 3. associative property of addition in R, additive inverses in R, identity property of addition in R

4. $p \leq \frac{7}{5}$ 4. associative property of multiplication in R, multiplicative inverses in R, identity property of multiplication in R

Thus $S = \{p \mid p \in R \text{ and } p \leq \frac{7}{5}\}$.

Note that multiplying both members by a negative quantity changes the order relation from $>$ to $<$.

▲ ■   Graphically, the solution set is

As a check, consider $16 - 5p \geq 9$. A replacement of 1 for p makes the sentence true because $16 - 6 > 9$; a replacement of 7/5 makes the sentence true because $16 - 7 = 9$; and a replacement of 2 makes the sentence false because $16 - 10 > 9$.

■ **B.6**   **Example 3**
▲ **B.7** $\{p \mid p \in I \text{ and } 5p + 8 \leq 2p - 1\}$

SOLUTION

1. $5p + 8 \leq 2p - 1$ 1. given
2. $(5p + 8) + [-2p + (-8)]$ 2. cancellation laws of addition
    $\leq [2p + (-1)] + [-2p + (-8)]$ in R for inequalities
3. $3p \leq -9$ 3. commutative and associative properties of addition in I, inverse and identity property of addition in I

4. $\frac{1}{3}(3p) \leq \frac{1}{3}(-9)$ 4. cancellation law of multiplication in I for inequalities

5. $p \leq -3$

▲ ■   Thus, $S = \{p \mid p \in I \text{ and } p \leq -3\}$, or $\{\ldots -5, -4, -3\}$. Graphically,

To check, substituting -4 in $5p + 8 \leq 2p - 1$ produces $-12 \leq -9$, which is a true statement. Substituting -3 produces $-7 \leq -7$, which is also true. But substituting -2 produces $-2 \leq -5$, which is false.

EXERCISES

1. Find solution sets for the following in R.
    a. $3p - 5 < 7$ b. $2p - 4 \geq 5p - 3$
    c. $-3p - 7 \geq 12$ d. $\frac{5}{6}p + 12 = 24$

2. Find solution sets for the following in I.

a. $2p + 6 \leq 12$ b. $3p - 5 \geq 7$

c. $(-2p/3) - 3 \geq 9$ d. $3p + 7 \leq 4 - 2p$

3. Find solution sets for the following in W.

a. $3p + 6 \leq 12$ b. $(5p/3) + 7 \geq 22$

c. $3p > 9 + 2p$ d. $3p + 2 < 2 + 3p$

4. Graph the solution sets for each of the following if the universe consists of the real numbers.

a. $3p + 6 \leq 12$ b. $3p < 11 + 2p$

c. $3p < 3p$ d. $3p > 12p + 4p$

SOLUTION SETS FOR COMPOUND SENTENCES

Throughout this text we have made a point of the meaning of the words "and" and "or" in mathematics. A sentence consisting of two clauses connected by "and" is true only if *both* the sentences are true. With sets, "and" means intersection.

When dealing with sets, we understand the word "or" to mean union. In the following examples we illustrate how to find solution sets for compound sentences where the universe is the set of reals. Let us translate the following sentences by substituting an operation for the words "and" and "or."

Example 1

■ **B.7**
■
$\{p \,|\, p > 6 \text{ and } p < 12\} = \{p \,|\, p > 6\} \cap \{p \,|\, p < 12\} = \{p \,|\, 6 < p < 12\}$, which means that p is between 6 and 12.

Example 2

■ **B.9**
$\{p \,|\, p < 5 \text{ and } p \geq 6\} = \{p \,|\, p < 5\} \cap \{p \,|\, p \geq 6\}$.

SOLUTION   If you consider the graphs of $p < 5$ and $p > 6$, you can see that the sets have no elements in common; therefore, the solution set is empty.

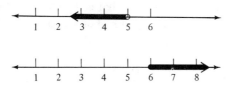

■   The graph of the solution is an empty graph or an empty set.

Example 3

▨ **B.9**
$\{p \,|\, p > 6 \text{ or } p < 12\}$

SOLUTION $\{p \,|\, p > 6\} \cup \{p \,|\, p < 12\}$. If we consider the graph of these sets, we find that their union is the set of reals.

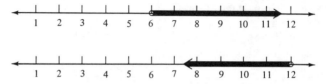

Therefore, our solution set contains everything to the left of 12 and everything to the right of 6. Together this takes in the whole set of reals.

Example 4

B.9 $\{p \mid p < 5 \text{ or } p \geq 6\}$

SOLUTION $\{p \mid p < 5\} \cup \{p \mid p \geq 6\}$. By unioning the sets shown, we produce the graph below.

Translate the graphs below into sentences using the connectives "and" and "or."

Example 1

SOLUTION   This graph shows that $2 \leq p < 5$. Therefore, we could write $\{p \mid 2 \leq p < 5\}$, or $\{p \mid p \geq 2\} \cap \{p \mid p < 5\}$, which translates as $\{p \mid p \geq 2 \text{ and } p < 5\}$.

Example 2

SOLUTION   This reads $p < 4$ or $p > 6$, that is, $\{p \mid p < 4\} \cup \{p \mid p > 6\}$, and translates as the set $\{p \mid p < 4 \text{ or } p > 6\}$.

Example 3

SOLUTION   This graph reads $p = 4$, which also means $\{p \mid p \leq 4\} \cap \{p \mid p \geq 4\}$. Both interpretations are quite correct. Since this is the case, we write either $\{p \mid p = 4\}$ or $\{p \mid p \leq 4 \text{ and } x \geq 4\}$. These are equal sets because they contain the same elements.

In the above examples, the universe was considered to be the real numbers. In the case that the universe is not the reals, there is no real convention as to how the set is described. For instance, let us describe the set whose graph is

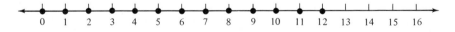

B.8 This set, S, is $\{-3, -2, -1, 0, 1\}$, a subset of the integers. We now describe the set in several ways.

1. $S = \{p \,|\, p \in I \text{ and } -3 \le p \le 1\}$
2. $S = \{p \,|\, p \in I \text{ and } p \ge -3\} \cap \{p \,|\, p \in I \text{ and } p \le 1\}$
3. $S = \{p \,|\, p \in I \text{ and } p > -4\} \cap \{p \,|\, p \in I \text{ and } p < 2\}$
4. $S = \{p \,|\, p \in I \text{ and } -4 < p < 2\}$
5. $S = \{p \,|\, p \in I \text{ and } p \ge -3 \text{ and } p \le 1\}$
6. $S = \{p \,|\, p \in I \text{ and } p > -4 \text{ and } p < 2\}$

If we specify that the universe is I, the integers, then we can also describe the set graphed above by the following:

7. $S = \{p \,|\, p \ge -3 \text{ and } p \le 1\}$
8. $S = \{p \,|\, p > -4 \text{ and } p < 1\}$
9. $S = \{p \,|\, -3 \le p \le 1\}$
10. $S = \{p \,|\, -4 < p < 1\}$

Note that for the last four descriptions we assumed that the universe was I, whereas the other descriptions specified the universe. You should verify that each set described above actually is $\{-3, -2, -1, 0, 1\}$. Needless to say, there seldom is *one* right way to describe a set, there are usually many correct ways.

As the above examples indicate, we need not be concerned with the grouping of statements within a compound sentence which involves a universe and two conditions. However, we must be concerned with grouping if there are more than two conditions on the universe. For example, we might have difficulty interpreting the problem below.

Example

B.9
▲ **B.8**
Graph the set $S = \{p \,|\, p \in W \text{ and } p \le 12 \text{ and } p \ge 5\} \cap \{p \,|\, p \in W \text{ and } 4 \le p \le 15\}$.

SOLUTION   Rewriting S we have $S = \{p \,|\, p \in W \text{ and } p \le 12\} \cap \{p \,|\, p \in W \text{ and } p \ge 5\} \cap \{p \,|\, p \in W \text{ and } 4 \le p\} \cap \{p \,|\, p \in W \text{ and } p \le 15\}$. Now graph each of these sets whose intersection is S.

▲

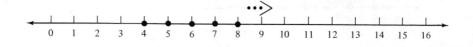

The graph of S consists of all those points that are common to the four graphs above, that is, as shown below.

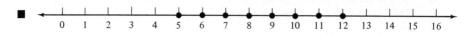

EXERCISES

1. Graph the set of real numbers that satisfy the statements below.
  a. $S = \{p \mid p \geq 2 \text{ or } p \geq 6\}$ b. $S = \{p \mid p \geq 2 \text{ and } p \geq 6\}$
  c. $S = \{p \mid p \leq 2 \text{ or } p \leq 6\}$ d. $S = \{p \mid p \leq 2 \text{ and } p \geq 2\}$

2. For the statements in Exercise 1, find the graphs if the universe is the set of
  a. integers   b. natural numbers

3. If $a < b < c < d$, graph the union and intersection of the pairs of sets given below. Assume that the universe is the set of reals.
  a. $\{p \mid a < p < b\}, \{p \mid c \leq p \leq d\}$
  b. $\{p \mid a \leq p < c\}, \{p \mid b \leq p \leq d\}$
  c. $\{p \mid b < p < d\}, \{p \mid b < p \leq c\}$

4. Describe the sets whose graphs are the following.

  a.

  b.

  c.

CHAPTER 10 **LOGIC**

SECTION A **STATEMENTS**

BEHAVIORAL OBJECTIVES

A.1 Be able to define a statement.

A.2 Be able to write a statement.

A.3 Given sentences, be able to differentiate between statements and non-statements.

A.4 Be able to define a statement which is a conjunction.

A.5 Given simple statements, be able to write conjunctions using the given statements.

A.6 Be able to define a statement which is a disjunction.

A.7 Be able to differentiate between statements which are conjunctions and statements which are disjunctions.

A.8 Given simple statements, be able to write disjunctions using the given statements.

A.9 Be able to define compound statement.

A.10 Be able to translate statements written in English to symbolic statements which are conjunctions or disjunctions.

A.11 Given a symbolic statement which is a conjunction or a disjunction, be able to translate the statement into an English sentence.

A.12 Be able to make a truth table showing the logical possibilities given 2 simple statements.

A.13 Be able to make a truth table showing the truth values for a statement which is a conjunction.

A.14 Be able to make a truth table showing the truth values for a statement which is a disjunction.

A.15 Given the truth values of simple statements, be able to find the truth value of compound statements formed by conjunction or disjunction.

A.16 Given several simple statements and their truth values, be able to form true conjunctions or disjunctions.

A.17 Given several simple statements and their truth values, be able to form false conjunctions or disjunctions.

A.18 Be able to state and prove, by use of truth tables, the laws of excluded middle and contradiction.

A.19 Be able to translate a disjunction to a conjunction.

A.20 Be able to translate a conjunction to a disjunction.

STATEMENTS

The purpose of this chapter is to introduce you to a method of translating statements to symbols and then determining the truth or falsity of those statements under various circumstances. In addition, we would like to develop some method for determining the validity of an argument.

To carry out our goal, we must first make a few definitions. In order to discuss the truth or falsity of statements, we must first be able to recognize a statement when we see one. Some examples of statements are listed below.

■ **A.2** **Examples**

▲ **A.3** 1. It is raining. (T or F)

2. $2 + 2 = 5.$ (F)

3. Cows are people. (F)

4. The moon is blue. (F)

5. The moon is *not* made of cheese. (T)

■ 6. John is 6' tall. (T or F)

Each of the above statements is either true or false at a given time, in that we can determine their truth value. The truth of the first statement can be determined in a given time context. The other five are obviously true or false,

▲ but never both at once. Some sentences have no definite truth value, and these are *not considered statements*. Several examples of these are seen below.

Examples

■ **A.3** 1. Is Mary pretty?

2. Make it nice.

3. Close the door.

4. Is she pretty?

5. I wish I were in love!

6. Walk home.

■ 7. Did you eat dinner?

We cannot assign truth value to a question, because it would have no meaning to say that the question "Is it raining?" is true or false. Obviously, the answer to a question *could be a statement*. The other examples above might be characterized as commands. Again, we cannot consider the truth value of this type of sentence. We also exclude combinations of words which *do not form sentences* or sentences which require value judgments. We should be ready to define a statement at this point. Notice that all of the examples of statements had the following characteristics:

1. They were declarative sentences.

2. They were true or false.

3. They could not be true and false at the same time.

Definition

■ A.1   *A statement is a declarative sentence which is true or false (or to which*
■   *a truth value can be assigned).*

Notice that each of the examples of statements was a simple sentence rather than a compound sentence. Each expressed a single idea. If we wish to make a compound statement, we connect two simple statements with the words "and" or "or," as shown below.

Examples

■ A.8   1. 2 < 4 or 4 ≥ 2.

■   2. 2 + 2 = 5 or 3 < 6.

■ A.5   3. It is a triangle, and it has four sides.

■   4. The sun rises in the east, and the sun sets in the west.

■ A.4   We give special names to such compound sentences. Those compound
A.6 ■   sentences connected by the word "and" we call a *conjunction*. Those
        compound sentences formed by using the connective "or" are called
▲   *disjunctions*. The words "and" and "or" are called *connectives*, and the
        simple statements used to form the sentence are called *components*.

Definition

■ A.9   *A compound statement is two or more simple statements connected by*
■   *one or more connectives.*

EXERCISES

1. Which of the following are statements?
  a. The sun is shining.   b. John is tall.
  c. 6 is greater than 4.   d. Jan is the best athlete in class.
  e. 10 is a small number.   f. Max is ten years old.

2. From the simple statements above, write three conjunctions and three disjunctions.

3. Which of the following statements are conjunctions, and which are disjunctions?
  a. Bill is 10 years old and has blond hair.
  b. 6 is greater than 4, and the sum of 6 and 4 is 10.
  c. The temperature is over 70° or the temperature is over 80°.
  d. 2 − 2 < 2, or 2 − 2 > 2.
  e. The movie is $1\frac{1}{2}$ hours long, and Jon is in his room.

4. Classify the sentences and phrases below as simple statements, conjunctions, disjunctions, or non-statements.
  a. A red dog.
  b. Is six dollars enough?
  c. He is old, and it is snowing.
  d. The hen, or it is tall.
  e. 3 < 2, and 5 < 4.
  f. He is over 6 feet tall, or the flowers are dead.
  g. If he runs, he will be on time.
  h. 2 + 2 = 4

USING SYMBOLS AND FINDING TRUTH VALUES FOR STATEMENTS

It is customary to symbolize statements by using small letters. The letters most often used are p, q, r, and s. By symbolizing statements and then assigning truth values to the components, we can determine the truth or falsity of simple and compound statements more easily. We use the symbol $\land$ to represent the word "and" (conjunctions), and we use the symbol $\lor$ to represent the word "or" (disjunctions). Therefore, we can now translate some compound statements to symbolic statements, as shown below.

■ **A.7 Examples**

▲ **A.10** 1. It is raining and the sun is shining. $p \land q$

2. It is snowing and freezing. $p \land q$

3. It is snowing or raining. $p \lor q$

▲ ■ 4. It is raining or the windows are closed. $p \lor q$

In the examples above, we do not infer that p and q are variables in that they may represent different statements at different times. We could have translated several simple sentences into symbols and then formed compound statements from them. When this is done, the symbols represent the same statement throughout the discussion. Several such examples follow. Let us translate the following symbolic statements to English sentences. Let

p mean   It is raining outside.
q mean   The windows are closed.
r mean   It is freezing outside.

■ **A.11 Examples**

▲ **A.7** 1. $p \lor q$ It is raining outside, or the windows are closed.

● **A.5** 2. $p \land r$ It is raining outside and freezing outside.
              It is raining, and it is freezing outside.
              It is raining and freezing outside.

Notice that the last translation is more natural than the first two, but they mean the same thing.

▲ 3. $q \lor r$ The windows are closed, or it is freezing outside.

● ■ 4. $p \land q$ It is raining outside, and the windows are closed.

We must next consider the possible truth values that can exist in a simple or compound statement. Suppose we make the statement "It is raining"; this statement is either true or it is false. Suppose we consider the statement "It is freezing," which can also be true or false. Once we combine these statements, then we have four possibilities; namely, both true, first true and second false, first false and second true, or both false. We summarize these four possibilities in Figure 10.1

■ A.12

p	q
t	t
t	f
f	t
f	f

■

Figure 10.1

The order of the possibilities in Figure 10.1 is not fixed. When we consider p and q together, we see that when p is true, q can be true or false; and when p is false, q can be true or false. Figure 10.1 is called a *truth table*, and the values assigned are called *truth values*. If we simply assign all possible values, we find that for a single statement there are two possibilities, for two statements there are four possibilities, and for three statements there will be eight possibilities. What would you predict for four statements?

Number of Statements	Number of Truth Values
1	2
2	4
3	8
4	?

If you predicted 16, then you can see a rule: 2 raised to the power of the number of statements produces the number of possible truth values. Therefore, if a compound statement had 4 different symbols, p, q, r, and s, we would need to make a truth table with 16 horizontal lines of values. This, of course, becomes tedious and not very creative work. Later we will be able to determine truth as well as validity without referring to such tables.

Now let us consider our reasons for going to the trouble of constructing truth tables. To consider the truth value of a statement, we must know all the possible values. Then we can determine whether a particular compound statement is true or not. We need one more piece of information: We must have a definition of the truth value of conjunctions and disjunctions.

Before we define the truth value of conjunction and disjunction, let us consider what common sense might dictate regarding these concepts. Suppose we state, "It is raining and the sun is shining." It is possible for one or both of the simple statements within the compound statement to be true or false. If it is raining and if the sun is shining as in a summer shower, we would judge the statement to be true. However, if it is not raining or if the sun is not shining, we would judge the statement false. This should imply that a conjunction is only true if both components of the conjunction are true. We show this in Figure 10.2. Notice that in the first row we read "p is true," "q is true," then "p and q is true."

■ A.13

p	q	$p \wedge q$
t	t	t
t	f	f
f	t	f
f	f	f

Figure 10.2

Consider now the disjunction "It is raining or the sun is shining." If it is raining, we consider that true. If the sun is shining, we also consider that true. If it is raining and the sun is also shining, we consider this to be logically true. In English, we do not always define the word "or" in this manner. As was mentioned in the previous chapter, the word "or" in mathematics takes on an inclusive meaning. The only time the disjunction is considered false is when it is not raining and the sun is not shining. Therefore, we define the disjunction as false when the components are both false; otherwise, we consider the disjunction to be true. The truth table in Figure 10.3 defines the disjunction. Notice in the first row that we read "p is true," "q is true," then "p or q is true."

■ A.14

p	q	$p \vee q$
t	t	t
t	f	t
f	t	t
f	f	f

Figure 10.3

Now that we have truth tables for conjunction and disjunction, we can evaluate statements in terms of the truth value of the components. Consider some examples below.

Example 1

■ A.15   Maine is in New England *and* rain is wet.

SOLUTION   Since both components are true, the conjunction is true.

Example 2
    $2 < 4$ *and* $3 + 2 = 6$.

SOLUTION   Since $3 + 2 \neq 6$, the conjunction is false.

Example 3
Lemons are sour *or* snow is warm.

SOLUTION Since "lemons are sour" is true, the disjunction is true.

Example 4
$2 \times 37(1 + 1) < 3$ *or* $\frac{1}{2} \times \frac{1}{2} = 1$.

SOLUTION Since both components are false, the disjunction is false.

Example 5
Christmas is a Buddhist holiday *and* Thanksgiving is celebrated all over the world.

SOLUTION Since both components are false, the conjunction is false.

Example 6
$3 + 4 = 6$ *or* $\frac{2}{3} \times \frac{3}{4} = \frac{1}{2}$

■ SOLUTION Since $\frac{2}{3} \times \frac{3}{4} = \frac{1}{2}$, the disjunction is true.

It is quite often the case that we can decide the truth or falsity of a statement only by assigning a truth value to the components. For example, the statement "It is raining and Maine is in New England" cannot be given a truth value except at some instance in time. If we wished, though, we could assign a truth value to such a statement. Many times in an argument, we start out by assuming that something is true so that we can better defend our case. Consider the examples below, which all rely on given truth values.

p means　It is raining. (let p be true)
q means　The sun is shining. (let q be true)
r means　The train is late. (let r be false)
s means　It is snowing. (let s be false)

Example 1
■ A.16　Construct a true conjunction and symbolize it using p, q, r, and s.

SOLUTION　It is raining and the sun is shining ($p \wedge q$). Since both components are true, this conjunction is true.

■

Example 2
■ A.17　Construct three different false conjunctions and symbolize them using p, q, r, and s.

SOLUTIONS　1. It is raining and the train is late ($p \wedge r$).
2. The sun is shining and it is snowing ($q \wedge s$).
3. It is snowing and the train is late ($s \wedge r$).

The first two statements are false because one component is false. The last one is false because both components are false.

Example 3

Construct a false disjunction and symbolize it.

SOLUTION It is snowing or the train is late ($s \lor r$). Since both components are false, the disjunction is false.

■

Example 4

■ **A.16** Construct three true disjunctions and symbolize them.

SOLUTIONS 1. It is raining or the sun is shining. ($p \lor q$)
2. It is snowing or it is raining. ($s \lor p$)
3. The sun is shining or the train is late ($q \lor r$).

The first statement is true because both components are true. The last two statements are true because one component is true.

■

Notice in the examples above that whenever there is a false statement in a conjunction, the conjunction is automatically false. Whenever a statement in a disjunction is true, the disjunction is automatically true. This can be most easily seen by looking at the truth table in Figure 10.4. Placing the truth tables for conjunction and disjunction side by side in Figure 10.4, we see that conjunction and disjunction have the same truth value in two instances only.

p	q	$p \lor q$	$p \land q$
t	t	t	t
t	f	t	f
f	t	t	f
f	f	f	f

Figure 10.4

EXERCISES

1. Consider the following statements, and find their truth value.
   a. The sky is blue and snow is cold.
   b. $3 < 2$ and $-3 < -2$.
   c. Fish are plants or people are cats.
   d. $2 \times 4 = 8$ and $2 \times 4 = 2 \times (2 \times 2)$.
   e. $3 \times (4 + 1) = (3 + 4) \times (3 + 1)$ or $3 \times (4 + 1) = 12 + 3$.
   f. Dogs fly or people breathe under water.

2. Make a truth table for the following:
   a. $p \lor q$ b. $p \land q$ c. $p \lor (q \land r)$

3. From the truth values given, evaluate the truth value of the symbolic statements shown below: p is true, q is false, r is false, and s is true.

a. $p \lor q$ b. $(p \lor s) \lor r$ c. $(p \lor r) \land s$
d. $q \lor r$ e. $(p \land q) \lor r$ f. $(p \land r) \land s$
g. $p \land s$ h. $p \land (q \land r)$

4. From the symbolic statements and truth values given below, write sentences in English which have corresponding form and truth values. Be sure that your translations are consistent throughout. Once you decide on a statement for p, q, etc., use it throughout the exercise.

a. $p \lor q$ (true) b. $p \lor r$ (true)
c. $p \land q$ (false) d. $q \land s$ (true)

5. In Exercise 4, what were the truth values of p, q, r, and s?

COMPOUND STATEMENTS WHICH INVOLVE NEGATION

We will introduce a new symbol at this point which will be used to negate a statement. The symbol minus, $-$, translates as "it is not the case." By doing this, we will be able to make a statement p and then make a statement $- p$, which will translate as "it is not the case that p" or "the opposite of p." If this statement is used together with "and" and "or," we can write $p \lor - p$ as well as $p \land - p$. Each of the previous statements is quite interesting, because one is always true and the other is always false. Consider the disjunction $p \lor - p$. If p is true, then $- p$ is always false, and if p is false, $- p$ is always true. Thus, the disjunction is always true. In the case of the conjunction, since one of the statements p or $- p$ is always false, the conjunction $p \land - p$ is always false. We summarize these results in Figure 10.5 along with the statement $- (p \land - p)$.

p	$- p$	$p \lor - p$	$p \land - p$	$- (p \land - p)$
t	f	t	f	t
f	t	t	f	t

Figure 10.5

■ A.18 Notice that $p \lor - p$ is always true, and so is $- (p \land - p)$. The first of these statements, $p \lor - p$, is called the *law of excluded middle*. The ■ other statement, $- (p \land - p)$, is called the *law of contradiction*. Later we will show that the law of contradiction and the law of excluded middle are equivalent statements. For now, we simply point out that $p \lor - p$ has the same truth values as does $- (p \land - p)$.

Let us now find the negation of a compound statement. We will first find a statement equivalent to $- (p \lor q)$. This is a very practical problem, because a solution means that we will have a statement equivalent to "it is not the case that p or q." Thus, we will be able to negate statements such as "$p \geq 5$." That is, we will have a statement equivalent to "it is not the case that p is greater than 5 or p is equal to 5."

To find a statement with the same truth values as $- (p \lor q)$, we first determine the truth values of $- (p \lor q)$, depending upon p and q, as in Figure 10.6.

p	q	$p \vee q$	$-(p \vee q)$
t	t	t	f
t	f	t	f
f	t	t	f
f	f	f	t

Figure 10.6

Now look at a truth table in Figure 10.7, which involves p, q, $-p$, and $-q$, the conjunction $-p \wedge -q$, and the disjunction $-p \vee -q$.

p	q	$-p$	$-q$	$-p \wedge -q$	$-p \vee -q$
t	t	f	f	f	f
t	f	f	t	f	t
f	t	t	f	f	t
f	f	t	t	t	t

Figure 10.7

Now compare Figures 10.6 and 10.7. We see that the statement
■ A.19 $-p \wedge -q$ has the same truth values as $-(p \vee q)$. Thus, we say that
■ the statement $-(p \vee q)$ is equivalent to $-p \wedge -q$.

Before we give examples using the above equivalent statements, we will develop another useful pair of equivalent statements.

Let r be the statement $-p$, that is, $r = -p$, and let $s = -q$. Then $-r = p$ and $-s = q$. Now look at the previous pair of equivalent statements: $-(p \vee q)$ is equivalent to $-p \wedge -q$. Replace each of the statements by their equal statements in terms of r and s. That is,

$$-(p \vee q) \qquad \text{is equivalent to } -p \wedge -q$$
$$-(-r \vee -s) \quad \text{is equivalent to } r \wedge s$$

Thus, we have that $-(-r \vee -s)$ is equivalent to $r \wedge s$. If two statements are equivalent, their negations are equivalent. Thus, $-[-(-r \vee -s)]$
■ A.20 ■ is equivalent to $-(r \wedge s)$, or $-r \vee -s$ is equivalent to $-(r \wedge s)$. Why?
Let us summarize the equivalent statements that we will use in the following examples.

■ A.19 ■ 1. $-(p \vee q)$ is equivalent to $-p \wedge -q$.
■ A.20 ■ 2. $-(r \wedge s)$ is equivalent to $-r \vee -s$.

Example 1
Translate the following conjunctions to disjunctions.

■ A.20 1. $-p \wedge -q$ is equivalent to $-(p \vee q)$
       2. $-(-p \wedge q)$ is equivalent to $p \vee -q$
■ 3. $-p \wedge q$ is equivalent to $-(p \vee -q)$

Example 2

■ **A.19**   Translate the following disjunctions to conjunctions.

    1. $-(-p \lor q)$ is equivalent to $p \land -q$

    2. $-(p \lor q)$ is equivalent to $-p \land -q$

■   3. $p \lor -q$ is equivalent to $-(-p \land q)$

Besides finding the truth value of statements which involve negation, conjunction, and disjunction, you should be able to translate statements into symbolic forms and symbolic forms into statements. Consider the examples below.

Example 1

From the given statements, translate the compound statements below to symbolic statements.

    $p = $ "x is greater than 5"
    $q = $ "x is an odd number"
    $r = $ "x is less than 10"
    $s = $ "x is a multiple of 5"

■ **A.10**   SOLUTIONS   1. x is an even number less than 10.

                      We negate q to get the statement "x is an even number." Since x is both even *and* less than 10, the solution is $-q \land r$.

                2. x is less than 10 or it is not.

                      This translates as $r \lor -r$.

                3. It is not the case that x is a multiple of 5 and is not greater than 5.

                      $-(s \land -p)$. Notice that the first negation in this case applies to the whole conjunction $(s \land -p)$.

Example 2

From the symbolic statements below, translate to a sentence in English.

■ **A.11**   SOLUTIONS   1. $p \land -q$

                      x is greater than 5 and *not* odd. It can be better stated as: "x is an even number greater than 5," which means $x \in \{6, 8, 10, \ldots\}$.

                2. $s \land r$

                      x is a multiple of 5 and is less than 10, which means that x is 0 or 5.

                3. $-(p \land -q)$

                      It is not the case that x is both greater than 5 and not odd. This statement can be further refined to read, "It is not the case that x is an even number greater than 5," which is equivalent to stating, "x is an odd number or x is less than 5."

EXERCISES

Use the given statements in the exercises below.

p = "x is less than 6"
q = "x is an even number"
r = "x is a multiple of 3"
s = "x is greater than 10"
t = "x is less than 10"

1. Translate the following statements into English statements.

   a. $p \lor q$ b. $p \land s$ c. $-(-q \lor -r)$
   d. $-t \land r$ e. $t \lor s$

2. Find the truth values of the statements in Exercise 1 when $x \in \{2, 4\}$.

3. Translate the following sentences to symbols.

   a. x is an even number greater than 10.
   b. x is greater than 10 or less than 6.
   c. x is a multiple of 3 which is not greater than 10.
   d. x is less than 10 or greater than or equal to 6.

4. Write conjunctions which have the same truth table as the given disjunctions.

   a. $p \lor -q$ b. $-q \lor r$ c. $-(-s \lor t)$ d. $-(-s) \lor q$

5. Write disjunctions which have the same truth table as the given conjunctions.

   a. $q \land r$ b. $-p \land -s$ c. $-(-r \land q)$ d. $-(r \land p)$

6. For any statements a, b, and c, is the relation "has the same truth value" an equivalence relation? Why or why not?

7. Make a truth table which shows $p \land -q$, $-p \land q$, $-(-p \lor q)$, and $-(p \lor -q)$.

8. Determine the truth value of the following statements.

   a. Animals drink water and money grows on trees.
   b. Triangles are squares and rectangles are triangles.
   c. Squares are rectangles or triangles are rectangles.
   d. It is snowing or it is not snowing.
   e. $6 - 3 = 3$ or $2 \times 3 = 5$.
   f. $2 \times 3 = 5$ and $2 + 3 = 5$.
   g. It rains in October or it snows in May.
   h. It is never true that today is not today.
   i. June follows September and June precedes September.

SECTION B   CONDITION AND BICONDITION

BEHAVIORAL OBJECTIVES

B.1   Be able to describe a conditional statement.

B.2   Be able to make a truth table for a conditional or biconditional statement.

B.3   Be able to describe antecedent and consequent.

B.4   Be able to define biconditional statement.

B.5   Be able to relate symbolically the biconditional to the conditional.

B.6   Be able to define implication.

B.7   Be able to define equivalence.

B.8   Be able to translate an English sentence of the conditional or biconditional form into a symbolic statement which is a conditional.

B.9   Be able to translate a symbolic biconditional and conditional statement into an English sentence.

B.10 Be able to give examples of English statements which illustrate equivalence or implication.

B.11 Be able to write symbolic statements of equivalence.

B.12 Be able to show that a statement is an implication or an equivalence by use of a truth table.

B.13 Be able to state and prove the commutative law of $\wedge$.

B.14 Be able to state and prove the commutative law of $\vee$.

B.15 Be able to state and prove the associative law of $\wedge$.

B.16 Be able to state and prove the associative law of $\vee$.

B.17 Be able to state and prove the distributive law of $\wedge$ over $\vee$.

B.18 Be able to state and prove the distributive law of $\vee$ over $\wedge$.

B.19 Be able to state and prove DeMorgan's theorem.

B.20 Be able to state and prove the law of double negatives.

CONDITIONAL AND BICONDITIONAL

■ **B.1**
■

We have been using the phrase "if ..., then ..." throughout the book. This is a very common form in normal speech and is very important in mathematics. Sentences of the "if-then" type are commonly called *conditionals*. Consider some examples below.

Examples
1. If it rains, the streets will be wet.
2. If it is a triangle, then it is a polygon.
3. If it is a square, then it is a quadrilateral or a triangle.

You will notice that the statements above are compound statements formed from two or more simple statements. The statements given above are true conditionals, but not all conditionals are true. We will symbolize this new connective by using an arrow. The statement "if p then q" is written $p \rightarrow q$. We will eventually define this new connective by using a truth table as has been our convention; however, it might serve us better to consider the possible values of p and q under certain circumstances so that we get an intuitive understanding of our definition.

It seems obvious that if both p and q are true, the statement "if p then q" should be true. In the same manner, if p is true and q is false, we might expect the statement $p \rightarrow q$ to be false. The two remaining cases do not seem very obvious; however, we consider them both to be true. If p is false and q is true, or if both p and q are false, we consider $p \rightarrow q$ to be true. Since p and q need not relate in any particular manner, it is difficult to defend this definition. We can explain this situation by stating that if p happens to be false, we cannot judge the conditional to be false; therefore, we judge it to be true. If we take one of the statements above and apply this set of values to it, our rationale may be clearer.

Consider this statement: "If it rains, the streets will be wet." We know by experience that this is true. Whenever it rains, the streets become wet. Therefore, if p and q are true, the conditional is true. Suppose that it rained (p is true) and the streets stayed dry (q is false); then our conditional would be false. In the third case, it did not rain (p is false) but the streets are wet from some other sources (q is true); this does not make the conditional false, so the conditional is considered true. In the final case, it does not rain (p is false) and the streets do not get wet (q is false), but this does not affect our conditional, which could still be true.

■ B.3 ■ Note that we call p the *antecedent* and q the *consequent* in the conditional $p \rightarrow q$.

Consider Figure 10.8, which shows the conditional $p \rightarrow q$ and its defined truth values. Notice that the conditional is only false when p is true and q is false. This means that the conditional is true in all cases but one. Later you will see that we can translate a conditional into a disjunction or the negation of a disjunction.

■ B.2

p	q	$p \rightarrow q$
t	t	t
t	f	f
f	t	t
f	f	t

Figure 10.8

Another common form of speech is called the *biconditional*. The word seems to imply its meaning, namely, two conditionals. When we use the phrase "if and only if" in mathematics, we are asserting that two objects
■ B.5 are equivalent, or that $a \rightarrow b$ and $b \rightarrow a$. The biconditional is such a form, although not all biconditionals imply that the statements connected are
▲ B.4 equivalent. We can define the biconditional by using the conditional as follows: "p if and only if q" means "if p then q *and* if q then p." Symbolically, we use a double arrow ($\leftrightarrow$), and $p \leftrightarrow q$ has the same truth table as
▲ ■ the compound statement $[(p \rightarrow q) \wedge (q \rightarrow p)]$. Rather than define a truth table for $p \leftrightarrow q$, we will construct a truth table (Figure 10.9) using the relationship we just discussed.

From Figure 10.9, you can see that $p \rightarrow q$ and $q \rightarrow p$ are both true only

p	q	$p \rightarrow q$	$q \rightarrow p$	$[(p \rightarrow q) \wedge (q \rightarrow p)]$	$p \leftrightarrow q$
t	t	t	t	t	t
t	f	f	t	f	f
f	t	t	f	f	f
f	f	t	t	t	t

Figure 10.9

when p and q are both true or both false. When this is the case, the biconditional is true.

It should be instructive for you to see how the conditional and biconditional are used in the English language as well as how they differ in meaning in the language context. Let us consider some examples.

Example 1

Which of the following statements would be considered true within the language context?

1. If it is raining, the streets will get wet.
2. The streets will get wet only if it is raining.

SOLUTION   The first statement is conditional, and experience tells us that it is true unless the streets are covered. The second statement is false, because it implies that the street's being wet is sufficient for us to know that it has rained.

Example 2

Which of the following conditionals could be written as biconditionals and still be true?

1. If it is a triangle, then it is a polygon.
2. If he is vice-president, then he presides over the Senate.
3. If he can vote in Pennsylvania, he is a citizen of the U.S.A.
4. If a rhombus has a right angle, then it is a square.

SOLUTIONS   1. This is a true implication, but the converse, $(q \rightarrow p)$, is not always true, since not all polygons are triangles.
2. In general, this is a true biconditional. However, if the vice-president is not able to attend or if he has died, then this does not hold. In this case, if $p \rightarrow q$ is true, then $q \rightarrow p$ will also be true.
3. This is a true implication, because one qualification for voting in Pennsylvania is that one be a citizen of the U.S.A. The converse is not true, because there are other requirements one must have to be a qualified voter in a state.
4. This conditional is true, and it is also a true biconditional. A rhombus is a quadrilateral with four equal sides. If it has a right angle, then it is a square. Therefore, $p \rightarrow q$ and $q \rightarrow p$. (See Figure 11.9, page 417.)

Again, in the examples above, we used statements which we could judge true or false from prior knowledge. This is not required of us in our discussion of logic. We could always use statements which have little or no meaning in the real world. For example, we could write statements like these:

1. If it is a blob, then it is a skurd.

2. All skurds are not blobs.

3. A skurd is a blob only if it is a blurb.

We will, from time to time, evaluate the truth of statements such as these.

At this point, we should make some distinctions within the set of conditional and biconditional statements. These distinctions have to do with the concept of truth. We stated earlier that both conditionals and biconditionals can be either true or false; however, when the conditional or biconditional is always true, we give it a special name.

Definitions

■ B.6   *A conditional that is always true is called an implication and is symbolized*
■   *by ⇒.*

■ B.7   *A biconditional that is always true is called an equivalence and is sym-*
■   *bolized by ⇔.*

From our definitions above, we are asserting that if the conditional $p \to q$ is logically true, we will call this an implication and read it "p implies q." In symbols, $p \Rightarrow q$. We will also read "p is equivalent to q", or $p \Leftrightarrow q$ if the biconditional is logically true. In the section which follows we will deal with this concept in more detail.

From what we have said thus far about the conditional and the biconditional, you should be able to judge statements of this type to be true or false if you are given truth values for the antecedents and consequents. You should also be able to translate from symbolic language. to the written word and from the written word to the symbolic language. Lastly, you should be able to construct truth tables for given statements. Consider some examples below.

Example 1

Using the given symbols, translate the sentences below into symbolic language. Let the following phrases mean p, q, r, and s, respectively: it is a blob, it is a gash, it is blue, it is a nono.

■ B.8   1. If it is a blob, then it is blue.
         2. It is a gash if and only if it is a blue nono.
         3. If it is a nono then it is not a blob.

SOLUTIONS   1. $p \to r$
■   2. $q \leftrightarrow (r \wedge s)$
             3. $s \to -p$

Notice that if all the statements above are true, we can also make a diagram which would describe the relationship between the terms blob, gash, blue, and nono. Consider Figure 10.10. Notice that there is no indication that all nonos are blue.

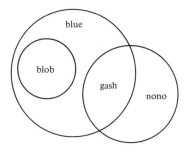

Figure 10.10

Example 2
Using the phrases given in Example 1, translate the symbolic statements below into sentences.

■ **B.9**

1. $r \rightarrow (p \lor q) \lor (r \land s)$
2. $(q \lor s) \rightarrow r$
3. $r \leftrightarrow (p \lor q)$

SOLUTIONS

1. If it is blue, then it is gash or a blob, or it is a blue nono.
2. If it is a gash or a nono, then it is blue.
3. It is blue only if it is a blob or a gash.
   Notice that these statements may be true or false, depending upon the situation defined. For example, in 3, it is obvious that other things may be blue. If we define blue as only those objects which are blobs or gashes, then 3 would be true.

EXERCISES

1. Construct truth tables for the following.
   a. $(p \lor q) \rightarrow q$ b. $p \rightarrow (p \lor q)$ c. $(p \land q) \rightarrow p$
   d. $(p \lor -p) \rightarrow r$ e. $(p \land r) \rightarrow (q \lor r)$ f. $-(p \land q) \rightarrow (-p \lor -q)$

2. From the given statements and symbols, translate the English sentences to symbolic statements.
   p = "it is a blurb" s = "it is a clop"
   q = "it is red" t = "it is a brib"
   r = "it is green" u = "it is a prop"

   a. If it is a prop, it is a blurb.
   b. If it is a clop, then it is a green blurb.
   c. No props are green.

d. If it is blue, it is not a blurb.

e. If it is a brib, it is not a blurb.

f. If it is a brib, then it is both blue and green.

3. Consider all the statements in Exercise 2 to be true. Try to make a single Venn diagram to illustrate the situation described above.

4. From the symbolic statements below, translate to English using the given statements.

p = "it is raining" r = "flowers are blooming"

q = "it is springtime" s = "3 and 4 is 7"

a. $r \rightarrow q$ b. $s \leftrightarrow (q \wedge r)$ c. $p \leftrightarrow q$

5. Write a compound statement which will always be true; use at least two connectives you have learned in this chapter.

6. For each of the given statements write a symbolic statement.

p = "Jon is honest"

q = "Bill is smart"

r = "Mary works hard"

s = "Bill will get a scholarship"

t = "They work hard"

u = "Mary will win a scholarship"

v = "Bill works hard"

a. Mary will win a scholarship only if she works hard.

b. Mary or Bill will get a scholarship if they work hard.

c. If Jon is honest and Bill does not win the scholarship, then they did not work hard or Bill is not smart.

d. If Mary does not work hard and Bill is not smart, then Mary and Bill will not win scholarships.

7. If a conditional is true in one instance, will the biconditional always be true in that instance? Why or why not?

8. If a biconditional is true in one instance, will the conditional always be true in that instance? Why or why not?

9. Let p be "It is snowing" and q be "It is sunny." Translate the following to symbolic form.

a. If it is snowing, the sun is shining.

b. If the sun is shining, then it is not snowing.

c. It snows only when the sun is shining.

d. When the sun is not shining, it never snows.

e. If it is not snowing, then the sun is shining.

10. Which of the following pairs of statements have the same truth table? Are they equivalent?

a. $p \vee q$ and $q \vee p$

b. $p \wedge q$ and $p \vee q$

c. $p \wedge q$ and $q \wedge p$

d. $(p \wedge q) \wedge r$ and $p \wedge (q \wedge r)$

e. $(p \wedge q) \vee r$ and $p \wedge (q \vee r)$

f. $p \vee (q \wedge r)$ and $(p \vee q) \wedge r$

g. $(p \vee q) \vee r$ and $p \vee (q \vee r)$

h. $p \wedge r$ and $q \wedge r$

11. From the pairs of statements above that have the same truth value,

a. do you believe that conjunction is commutative? Why?

b. do you believe that disjunction is commutative? Why?

c. do you believe that disjunction is associative? Why?

d. do you believe that conjunction is associative? Why?

12. Which of the pairs of statements below have the same truth table? Are they equivalent?

a. $p \lor (q \land r)$ and $(p \lor q) \land (p \lor r)$
b. $p \land (q \lor r)$ and $(p \land q) \lor (p \land r)$
c. $p \lor (q \land r)$ and $(p \lor q) \lor (p \lor r)$
d. $(q \land r) \lor p$ and $(q \lor p) \land (r \lor p)$
e. $(q \land r) \lor p$ and $q \land (r \lor p)$
f. $(q \lor r) \lor p$ and $(q \lor p) \lor (q \lor r)$

13. From Exercise 11 can you state the following symbolically? Use $\Leftrightarrow$ where it is possible!
a. a right distributive property of $\lor$ over $\land$.
b. a left distributive property of $\lor$ over $\land$.
c. a right distributive property of $\land$ over $\lor$.
d. a left distributive property of $\land$ over $\lor$.

IMPLICATION AND EQUIVALENCE

In this section we will further consider implication and equivalence, which were defined in the last section. We often say that something implies something else. For example, a politician may state that since his opponent quotes the *Radical News*, his opponent is a communist; or, if someone has friends who are known communists, they are unfit for office. Actually, since there is more than one possible set of truth values to a conditional, such statements are not logically true. *Such statements are conditionals but not logical implications.*

Even though implication takes on the same structure as the conditional, it should not be confused with the conditional. A simple way of differentiating between the two concepts is to state that the conditional is a compound statement of the if-then form, whereas implication is of the same form but there is a definite relationship between the antecedent p and the consequent q such that p implies q if and only if the conditional formed is logically true. Symbolically, we differentiate the conditional $\rightarrow$ from the implication $\Rightarrow$, and the biconditional $\leftrightarrow$ from equivalence $\Leftrightarrow$. Consider some examples of implication below.

Example 1

B.10   If it is a triangle, then it is a polygon. Let p mean "if it is a triangle," and let q mean "it is a polygon."

SOLUTION   There are four possibilities in the statement above.

1. Both are true and so is the conditional. (p and q are both true)
2. It is not a triangle and it is a polygon, that is, it is a square. (p is false and q is true)
3. It is not a triangle and it is not a polygon. (p and q are both false)
4. It is a triangle and it is not a polygon. (p is true and q is false)

The first three possibilities do not contradict the implication. If 4 can happen, then the given statement is not an implication. Actually, the figure should always be a polygon if it is a triangle. It is possible, as indicated, to have a figure which is a polygon but not a triangle. It is also possible to have a non-triangle which is also a non-polygon. Whenever a conditional is such that q is always true when p is true, we have an implication, $p \Rightarrow q$.

Example 2

■ B.10   At standard conditions, ice will form from water if the temperature is below 32°F. Notice that this can be read "if the temperature is below 32°F, then ice will form"; however, the "if" clause is the antecedent.

SOLUTION   Since it is never the case that ice will not form under these conditions, the statement given is an implication.

Example 3

■ B.10   If Jon is first in his class, then he will be valedictorian.

SOLUTION   If valedictorian is always chosen in this manner, then the statement is an implication. It is obvious that if Jon is not first, he will not be valedictorian. It will always be true that if he is first in his class, then he will be valedictorian. Actually, this statement could be reversed and would still be true. When such a situation occurs, we have a biconditional which is always true, and we call it an *equivalence*, or we say the statements are logically equivalent.

Let us consider some examples of equivalence.

Example 1

■ B.10   A person is a male *only if* he is not a female. Notice the use of "only if," which takes the place of "if and only if."

SOLUTION   Since we know that there is a partition on the set of people which produces males and females only, this statement is an equivalence. It is not possible to be a male and a female at once; therefore, if someone is a male, then that person is not a female, and if someone is not a male, then that person is a female. The original statement is a biconditional which is always true.

Example 2

■ B.10   A color is blue, yellow, or red if and only if it is a primary color.

SOLUTION   By definition, if we mention primary color, we think of blue, yellow, or red. If we ask the question, "Is a color a primary color?" the answer is always Yes if the color mentioned is blue, yellow, or red; and the answer is always No if the color mentioned is not one of the three. If we state that a color is *not* a primary color, we know that it is not blue, yellow, or red.

Example 3

■ B.10   A color is red if and only if it is not the case that it is not red.

SOLUTION   We are asserting that the statement "a color is red," or p, is equivalent to a statement which is the double negation of the statement "a color is red," or $-(-p)$. Symbolically this

B.11 ▲   reads "$p \Leftrightarrow -(-p)$." The simplest solution to this example would be to show by a truth table that whenever p is true, $-(-p)$ is true, and whenever $-(-p)$ is true p is true. We are asserting here that "$p \to -(-p)$" and "$-(-p) \to p$" are always true. You will notice in Figure 10.11 that both the antecedent and the consequent have *exactly the same* truth values. This is another way to show that two statements are equivalent.

■ B.12

p	$-p$	$-(-p)$	$p \Leftrightarrow -(-p)$
t	f	t	t
f	t	f	t

Figure 10.11

As a result of our definition of equivalence and the work done in Section A of this chapter, we are ready to state several theorems. Each of these has been proved earlier but not stated as a theorem. See Exercises 10–13 on pages 392–393 as well as Figures 10.6 and 10.7 on page 384 for proof of the statements below.

Theorems

1. *For all p and q statements*

B.13 ■ $p \wedge q \Leftrightarrow q \wedge p$ *commutative law of* $\wedge$
B.14 ■ $p \vee q \Leftrightarrow q \vee p$ *commutative law of* $\vee$

2. *For all p, q, and r statements*

B.15 ■ $(p \wedge q) \wedge r \Leftrightarrow p \wedge (q \wedge r)$ *associative law of* $\wedge$
B.16 ■ $(p \vee q) \vee r \Leftrightarrow p \vee (q \vee r)$ *associative law of* $\vee$

3. *For all p, q, and r statements*

B.17 ■ $p \wedge (q \vee r) \Leftrightarrow (p \wedge q) \vee (p \wedge r)$ *distributive law of* $\wedge$ *over* $\vee$
B.18 ■ $p \vee (q \wedge r) \Leftrightarrow (p \vee q) \wedge (p \vee r)$ *distributive law of* $\vee$ *over* $\wedge$

4. *For all p and q statements*

◄ B.19 $-(p \vee q) \Leftrightarrow (-p \wedge -q)$ *DeMorgan's law*
■ $-(p \wedge q) \Leftrightarrow (-p \vee -q)$ *DeMorgan's law*

5. *For all p statements*

.20 ■ $p \Leftrightarrow -(-p)$ *law of double negative*
   (see Example 3 above)

EXERCISES

1. Use truth tables or known properties to classify the following as conditionals, biconditionals, implications, or equivalences.

a. $(p \wedge q) \rightarrow (-p \vee q)$

b. $(p \rightarrow -q) \leftrightarrow (-q \rightarrow p)$

c. $(-p \vee q) \rightarrow -(-q \rightarrow p)$

d. $(p \rightarrow q) \leftrightarrow (-p \wedge q)$

e. $(p \vee q) \rightarrow (-p \vee -q)$

f. $(p \rightarrow -q) \leftrightarrow (q \rightarrow -p)$

g. $(-p \vee q) \leftrightarrow -(p \wedge q)$

h. $(p \vee -p) \rightarrow (q \vee -q)$

2. Let the statements below take on the symbolic representation given. Decide which symbolic statements are implications or equivalences, and explain.

p = "it is raining"

q = "the sun is shining"

r = "the ground is dry"

s = "the ground is wet"

a. $-r \rightarrow -q$ b. $s \rightarrow -r$ c. $r \leftrightarrow s$ d. $-q \leftrightarrow p$

e. $-r \rightarrow s$ f. $s \leftrightarrow -r$ g. $-r \leftrightarrow s$ h. $q \rightarrow r$

3. If q is logically false, which of the following statements are logically true? Why?

a. $q \rightarrow p$ b. $q \leftrightarrow p$ c. $(q \vee -q)$

d. $(q \wedge -q)$ e. $(p \vee -q)$ f. $(-q-p) \leftrightarrow p$

g. $(q \rightarrow -p) \rightarrow (p \rightarrow -q)$

4. Show by the use of truth tables which of the following are equivalent.

a. $-r$ and $-[-(-r)]$

b. $(-p \wedge q)$ and $[(p \vee (-q)]$

c. $-(p \rightarrow q)$ and $(p \rightarrow -q)$

d. $(p \rightarrow -q)$ and $(-p \vee q)$

5. Make a table and prove that the following are true or false.

a. $(p \rightarrow q) \Leftrightarrow (-q \rightarrow -p)$

b. $(-q \rightarrow p) \Leftrightarrow (-p \rightarrow -q)$

c. $(p \wedge q) \Rightarrow (p \vee q)$

d. $(p \vee q) \Rightarrow (p \wedge q)$

e. $(-p \rightarrow -q) \Leftrightarrow (q \rightarrow p)$

f. $p \Rightarrow p \wedge q$

6. Prove the following statements from the theorems listed in this section.

a. $p \wedge -p \Leftrightarrow -p \wedge p$

b. $(q \wedge r) \vee p \Leftrightarrow (q \vee p) \wedge (r \vee p)$

c. $-p \Leftrightarrow -[-(-p)]$

d. $(-p \vee q) \Leftrightarrow -(p \wedge -q)$

7. Complete the following statements.

a. $p \vee (q \wedge r) \Leftrightarrow$	distributive of $\vee$ over $\wedge$
b. $(p \wedge q) \vee (p \wedge r) \Leftrightarrow$	distributive of $\wedge$ over $\vee$
c. $(p \wedge q) \wedge r \Leftrightarrow$	associative of $\wedge$
d. $p \wedge q \Leftrightarrow$	commutative of $\wedge$
e. $-(p \vee q) \Leftrightarrow$	DeMorgan's theorem

SECTION C TAUTOLOGY

BEHAVIORAL OBJECTIVES

C.1 Be able to define tautology.

C.2 Be able to give an example of a statement that is a tautology.

C.3 Given a conditional statement, be able to form the converse, inverse, and contrapositive.

C.4 Be able to prove by using truth tables that a statement is or is not a tautology.

C.5 Be able to write tautological statements using statements and their inverses, converses, and contrapositives.

TAUTOLOGY AND EQUIVALENT CONDITIONALS

In the last section we dealt primarily with implications and equivalence. Both of these concepts were defined as statements which are always true. All statements which are implications and equivalences are *tautologies*. It should follow from this that a discussion of tautologies would deal with statements which are always true.

Before we define tautology, we will consider some examples of statements which are tautologies, or *tautological statements*.

Examples

■ C.2   1. The statement "$p \vee -p$" is a tautology. Since one of the statements is always true when the other is false, the disjunction is always true.

2. The statement "$-(p \wedge -p)$" is a tautology. You should see that this is equivalent to the statement above and can be derived from it. Again, the statement $-(p \wedge -p)$ must always be true, because the statement $(p \wedge -p)$ is always false.

3. The statement "$(p \wedge -p) \rightarrow r$" is a tautology. This statement is an implication also, and it is always true because the antecedent is always
■   false.

We should now be able to define tautology from the examples given and the work in the previous section.

Definition

C.1 ■   *A tautology is a statement which is constructed so that it is always true.*

In the previous section we discussed briefly the concept of a conditional statement. Let us consider a few examples of given conditional statements and their equivalent conditional statements.

Example 1

◀ C.3   Given the conditional $p \rightarrow -q$, find the inverse, converse, and contrapositive.

SOLUTION   The *converse* is a conditional in which the antecedent and consequent are reversed; therefore, $-q \rightarrow p$ is the converse of the given conditional, $p \rightarrow -q$.

The *inverse* is simply the negation of the antecedent and consequent without a change in order. Thus, the inverse in this instance is $-p \rightarrow q$.

The *contrapositive* is produced by negating and changing the order of the antecedent and consequent of the given statement. In this case we produce $q \rightarrow -p$.

It is left to you to show that the two statements below are tautological.

1. $(p \to -q) \Leftrightarrow (q \to -p)$
2. $(-p \to q) \Leftrightarrow (-q \to p)$

■ **C.3** Notice that if $(-p \to q)$ were the given statement, $(-q \to p)$ would be
■ its contrapositive.

Example 2

■ **C.3** Given the statement "$-p \to -q$," construct its contrapositive, and
show that the two statements are equivalent.

▲ **C.5** SOLUTION The contrapositive of $-p \to -q$ is $q \to p$. By a truth table
(Figure 10.12), we can show that these statements are equiv-
● **C.4** alent and that the equivalence formed must be a tautology.

p	q	$-p$	$-q$	$(-p \to -q)$	$(q \to p)$	$(-p \to -q) \Leftrightarrow (q \to p)$
t	t	f	f	t	t	t
t	f	f	t	t	t	t
f	t	t	f	f	f	t
f	f	t	t	t	t	t

Figure 10.12

Example 3

Given the statement "$-p \to q$," write the converse and inverse, and
show that they are equivalent.

▲ **C.5** SOLUTION The converse of $-p \to q$ is $q \to -p$. The inverse of $-p \to q$
is $p \to -q$. It is left to you to show that these statements are
equivalent and that $(q \to -p) \leftrightarrow (p \to -q)$ is a tautology.
Note: If the given statement, $-p \to q$, is true, then the converse
and inverse are not necessarily true. What about the contra-
▲ ■ positive?

EXERCISES

1. Construct three tautological statements from the given statements.
   a. $(p \lor -p)$ b. $(-q \to -p)$
   c. $(q \to p)$ d. $(p \lor -q)$
   e. $(p \to q)$ f. p

2. Which of the following statements are true? Explain.
   a. All implications are tautologies.
   b. All tautological statements are implications or equivalences.
   c. The set of equivalent statements in logic forms an equivalence relation.
   d. Any true statement is a tautology.
   e. The statement "$[(p \lor q) \land -p]$" is a tautology.

3. From the conditional $(p \lor q) \to q$, construct the converse, inverse, and contrapositive.

4. From the statements below, find pairs which are contrapositives. Explain.

a. $(p \land q) \to -r$ b. $-r \to (p \lor -q)$

c. $-(-p \land q) \to r$ d. $r \to (-p \land -q)$

e. $r \to -(-p \land q)$ f. $-(p \lor q) \to -r$

g. $r \to (-p \lor -q)$

5. Which of the following statements are tautologies? Explain.

a. $-q \to -p$ b. $(p \lor q) \to p$

c. $(p \land -p) \to q$ d. $(p \lor -r) \to q$

e. $(p \land q) \to (p \lor q)$ f. $(p \to q) \leftrightarrow (-p \lor q)$

g. $[(p \to q) \land (q \to p)] \leftrightarrow (p \leftrightarrow q)$ h. $[(p \to q) \land (q \to r)] \to (p \to r)$

SECTION D DISJUNCTION, SYLLOGISM, AND PROOF

BEHAVIORAL OBJECTIVES

D.1 Be able to state and prove the law of excluded middle.

D.2 Be able to state and prove the law of contradiction.

D.3 Be able to derive statements of disjunction from conditional statements.

D.4 Be able to state and prove the law of syllogism.

D.5 Be able to state and prove modus ponens.

D.6 Be able to state and prove modus tollens.

D.7 Be able to state and prove the law of simplification for $\land$.

D.8 Be able to use the laws developed in this chapter for deductive proof.

D.9 Be able to state and prove the law of contrapositive.

SOME LAWS OF LOGIC AND PROOF

In previous sections of this chapter we developed some laws of logic; several of these laws were stated as Theorems (see Theorems 1–5). We have also developed several laws, which have not yet been stated as theorems. We will now state these as theorems and leave the proofs to you.

The first of these laws we called the law of excluded middle, which stated that "$p \lor -p$" is always true (Fig. 10.5). From this we derived the law of contradiction, which stated that "$-(p \land -p)$" is always true. We also showed that from these two laws, the statement "$p \land -p$" is always false and also named this the law of contradiction (Fig. 10.5).

Theorem 6 *(law of excluded middle)*

■ **D.1**   *For all p statements*

■ $p \lor -p$ *(is true)*

Theorem 7 *(law of contradiction)*

■ **D.2**   *For all p statements*

■   1. $- (p \wedge - p)$ *(is true)*
2. $p \wedge - p$ *(is false)*

When dealing with compound statements involving negation, we showed equivalent statements, one of which was a conjunction and the other a disjunction. It might be of value for you to be able to derive the relationship between the conditional statement and statements of conjunction or disjunction.

■ **D.3**   Consider Figure 10.13. What can you replace □ and ○ with?

p	q	$p \to q$	□ ∨ ○	$(p \to q) \leftrightarrow (□ ∨ ○)$
t	t	t	t	t
t	f	f	f	t
f	t	t	t	t
f	f	t	t	t

Figure 10.13

You should be able to fill the square and circle in the table above with p, q, $- p$, or $- q$ so that the table is correct. Then, you should be able to relate conditional statements to disjunction. Having done this, you should then be able to relate conditional statements to conjunction, since it is possible to translate conjunctions to disjunctions.

If you had trouble doing this task, consider that the conditional is false in one instance only. What pair of statements taken from p, q, $- p$, or $- q$ can be joined by disjunction so that the conditional is false when p is true and q is false? This should help to answer the question for you.

You probably noticed in the last exercise set that the statement in Exercise 5h was a tautology. In fact, we call this the *law of syllogism*. Notice that it looks much like the transitive property of relations, because it states that "if $p \Rightarrow q$ and $q \Rightarrow r$, then $p \Rightarrow r$." This law can be extended to form a chain of logical implications. If we are given several statements which in turn imply other statements, then we can imply a statement which we plan to prove. This law may be the most important law of logic, because deductive proof is based upon it. Before we extend this law, we will state it as a theorem. The law is named the *law of syllogism* or the *transitive law of implication*.

Theorem 8 *(law of syllogism, or transitive law of implication)*
    For all p, q, r statements

■ **D.4** $[(p \Rightarrow q) \wedge (q \Rightarrow r)] \Rightarrow (p \Rightarrow s)$

Quite often this theorem is extended to form a finite chain of implication.

It is then called the *chain law*, or *extended* transitive law of implication. For example,

$$[(p \Rightarrow q) \wedge (q \Rightarrow r) \wedge (r \Rightarrow s) \wedge (s \Rightarrow t)] \Rightarrow (p \Rightarrow t)$$

You should be able to prove Theorem 8 with truth tables and the extended law by applying the theorem several times. We leave these proofs as exercises.

EXERCISES

1. State a rule, law, or relationship which explains why each of the statements below is true.

a. $[(p \wedge q) \vee -(p \wedge q)]$

b. $[(p \rightarrow q) \Leftrightarrow -(-p \vee q)]$

c. $[(p \vee -q) \Leftrightarrow -(-p \wedge q)]$

d. $[(p \wedge q) \Leftrightarrow -(-(p \wedge q))]$

e. $\{[(q \Rightarrow r) \wedge (r \Rightarrow q)] \Rightarrow (q \Rightarrow q)\}$

2. Given that $p \rightarrow r$ and $r \rightarrow s$ are true, prove by stating a law that $p \rightarrow s$.

3. Prove without the use of a truth table that the statement below is always true.

$$[(q \Rightarrow r) \wedge (r \Rightarrow s) \wedge (s \Rightarrow p)] \Rightarrow (q \Rightarrow p)$$

(*Hint*: If $q \Rightarrow r$ and $r \Rightarrow s$, then what is true?)

4. Fill in the blank spaces with statements p, q, r, or the negation of the statements given such that each statement below will be an illustration of a rule, law, or relationship. Name the rule, law, or relationship, or explain it.

a. $(p \vee \square) \leftrightarrow -(\square \vee \bigcirc)$

b. $(q \rightarrow p) \leftrightarrow (\square \vee \bigcirc)$

c. $(p \wedge \square) \leftrightarrow (\bigcirc \rightarrow -r)$

d. $[(p \rightarrow \bigcirc) \wedge (\bigcirc \rightarrow \square)] \rightarrow (\diagdown \rightarrow q)$

5. State Theorems 7, 8, and 9 symbolically.

DEDUCTIVE PROOF

A statement to be proved consists of some given statements which we call *hypotheses* and a desired conclusion which we are asked to logically reach. To enable us to reach a logical conclusion, we may sometimes state in one line a law, theorem, or property. A conclusion can be reached in several ways, and often one person does not see the logic that another person may use.

Since children often prove by the use of intuition they can be stilted by the teacher's refusal to recognize the child's logic. It is important for the teacher to be able to understand, as well as think, in such a way that many possible routes of proving are open to her students. Too often, a student's lack of verbal ability does not allow communication between student and teacher. With children, proof by drawing, writing, and describing are all acceptable means to conclusion. Teachers should leave all possible routes open to the student and should not force verbal or written proof too quickly. Intuition should be allowed and encouraged.

The method of deductive proof may not always be the best route to a conclusion, but it is one route that the mathematician, scientist, and logician use quite often. The deductive method requires establishing a chain of logical statements, each of which implies the next. Therefore, most deductive proofs use the chain rule, or extended transitive property. Within the chain, we justify step after step by stating rules, laws, definitions, and properties. Some examples of this type of proof are given below.

Example 1

Given that the door is red. If the door is red, then it is not pretty. If the door is red, Jon will like it. Prove that Jon will like it.

p = "the door is red"
q = "it is not pretty"
r = "Jon will like it"

SOLUTION   We are given these hypotheses:
1. p
2. $p \to q$
3. $p \to r$

The proof goes this way.

1. p	1. given
2. $p \to r$	2. given
3. r	3. definition of true conditional

In the proof above we used the definition of the conditional, which states that a conditional is false only when the antecedent is true and the consequent is false. Since both p and $p \to r$ were given as true, we concluded that r must be true.

The proof above could have been done simply by stating a law named *modus ponens*. This law states that if p is true and $p \to q$ is true, then q is true.

Theorem 9 *(modus ponens)*

D.5   *For p and q statements*

if p (is true)

and

if p → q (is true)

then

q (is true)

We will use this law in the future, and the proof will be more formal and less intuitive. It is easily shown that this is a law by a truth table; however, this proof is left as an exercise.

■ **D.6**   Now, as a result of this we can show that if $p \rightarrow q$ is true, its logical equivalent is true; namely, $\sim q \rightarrow \sim p$. If $\sim p$ is true, then we have the following chain of logic.

$p \rightarrow q$ (is true)

therefore,

($\sim q \rightarrow \sim p$) (is true)   Why?

If

$\sim q$ (is true)

then

■ $\sim p$ (is true)   Why?

We state this line of reasoning in a simpler form and call it *modus tollens*.

Theorem 10 *(modus tollens)*
■ **D.6**   *For p and q statements*

if $p \rightarrow q$ *(is true)*

and

if $\sim q$ *(is true)*

then,

■ $\sim p$ *(is true)*

Notice that we could have stated Theorems 9 and 10 as follows:

◀ **D.5** ■ 1. $[p \wedge (p \rightarrow q)] \rightarrow q$ modus ponens
◀ **D.6** ■ 2. $[(p \rightarrow q) \wedge \sim q] \rightarrow \sim p$ modus tollens

Since the antecedents are conjunctions, the order of statements could be changed with respect to the $\wedge$ symbol. Why?

Example 2
■ **D.8**   Given that r is true, $-q \rightarrow s$ is true, and $q \rightarrow -r$ is true, prove that s is true.

PROOF 1

1. $q \rightarrow -r$	1. given
2. $r \rightarrow -q$	2. contrapositive
3. $-q \rightarrow s$	3. given
4. r	4. given
5. $-q$	5. theorem 9 (modus ponens)
6. s	6. theorem 9 (modus ponens)

Now since we have several theorems, we could shorten the proof given above.

PROOF 2

1. $q \rightarrow -r$	1. given
2. r	2. given
3. $-.q$	3. theorem 10 (modus tollens)
4. $-q \rightarrow s$	4. given
■ 5. s	5. theorem 9 (modus ponens)

Example 3

Choose from the following hypotheses those which are necessary for the proof and indicate why the others are not needed: p, $-q \rightarrow r$, $-s \vee -r$, $p \wedge -q$, $p \vee -q$, and $s \rightarrow -r$ are true. Prove that $-s$ is true.

There are several hypotheses that can easily be deleted. Since we know that $p \wedge -q$ is true, we know that $-q$ is true by definition. Since we know that $p \wedge -q$ is true, we need not be given that $p \vee -q$ is true. Since $-s \vee -r$ and $s \rightarrow -r$ are equivalent, only one is needed. We will use the latter in the proof below; this is simply a matter of choice and either would do.

PROOF

■ D.8 1. $-q \rightarrow r$	1. hypothesis
2. $s \rightarrow -r$	2. hypothesis
3. $r \rightarrow -s$	3. contrapositive of 2
4. $[(-q \rightarrow r) \wedge (r \rightarrow -s)]$	4. definition of conjunction
5. $-q \rightarrow -s$	5. law of syllogism (transitive law)
6. $p \wedge -q$	6. hypothesis
7. $-q$	7. definition of conjunction
8. $-s$	8. modus ponens (5 and 7)

Notice that we can again use our theorems to shorten this proof. This also allows you to consider alternative routes in proving statements.

PROOF

1. $p \wedge -q$	1. given
2. $-q$	2. definition of conjunction
3. $-q \rightarrow r$	3. given
4. r	4. theorem 9 (steps 2 and 3)
5. $s \rightarrow -r$	5. given
■ 6. $-s$	6. theorem 10 (steps 4 and 5)

In steps 1 and 2 we used the definition of conjunction. This is often stated as a theorem. We will now state this as simplification of $\wedge$.

Theorem 11 *(simplification of $\wedge$)*
■ D.7   *For all p and q statements*
     1. *if p $\wedge$ q*   *(is true)*
       then p   *(is true)*

2. *if p ∧ q*   (*is true*)

■   *then q*   (*is true*)

Before offering some exercises we summarize the laws and theorems below.

Theorem 1   commutative laws for ∧ and ∨
Theorem 2   associative laws for ∧ and ∨
Theorem 3   distributive laws of ∧ over ∨ and ∨ over ∧
Theorem 4   DeMorgan's law
Theorem 5   law of double negative
Theorem 6   law of excluded middle
Theorem 7   law of contradiction
Theorem 8   law of syllogism (transitive law)
Theorem 9   modus ponens
Theorem 10   modus tallens
Theorem 11   simplification of ∧

We also showed that a statement and its contrapositive are equivalent. We will now state this as a theorem.

Theorem 12 *(law of contrapositive)*

■ **D.9**   *For all p and q statements*

■ $(p \rightarrow q) \Leftrightarrow (-q \rightarrow -p)$

EXERCISES

1. Name a theorem that the statements below illustrate.
   a. If he is nice, then he is happy. He is nice. Therefore, he is happy.
   b. If 3 is greater than 4, then 4 is greater than 5. But 4 is *not* greater than 5. Therefore, 3 is not greater than 4.
   c. If $2X + 5 = 7$, then $2X = 2$. If $2X = 2$, then $X = 1$. Therefore, if $2X + 5 = 7$, then $X = 1$.
   d. If a snake is a reptile and a whale is a mammal, then a whale is a mammal.

2. Given: q and $-p \wedge q$
   Prove: $-p$

3. Given: If it is a square, then it is a rhombus and a rectangle. If it is a rhombus, then it is not a circle. It is a circle.
   Prove: It is not a square.

4. Given: If it is a rhombus, then it is a parallelogram. If it is a parallelogram, then it is a trapezoid. It is not a trapezoid.
   Prove: It is not a rhombus.

5. Given: If it is a quadrilateral, then it is a polygon. If it is a polygon, then it is not a circle. It is a circle.
   Prove: It is not a quadrilateral.

6. Given: If Jon does not love Mary, then he loves Paula.
  When Jon loves Paula, Mary stays home.
  Mary does not stay home.
  Prove: Jon loves Mary.

7. Given: $p, p \rightarrow q, q \rightarrow r, -s \rightarrow -r$
  Prove: s

8. Given: $p \vee q, p \rightarrow q, q \rightarrow r$
  Prove: $q \vee r$

9. Given: It is a tall tree.
  If it is a tree, then it is green.
  If it is green, then it is not red.
  Prove: It is not red.

10. Write a conclusion that follows from the given statements. State a law as your reason.
  a. Pennsylvania is in the United States. If Pennsylvania is in the United States, then Philadelphia is in the United States.
  b. If Pennsylvania is in the United States, then New York and Atlantic City are in Pennsylvania. New York and Atlantic City are not in Pennsylvania.
  c. May loves to write. If May loves to write, then her dinner can be improved by making it warmer.
  d. If the manual is easy to read, then Daniel can work on his car. If Daniel can work on his car, then most people could travel easily.
  e. A good detergent will clean clothes. If a good detergent will clean rugs, then Adam's clothes can be cleaned. If Adam's clothes can be cleaned, then his mother will be happy.
  f. David and Sara go. If David goes, then Sara is happy.

CAPSULE VI GAMES

We might characterize educational games as a method of packaging concepts. The introduction of skills and concepts in a game form is intrinsically motivating, due to the fact that the subject matter seems secondary to the game while the role which the learner takes in dealing with this material is an active one.

Using games as a mode of instruction has several advantages. It seems obvious that children enjoy playing games, and, in fact, playing games is part of the child's existence. This makes game playing a very natural thing which draws on the child's experience. Second, games provide an opportunity for children to deal with very complex problems in a concrete way. Many games are created to simulate real-life situations, thus being more real than many of the strategies used in traditional text-book-workbook teaching. Also, some children may not feel as badly about losing a game as being told by a teacher that they are wrong. This last aspect allows the student to make and correct errors in strategy, to strengthen skills, and to freely interact with others without the usual judgmental atmosphere found in many school situations. Another positive aspect of learning through educational games is that attention span is often greatly increased. Children who cannot sit through a short lesson can often play a game for hours.

In Paschal's article, it is pointed out that one of the best ways of working with disadvantaged students in the elementary school is through the use of games. On the other hand, Glenadine Gibb points out that bright children need motivation to think more analytically, and one approach to this problem is that of employing games. In the Nova Academic Games project directed by James Coleman at Johns Hopkins University, the goal is to integrate academic games into the curriculum. In one of the studies in this project, a group of children from the fourth, fifth, and sixth grades participated in playing the game of *Equations*. The group's increase in both skills and concept development was almost double their usual average increase in a nine-week period. All this need not be conclusive evidence for using games, but it certainly is impressive.

The use of card games has often been prohibited in schools because of the negative connotation of such games. Yet we all know that most card games require decision making and call on numerical computation, probability, memory, and learning rules. Peter Gaurau from Springfield College has developed a curriculum for beginning mathematics by using a series of card game lessons. He uses the first ten cards, ace through ten, of every suit in most of his early lessons. These games are designed to develop counting, ordering, naming, and recognizing number size and relationships. He employs games such as *War* and *Concentration*. We also suggest for older children the game *Cribbage*, which involves counting, sequencing, and decision making.

Teachers should become familiar with commerical games, read about game theory, and try creating games for classroom use. The process of creating games and then playing them is an exciting one. Children as well as teachers should actively participate in this process. It is important to note that the use of games does not replace the well-planned instructional curriculum, but it enriches and supplements it. The teacher must be able to choose the right games and present them in a stimulating way. There should be a follow-up procedure for most games, particularly simulation games. This procedure can lead to strengthening concepts developed and clarifying the decision-making aspect developed in the game. Children may want to replay games or portions of games and apply what they have learned during the game and in the follow-up discussion. Many games do not require a follow-up discussion on the part of the teacher, and, in fact, such a follow-up may in these cases stilt the excitement and motivation of the game for the students.

As a summary, we repeat the following major points regarding games. Games can increase student motivation, clarify difficult concepts and processes, help verbal interaction, increase attention span, and reduce the need to produce the "right answer." Games are most effective as a supplement to the other teaching modes used. Playing and creating games allow children to participate in what Bruner calles the "methodology of the discipline." Play some games!

SUGGESTED READINGS

Allen, Ernest E. "Bang, Buzz, Buzz—Bang, and Prime," *The Arithmetic Teacher*, June, 1969, pp. 494–495.

Deans, Edwina. "Games for the Early Grades," *The Arithmetic Teacher*, February, 1966, pp. 140–141.

———. "More Games for the Early Grades," *The Arithmetic Teacher*, March, 1966, p. 238.

Doffer, Dora. "The Role of Games, Puzzles, and Riddles in Elementary Mathematics," *The Arithmetic Teacher*, November, 1963, pp. 450–452.

Gibb, Glenadine. "Basic Objectives of the Program," *Mathematics in Elementary Education*, edited by Nicholas Vigilante. London: Macmillan, 1969, p. 50.

Golden, Sarah. "Fostering Enthusiasm Through Child-Created Games," *The Arithmetic Teacher*, February, 1970, p. 113.

Gordon, Alice. *Games for Growth*. Palo Alto, California: Science Research Associates, 1970.

Gurau, Peter. "A Deck of Cards, a Bunch of Kids, and Thou," *The Arithmetic Teacher*, February, 1969, pp. 115–117.

Massey, Tom. "Dominoes in the Mathematics Classroom," *The Arithmetic Teacher*, January, 1971, pp. 53–54.

Overholster, Jean S. "Hide-a-Region . . . $N \geq 2$ Can Play," *The Arithmetic Teacher*, June, 1969, pp. 496–497.

Paschal, Billy. "Teaching the Culturally Disadvantaged Child," *The Arithmetic Teacher*, May, 1966, pp. 369–374.

Stark, George. "A Games Theory in Education," *School and Society*, vol. 96, no. 2301, p. 43.

CHAPTER 11 **GEOMETRY**

SECTION A **POINTS, LINES, AND PLANES**

BEHAVIORAL OBJECTIVES

A.1 Describe a point by using physical examples.

A.2 Explain the notation for a point on a plane.

A.3 Describe a plane by using a physical model.

A.4 Describe a line by using a physical example.

A.5 Be able to name a line by using the standard notation.

A.6 Define a line segment.

A.7 Be able to describe a given line segment by using the standard notation in the text.

A.8 Define a ray.

A.9 Be able to name a ray by using the standard notation.

Much of the world around us is described geometrically. Traffic signs are made of familiar geometric shapes. Boxes, bricks, and homes can be described in terms of geometric solids. We often use physical models of geometric figures to illustrate the "ideal figures" of geometry. Children become familiar with the physical models early. For example, a line segment may be modeled by a piece of string, the edge of a page in a book, or the intersection of two walls in a room.

In this chapter we introduce mathematical objects and notation that will enable us to deal with geometry in the "ideal" sense. We will first summarize some of the basic notions required to do geometry. Some of these notions may be familiar to you if you have had some dealings with geometry; others will be completely new.

■ A.1 The concept of a number line was discussed in Chapters 5, 7, and 9 along with the notion of a point on the line. However, we must be sure to note that the physical representation of a point as a dot or the tip of a pencil is nothing more than an illustration to give an intuitive idea of

▲ A.2 the concept "point." *We think of a point as a location in space and usually name a point by a capital letter.* In Figure 11.1 the set of points symbolized

▲ ■ by the dots is named $\{A, B, C\}$.

■ A.3 It seems that the points A, B, and C all lie on the same flat surface. Such a flat surface is often described as a *plane*. A plane can be illustrated by employing a physical model such as a window, wall, or blackboard. This type of physical example could cause some basic misunderstandings, because a plane is not bounded as are the objects mentioned above.

Again, we should make it clear that these physical representations serve only as intuitive models for a plane. *A plane can best be described as a set of points, all of which lie on a flat surface that continues infinitely in all*

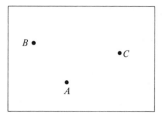

Figure 11.1

directions. In this way, if we consider a blackboard as a physical example, we must point out that this physical model only serves as a portion or subset of the plane. The plane itself continues in all directions. A plane is usually named by three points such as those in Figure 11.2, which we
■ might describe as *plane ABC.*

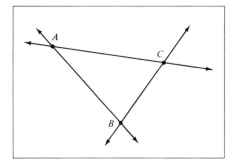

Figure 11.2

■ **A.4**   A *line* is another idea we describe rather than define. (Notice that we did not define point or plane.) A line is a subset of a plane. *We say that two points determine a line, and that this line is a subset of every plane that*
▲ **A.5** ■ *contains the two points.* We name lines by using the names of two points of the line. For example, the line through the points A and B is symbolized by $\overleftrightarrow{AB}$. The double-headed arrow indicates that the line extends infinitely
▲ in both directions. Figure 11.2 shows the lines $\overleftrightarrow{AB}$. $\overleftrightarrow{BC}$, and $\overleftrightarrow{AC}$.
■ **A.6**   A *line segment* is a subset of a line. *It consists of two points on a given*
■ *line and all the points in between.* We describe the line segments in Figure 11.3 as $\overline{AB}$, $\overline{BC}$, and $\overline{AC}$. Notice that we do not place arrows on the ends of these segments, because these segments end. Figure 11.3 shows a line
■ **A.7** with several points on it. Can you name the line segments represented? If you listed $\overline{AB}$, $\overline{AC}$, $\overline{AD}$, $\overline{BC}$, $\overline{BD}$, and $\overline{CD}$, you were correct. *Note:* The

Figure 11.3

points used to name the line segments are called endpoints. We should also note that any pair of points of a line segment could be used to name the line that contains the segment. Thus, in Figure 11.3, while $\overline{AB}$ and $\overline{CD}$ are different segments, $\overleftrightarrow{AB}$ and $\overleftrightarrow{CD}$ are the same line. Since we assumed in Chapter 9 that there is a one-to-one correspondence between the points of a line and the set of real numbers, we have that the endpoints of a segment correspond to two real numbers and the segment consists of all points corresponding to the real numbers between and including the two given real numbers.

■ **A.8**   *A ray is another subset of a line which includes an endpoint and the set of all points to the right of that endpoint or the left of that endpoint, but not* ■ *both.* Another way of describing a ray is as a subset of a line that consists of all the points corresponding to the real numbers greater than or equal ■ **A.9** to some given real number. What does a ray look like? It's half of a line, *including an endpoint*, as in Figure 11.4! We symbolize the ray in Figure 11.4 by $\overrightarrow{AB}$.

Figure 11.4

In Figure 11.4 the ray $\overrightarrow{AB}$ consists of all the points corresponding to numbers greater than or equal to the real number which corresponds ■ to the point A.

EXERCISES

1. Name all of the line segments on $\overleftrightarrow{AB}$ below.

2. In the figure in Exercise 1 which rays are subsets of the ray $\overrightarrow{AB}$? Which rays are not subsets of the ray $\overrightarrow{AB}$?

3. a. Is $\overline{AB} = \overline{BA}$?   Why?
   b. Is $\overrightarrow{AB} = \overrightarrow{BA}$?   Why?
   c. Is $\overleftrightarrow{AB} = \overleftrightarrow{BA}$?   Why?
   d. Is $\overline{AB} = \overline{AY}$?   Why?
   e. Is $\overrightarrow{AB} = \overrightarrow{AY}$?   Why?
   f. Is $\overleftrightarrow{AB} = \overleftrightarrow{AY}$?   Why?

4. a. If A, B, C are on a plane and D is a point not on the plane, how many planes can you name using A, B, C, D?
   b. How many lines?
   c. How many rays?

5. a. Given lines $\overleftrightarrow{AB}$ and $\overleftrightarrow{CD}$, how many points at most could they have in common?
   b. Given lines $\overleftrightarrow{AB}$, $\overleftrightarrow{CD}$, $\overleftrightarrow{EF}$?
   c. Given lines $\overleftrightarrow{AB}$, $\overleftrightarrow{CD}$, $\overleftrightarrow{EF}$, and $\overleftrightarrow{GH}$?
   d. Can you generalize your findings?

SECTION B **ANGLES**

BEHAVIORAL OBJECTIVES

B.1 Be able to describe the intersection of points, lines, and rays on a plane.

B.2 Be able to describe the union of points, lines, and rays on a plane.

B.3 Define an angle by using the operation of union with sets of points previously defined.

B.4 Be able to name a given angle and its vertex.

B.5 Describe what we mean by measure of an angle.

MORE GEOMETRIC FIGURES

Since we have described the geometric figures thus far as sets of points, we can apply the operations of set union and intersection to these sets and produce new sets. In this manner, we can define other objects with which you should become familiar. Notice in Figure 11.5 that we show the intersection of two line segments in several positions on the plane.

 B.1

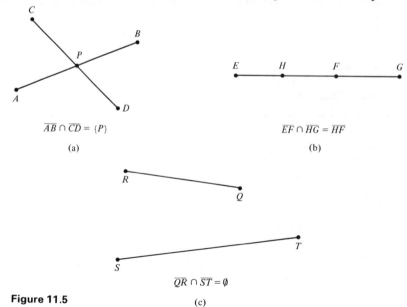

$$\overline{AB} \cap \overline{CD} = \{P\}$$

(a)

$$\overline{EF} \cap \overline{HG} = \overline{HF}$$

(b)

$$\overline{QR} \cap \overline{ST} = \emptyset$$

Figure 11.5

(c)

In Figure 11.5 we have shown three different types of intersections: a point, a line segment, and an empty set. Suppose the diagrams above were those of lines intersecting rather than line segments. What would ■ the possible intersections be?

We can also illustrate set union as we did set intersection. Suppose we consider the union of two rays, as shown in Figure 11.6. Notice that in a, the union of two rays with a common endpoint is a line; in b, the

■ B.2

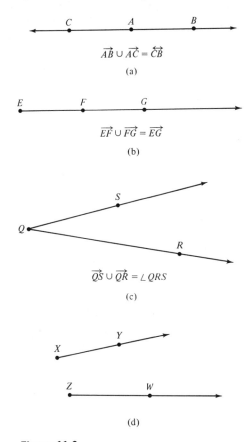

$$\overrightarrow{AB} \cup \overrightarrow{AC} = \overleftrightarrow{CB}$$

(a)

$$\overrightarrow{EF} \cup \overrightarrow{FG} = \overrightarrow{EG}$$

(b)

$$\overrightarrow{QS} \cup \overrightarrow{QR} = \angle QRS$$

(c)

(d)

Figure 11.6

union is a ray; in c, the union is an angle; and in d, the union is simply ■ two disjoint rays. Can you think of any other possibilities for two rays? We mentioned earlier that we can produce new sets by operating on sets already described. Notice that we have done this in Figure 11.5 and 11.6; however, in every case but one, we have simply produced sets of points which we could describe based on earlier description. In Figure 11.6c we produced a new set which we called an angle; its description is based ■ B.3   on our earlier descriptions and the operation of union. In this way, *we can define an angle as the geometric figure formed by the union of two rays*

■   *with a common endpoint.* Notice that Figure 11.6a falls into this same set of objects. Actually, a straight line is an angle with a special name. We call it a *straight angle*, and we say that its measure is 180°. Does Figure 11.6b fit the definition of an angle? Why is Figure 11.6d *not* an angle?

■ **B.4**   Before we discuss the measure of an angle, let us consider the notation employed in Figure 11.6 to describe an angle. We named the angle ∠ *QRS*. The symbol ∠ represents the word "angle," and the first and last letters, *Q* and *S*, are points on different rays. The letter in the middle, *R*, is the common endpoint of each ray, which is also called the *vertex* of the angle. In some cases we use the less precise name for an angle, ∠ *R*, using the

■   vertex only.

■ **B.5**   To find the measure of an angle, we place a protractor on the angle, as shown in Figure 11.7. We put the center point of the protractor on the vertex of the angle and the base of the protractor along one of the rays.

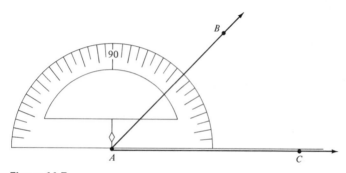

Figure 11.7

We see that the other ray extends to mark a number on the protractor. This number is called the *measure of the angle at A*, or the *measure of angle A*. It is important to note that the lengths of the sides of an angle are not related to an angle's measure, because the sides are rays, which have

■   infinite length.

EXERCISES

1. From the statements below draw a diagram to describe the situation.

a. $\overleftrightarrow{AB} \cap \overleftrightarrow{CD} = P$ b. $\overline{AB} \cap \overline{CD} = \overline{CB}$
c. $\overline{AB} \cap \overline{CD} = \varnothing$ d. $\overrightarrow{AB} \cap \overrightarrow{CD} = E$
e. $\overrightarrow{AB} \cap \overrightarrow{BC} = B$ f. $\overrightarrow{BC} \cap \overrightarrow{CD} = \overrightarrow{CD}$
g. $\overline{AB} \cap \overrightarrow{AB} = \overline{AB}$ h. $\overline{AB} \cap \overleftrightarrow{CD} = X$
i. $\overline{AB} \cap \overleftrightarrow{CD} = \overline{AB}$

2. From the statements in Exercise 1 complete the statements that follow.

a. From b, name $\overline{AB} \cup \overline{CD} = $?
b. From e, name $\overrightarrow{BA} \cup \overrightarrow{BC} = $?
c. From f, name $\overrightarrow{BC} \cup \overrightarrow{CD} = $?
d. From g, $\overline{AB} \cup \overrightarrow{AB} = $?
e. From i, $\overrightarrow{AB} \cup \overrightarrow{CD} = $?

3. Write symbols for the figures shown below.

(a)

(b)

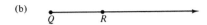

(c)

(d)

4. Using symbols, describe the unions and intersections in each figure below.

(a)

(b)

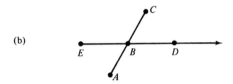

(c)

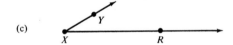

(d)

SECTION C **CURVES AND POLYGONS**

BEHAVIORAL OBJECTIVES

C.1 Define curve, and give an example of a curve.

C.2 Define closed curve, and give an example of a closed curve.

C.3 Define simple closed curve, and give an example of one.

C.4 Given several curves, be able to identify closed curves, simple closed curves, and curves which are not closed.

C.5 Define polygon, and give an example of a polygon.

C.6 Be able to name and describe polygons with 3, 4, and 5 sides.

C.7 Be able to define special types of quadrilaterals.

C.8 Given a description of a quadrilateral with special characteristics, be able to name the quadrilateral.

C.9 Be able to explain the concept of distance as it relates to points on the real number line.

Before we discuss some familiar geometric figures, we will consider the idea of a simple closed curve. The word "simple" is often misleading. If one were to look at Figure 11.8 he would consider the first drawing to be anything but a simple curve. Yet our definition of simple closed curve will be such that it will include the first figure. *We will define curve as a path from one point to another. If the starting point and ending point of a curve are the same, then we call the curve closed.* Thus, Figure 11.8 shows six different curves, three of which are closed (a, d, and f). *If a closed curve does not cross over any point, excluding the starting point, it is called a simple closed curve.* This means that simple curves do not cross or intersect themselves. Therefore, curves a and d of Figure 11.8 are simple closed curves, but f is not a simple closed curve.

■ C.1
▲ C.2
●C.4▲■
■ C.3

● ■

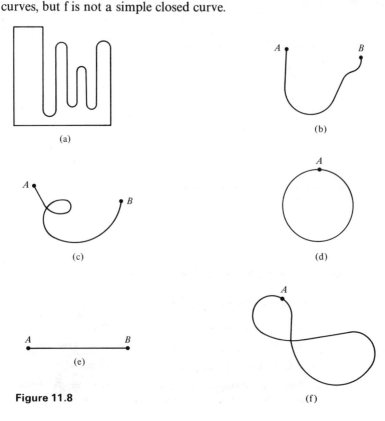

(a)

(b)

(c)

(d)

(e)

Figure 11.8

(f)

■ C.5   The simple closed curves which we deal with most often are called *polygons*. Polygons are formed by the union of three or more line segments. In this way, the simplest polygon would be a three-sided polygon,
▲ C.6   which we call a triangle. The word "triangle" refers to the three angles found in the figure. In this manner, we name polygons (or many-sided figures) so that a four-sided polygon is called a *quadrilateral*, a five-
▲ ■   sided polygon is called a *pentagon*, and a six-sided figure is called a *hexagon*.

    The set of quadrilaterals has several subsets with special characteristics. We will define them below and then diagram them in Figure 11.9.

■ C.8

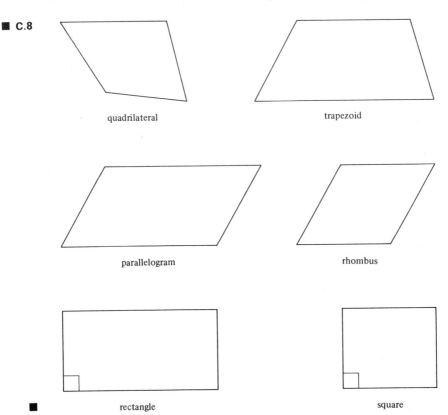

quadrilateral   trapezoid

parallelogram   rhombus

■   rectangle   square

Figure 11.9

■ C.7   A *trapezoid* is a quadrilateral with one pair of parallel sides.

    A *parallelogram* is a quadrilateral with two pairs of parallel sides.

    A *rhombus* is a parallelogram with adjacent sides equal.

    A *rectangle* is a parallelogram with at least one right angle.

■   A *square* is a rectangle with adjacent sides equal.

    Let us now use the line segment to introduce the concept of distance.
    If *A* and *B* are two points in the plane, then we know that there is a line
■ C.9   in the plane which contains these two points. We also know that there

is a 1–1 correspondence between the points on line *AB* and the real number line. Thus, we have two real numbers *a* and *b* which correspond to the points *A* and *B*, respectively. Consider the numbers $a - b$ and $b - a$. One of these numbers is positive. Suppose $b - a$ is positive. We then define the distance between the points *A* and *B* to be the positive number $b - a$. To show that the definition of distance between points is not unrealistic, consider the following example.

Example
What is the distance between the points which correspond to 4 and 15?

SOLUTION   Since $15 - 4 = 11$ and $4 - 15 = -11$, we say the distance between these points is 11.

It should be clear that if we lay the line segment *AB* against a number line, we should be able to find its measure. Is it clear that there is a number that we call the distance from *A* to *B*? We call this number the *measure of line segment AB*. In general, the distance between two points is considered to be positive and is defined as the absolute value of $a - b$, written $|a - b|$. If $a - b > 0$, then $|a - b| = a - b$, and if $a - b < 0$, $|a - b| = -(a - b)$. Therefore, if $a = 15$ and $b = 4$, $|15 - 4| = 11$ and $|4 - 15| = -(-11)$, or 11.

EXERCISES

1. Illustrate the following terms with a sketch.
   a. line segment $\overline{AB}$
   b. ray $\overrightarrow{XY}$
   c. point *B*
   d. line $\overleftrightarrow{AB}$

2. How many ways can a pair of lines intersect on a plane? Explain.

3. How many ways can a ray and a line intersect on a plane? Explain.

4. How many ways can a line segment and a ray intersect on a plane? Explain.

5. Which of the following statements are true? Why?
   a. All triangles are polygons.
   b. All squares are rhomboids.
   c. Every rectangle is a trapezoid.
   d. Every square is a parallelogram.
   e. Every three-sided polygon is a triangle.

C.8   **6.** Draw and name the following.
   a. a four-sided polygon with all sides equal
   b. a quadrilateral with all right angles
   c. a five-sided polygon
   d. a quadrilateral with all sides different in length

7. Use two rays to draw the following angles
   a. with measure 0°
   b. with measure 90°
   c. with measure 180°
   d. with measure 270°
   e. with measure 360°

8. Classify the curves below as closed, simple, complex.

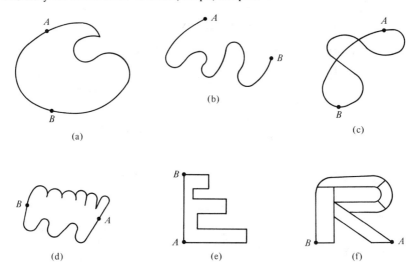

(a)

(b)

(c)

(d)

(e)

(f)

■ C.8

9. Classify the polygons below by naming them.
   a. a four-sided polygon with four right angles
   b. a four-sided polygon with a pair of parallel sides
   c. a four-sided polygon with two pairs of parallel sides
   d. a five-sided polygon

10. Trace the figures below without raising your pencil or going over any line twice. In which cases can this be done? Why? (*Hint*: Count the number of segments which touch at a given point. How many are odd numbered, how many are even numbered?)

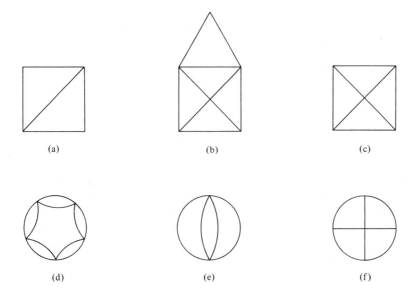

(a)

(b)

(c)

(d)

(e)

(f)

SECTION D **RIGID MOTIONS**

BEHAVIORAL OBJECTIVES

D.1 Be able to give a physical model of a rigid mapping.

D.2 Be able to define a mapping of a plane.

D.3 Be able to explain how a rigid mapping affects the distance between two points.

D.4 Be able to define and give examples of the composition of mappings.

D.5 Be able to define equality of mappings.

D.6 Be able to find the group properties of symmetries.

D.7 Given two symmetric designs, be able to find a rigid motion that will set up a mapping between the designs.

D.8 Be able to decide whether a rigid motion corresponding a set to its image is a proper or an improper rigid motion.

D.9 Be able to define a translation by a physical model.

D.10 Be able to describe the characteristics of a translation.

D.11 Be able to define a rotation by a physical model.

D.12 Be able to describe the characteristics of a rotation.

D.13 Be able to use the properties of a rotation to determine when the composition of two rotations is a rotation.

D.14 Be able to compose rotations and translations.

GEOMETRY BY TRANSFORMATIONS; PROPER RIGID MOTION

It should not be surprising to find that wherever there occurs in nature some regularity, mathematics enters and soon finds a way to describe that regularity in a very exact fashion. We will illustrate this characteristic of mathematics by first examining two-dimensional repetitive designs such as are commonly found in a wallpaper or a floor covering. We shall consider that some particular figure is repeated infinitely many times to make up the design, which covers the Euclidean plane. That is, the figure repeats itself over and over in some regular fashion so as to fill a flat surface that extends infinitely far in all directions.

Imagine now a large (as large as the plane) sheet of glass laid upon the plane and upon which the design has been etched in perfect detail. Thus, we have two copies of the design, one on the plane and another on the glass. We now assume that the piece of glass can be moved to other positions on the plane in such a manner that each figure on the glass fits exactly over a figure on the plane. We also assume that, because of the regularity of the figure in the design, when the glass is moved to a new position as described, every figure on the glass is exactly above a figure in the design.

■ **D.1** The movement of the glass, as described above, is a physical model of what is called a *rigid motion*, or *rigid mapping*, of the plane onto itself. The movement of the sheet of glass, or rigid motion, determines a correspondence between the initial points of the design and the second points of the design.

Suppose that the design is a triangle. On the plane the figure might be labeled *ABC*, as shown in Figure 11.10. The design on glass might be

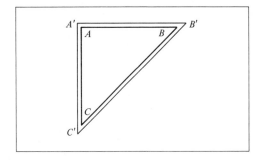

Figure 11.10

labeled *A'B'C'*. In Figure 11.10 we see two triangles *A'B'C'* and *ABC*, but we can imagine that *ABC* is directly below the plane. In Figure 11.11, we have moved the glass, while the plane remains in the same position. Consider now the distance from *A* to *A'*, as well as the distances *B* to *B'* and *C* to *C'*. If we were able to carefully measure these distances, we would find that they were all equal. Measure the distances between *A* and *A'*, *B* and *B'*, and *C* and *C'* to satisfy yourself that the distances are constant. Note that we took the license of making the plane look as if it were finite, although this is not the case.

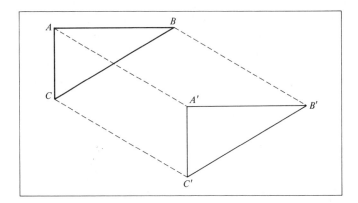

Figure 11.11

To picture what we are doing, cut a triangle out of cardboard and use it to follow the discussion. Trace the outline of the triangle, as in Figure 11.11, on a piece of paper, and then without turning it slide the triangle across the paper. What do you notice about the distance from A to A', B to B', and C to C'? *Remember*, A, B, and C represent the triangle on the paper, while A', B', and C' represent the cardboard figure, or figure on the glass.

We are now ready to define this type of movement on the plane, which we will refer to as a *mapping*.

Definition

■ **D.2**
■ *A mapping of a plane is a correspondence M which associates with every point P of the plane a point P' of the plane.*

This correspondence is symbolized by $P' = M(P)$, or $P \rightarrow P'$, or $M:P \rightarrow P'$, and is really a description of a movement that carries a given point P into another point P' or $M(P)$. It may be that the mapping sends every part into itself, in which case the figure does not move, or a mapping could move every point one inch above the initial point. Try tracing your triangle and moving one point, A, for example, to a point one inch above it. What happens to B and C?

A mapping M is called a *rigid mapping* if for any pair of points P_1 and P_2 the distance between $M(P_1)$ and $M(P_2)$ is the same as the distance

■ **D.3** between P_1 and P_2. Thus, a mapping is called a rigid mapping if it preserves
■ the distance between points. It should be obvious that the rigid motion of the glass described earlier is an intuitive or physical example of a rigid mapping of the plane onto itself.

Not all motions of the sheet of glass result in a perfect correspondence between the design on the plane and the design on the glass. Any movement of the glass that results in perfect match of the design will be called a *symmetry* of the design. Thus, under a symmetry every point of the figure on the glass is directly above the corresponding point of the figure on the plane, and we say that the design is *invariant* under the symmetries of the design. By invariant we simply mean that the mapping has not changed the figure on the glass in any way other than distance from the original figure that was on the plane. A good way of illustrating this concept to children is to use tracing paper to produce two figures which are alike in shape and size but different in position. If the child slides the tracing copy across the original and stops at some point to measure these distances, he will find that the distances remain constant.

The teacher might also try using spirit duplicators to produce several copies of a figure ABC and then on the same spirit master mark the figure $A'B'C'$ (as shown above) and have the children move their duplicate copies in any way they please and measure distance. Since children might have trouble keeping both copies in place while measuring, the teacher could have them staple the copies together after movement. Cutout figures could also be used so that children could see that the shapes fit over each other easily.

It also helps children to visualize the movement if each side of the figure is colored; that is, the sides of the triangle could be painted yellow, red, and blue. In this way, the sides of the triangle that is moved can be easily matched to the triangle that is in the original position.

If we now consider the set of all symmetries of the design, we find some characteristics that should be reminiscent of other mathematical systems. For two symmetries M_1 and M_2, consider the result of performing M_1 and then M_2. Since the design is invariant under both M_1 and M_2, the design will be invariant under the sequence of performing M_1 and M_2 in order. We call the mapping that results from performing first M_1 and then M_2 the *composition of M_1 by M_2*, and we write the composition M_2M_1. Thus, the result of composing two symmetries is again a symmetry of the design. We will state without formal proof that the set of symmetries of a design is closed under composition.

Consider an example of composition, as illustrated in Figures 11.12–11.14. Notice that we perform M_1 first and then M_2. We move a triangle

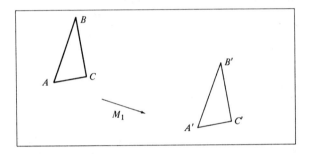

Figure 11.12

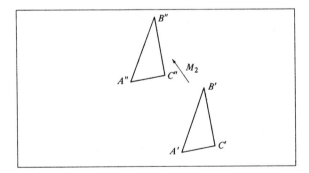

Figure 11.13

ABC by M_1 to $A'B'C'$ and then move $A'B'C'$ by M_2 to $A''B''C''$. Notice that the distances between corresponding points of ABC and $A''B''C''$ are still constant. Notice also that the figures remain in the same shape. Figure 11.14 shows the composition M_2M_1 without the intermediate step. To consider another example, cut out a triangle and trace it on a piece of paper. Label the tracing ABC as in Figure 11.12. Now, move the triangle one inch to the right and trace it again. Mark this $A'B'C'$. Now move it up two inches trace and mark it $A''B''C''$. If M_1 was the first move

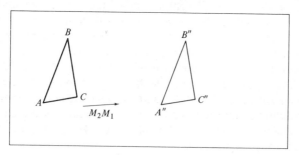

Figure 11.14

and M_2 the second move, then M_2M_1 was a composition of a mapping which sent ABC to $A''B''C''$. We could say that we performed M_1 on ABC to produce $A'B'C'$ then we performed M_2 on $A'B'C'$ to produce $A''B''C''$. This can be written as follows:

1. $A'B'C' = M_1(ABC)$
2. $A''B''C'' = M_2(A'B'C')$
■ 3. $A''B''C'' = M_2M_1(ABC)$

This can be done by children in the same way as you did with the cutout triangle. Exercises like these are excellent ways to develop a readiness for the concept of composition of mappings.

■ **D.6**   If a mapping is such that every point in the plane goes to itself, it is called an *identity mapping*, and we symbolize it by I. Such a mapping would simply send triangle ABC to triangle ABC such that $I(A) \rightarrow A$, $I(B) \rightarrow B$, and $I(C) \rightarrow C$ under the mapping I. The properties of identity mapping are related to the usual properties of identity:

1. For every point P, $I(P) = P$.
2. For every symmetry M, $IM = MI = M$.
■ 3. For every symmetry M, $MM^{-1} = I = M^{-1}M$.

The use of the equals sign in properties is really an equality between mappings.

We now formally define equality between mappings.

Definition

■ **D.5** ■   *Two mappings M_1 and M_2 are equal iff for every point P, $M_1(P) = M_2(P)$.*

(An example of two equal mappings is the mapping to the right one inch and up one inch, and the mapping up one inch and to the right one inch.)

Notice in property 3 above that we used the notation MM^{-1}. This notation is similar to that used in previous chapters to indicate inverses. Thus, we are saying that if there is a mapping M^{-1} which relates in this manner to M, then it produces I. Remember that I is a necessary but not sufficient condition for M^{-1}. We must be able to describe an identity mapping before we can discuss inverse. Let us consider the mapping M^{-1} in the set of symmetries.

Consider what would happen if we mapped ABC to $A'B'C'$ and then moved $A'B'C'$ back to position ABC. The resulting mapping would also be a composition of M_1 with another mapping, which we will call M^{-1}.

■ **D.6** It is not necessary for clarity to draw this. You might lay a book of matches on a piece of paper and trace around it, then move it to a second position and again trace it. If you then reverse the process, you have composed the function with M^{-1}, which we write $M^{-1}M = I$.

The set of symmetries has the property that for symmetry M there is another symmetry M^{-1} which is simply the reverse mapping of M. That is, if M maps P into P', M^{-1} maps P' into P. Thus, $M^{-1}M(P) = P$, because $M(P) = P'$ and $M^{-1}M(P) = M^{-1}(P') = P$, so the set of symmetries remains closed under composition. We call the mapping $M^{-1}M$ a symmetry. Since $M^{-1}M(P) = P$, this mapping corresponds each point to

■ itself; that is, it leaves every point fixed.

EXERCISES

■ **D.6** 1. Show that the set of symmetries of a design where the operation is composition form a
■   group.

2. Let L be some number line in the plane. If T_5 is the mapping that carries 0 into 5, that is, $T_5(0) = 5$, then
   a. $T_5(1) = ?$ b. $T_5(2) = ?$
   c. $T_5 T_5(2) = ?$ d. $T_5(-5) = ?$
   e. the image of the segment 12 is?

3. If T_p is the mapping of the number line into itself that carries 0 into p, that is, $T_p(0) = p$, and $S = \{T_p | p \text{ is real}\}$, then
   a. $T_3 T_2(5) = ?$ b. $T_2 T_3(5) = ?$
   c. $T_0(5) = ?$ d. $T_{-2} T_2(5) = ?$
   e. $T_0 T_2(5) = ?$

4. Let S be the set of mappings in Exercise 3, and let ∘ be the operation of composition of mappings.
   a. Is S closed under ∘?
   b. Is S commutative under ∘?
   c. Is there an identity element in S under ∘?
   d. Do inverses exist in S under ∘?

5. For each of the following mappings, describe the inverse mapping.
   a. up one inch
   b. right one inch
   c. down one inch
   d. one inch at an angle of 45°
   e. down one inch and left two inches

6. For each mapping in Exercise 5, describe an equal mapping.

TYPES OF RIGID MOTION

There are several types of rigid motions of the sheet of glass that we wish to clarify. Let us first attempt to differentiate between these rigid motions by an example. The figure that will make up the design will consist of a square with the letter N superimposed. Before we discuss this design, you should make the letter N on a small piece of paper. Now slide the N to the right. What does it look like? Turn the N clockwise 90°, 180°, 270°, and 360°. Can you draw the figure after each turn? Now consider the drawings you have made and those in Figure 11.15.

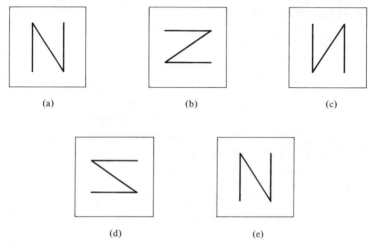

(a) (b) (c)

(d) (e)

Figure 11.15

■ **D.7** If Figure 11.15a were copied on the sheet of glass, the glass could be moved along the paper so that the design would match that of Figure 11.15b. To do this, we would slide the glass along the paper and give it a quarter turn clockwise. Note that the glass could remain in contact with the paper throughout the operation; that is, the glass need not be picked up and flipped over.

Now try writing the N on both sides of the paper in such a way that each point on one side corresponds to a point on the other side. What happens to the figure if you turn it over around a vertical axis through the N? What happens if you rotate it as before? How many different draw-
■ ings can you produce in this way?

As you can see from what you have done, it is not possible to slide the glass imprinted with Figure 11.15a so that it will coincide with Figure 11.15c or with Figure 11.15d with the restriction that the glass remain in contact with the paper throughout the operation.

The point to be taken here is that we can class rigid motions into at least two categories. The first is proper rigid motions, or motions of the glass that will make the figures coincide without lifting the glass from the paper. The second category, called improper rigid motions, is those motions that require the glass to be flipped over in order to make the figures coincide. We will consider in more detail both of these classes of motions in the following sections.

EXERCISES

1. What kind of movement would be needed to make Figure 11.15a coincide with Figure 11.15c? (*Hint*: You cannot just slide the piece of glass.)

2. What kind of movement would be necessary to make Figure 11.15a coincide with Figure 11.15d?

3. What kind of motion would be necessary to make Figure 11.15a coincide with Figure 11.15e?

4. Consider the figures below.

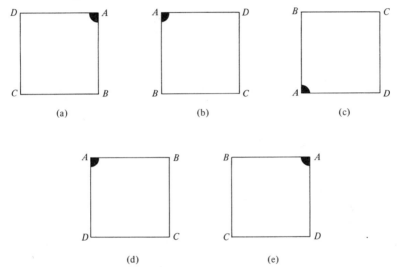

(a)   (b)   (c)

(d)   (e)

a. What kind of movement would make figures a and b coincide?
b. What kind of movement would make figures a and c coincide? a and d? a and e?
c. What kind of movement would make b and c coincide? b and d? b and e?

5. Which of the movements in Exercise 4 can be done without turning the figure over?

TRANSLATION AND ROTATION

D.8   A sliding motion of the glass which does not involve turning the glass over is called a *proper rigid motion*. Any motion of the glass that involves turning over the glass is called an *improper rigid motion*. Remember our description of a flip, or reflection, at the end of Chapter 6!

By visualizing the sheet of glass, can you classify the kind of motion that will result when two proper rigid motions are performed in succession? What this question is really asking is: "Is the set of proper rigid motions closed where the operation is performed one after the other?"

There are two kinds of proper rigid motions, *translations* and *rotations*. We will begin by considering translations.

■ **D.9**   Let us choose a particular direction and mark it on a sheet of paper, *DD′*. We will use an arrow to indicate direction. Also, let a segment of length *a* be given. Let *P* be any point on the paper. Put the sheet of glass on the paper and mark it over *P*. Then move the glass a distance *a* in the direction of *DD′*. The point under the mark on the glass is called *P′*, and we say that *P′* is obtained from the point *P* by translation through the distance *a* in the direction *DD′*. See Figure 11.16.

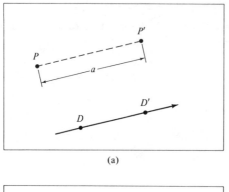

(a)

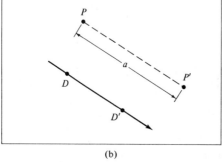

(b)

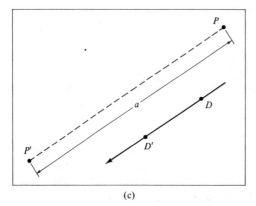

(c)

Figure 11.16

Another way of describing the translation which is determined by both the distance a and the direction DD' is that for a given point P there is a point P' such that its distance from P is a and its direction from P is DD'.

The point P' which is obtained from P by this translation is sometimes called the image of P under the translation which is determined by the distance a and the direction DD'.

■ **D.3** Cut a triangle and label it PAB. Place point P of your triangle on top of point P in Figure 11.16a and move the triangle along the ray $\overrightarrow{DD'}$. Notice how when you move triangle PAB along the line DD' in Figures 11.16a, b, and c, all points of the triangle move a distance, a, in the same direction. Now trace the triangle on paper. Mark points P, A, and B on the drawing and on the triangle. Draw a line DD' in any direction and move the triangle in such a way that the point P moves through a distance of two inches. In this manner $a = 2''$. Now trace the triangle and label it $P'A'B'$. Find the distance from A to A' and from B to B'. Try this several times, moving in several different directions. What did you find to be true of the distances? This idea is illustrated in Figure 11.17.

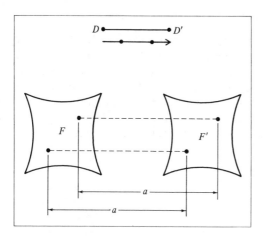

Figure 11.17

If we consider some figure F in the plane, then the figure F', consisting of the images of the points of F, is called the figure produced from F by a translation in the direction DD' through a distance a. (See Figure 11.17.) We see that the entire figure F has been "moved" or has undergone a "slide" a distance a in the direction of DD'. That is to say, all the points

■ **D.10** of F have been moved in the same direction for the same distance. This type of rigid motion is thus characterized by the fact that the line segments joining corresponding points have the same length and direction. That is, the line segments AA' and BB' are parallel and equal in length. This type of translation was illustrated earlier in Figures 11.10 and 11.11 and was also discussed in some detail in the first section.

Let us now look at some specific properties of translation. Can you show that the result of translating a circle will be another circle? To do this, cut out a circle about the size of a half dollar. Trace it on paper and mark four points, A, B, C, D, on the circle. Draw a line in some direction from A for a distance of three inches. Mark the endpoint A'. Draw a three-inch line from B parallel to the first line. Mark the endpoint B'. Do this for C and D, producing four points marked A', B', C', D'. What is the relative position of A' to B' to C' to D' with respect to the original positions, A, B, C, D? Now lay the cutout on the first circle and mark the cutout figure A, B, C, D. Now move the cutout so that you can lay it on A', B', C', D'. Are you convinced that if you could do this for all points of the circle, the new figure formed in this way would be a circle?

■ **D.10**
Note that a translation is really a rigid motion of the sheet of glass, so the translation T is a rigid mapping of the plane into itself. Thus, for the points A and B of a figure, the corresponding points $A'B'$ (i.e., $T(A)$ and $T(B)$) have the property that the distance between A and B is equal to
■
the distance between A' and B'. Let us illustrate this.

Example 1

What happens when a straight line is translated?

SOLUTION   Suppose that the line L is translated into the line L'. Will L and L' intersect? The answer to this question is No, which we will now try to justify; we will show that the only alternative is impossible. If the only alternative to our conclusion is impossible, then our conclusion must be correct. (Recall from Chapter 10 that $p \vee -p$ is always true.) The only alternative to our conclusion is that the lines intersect. See Figure 11.18.

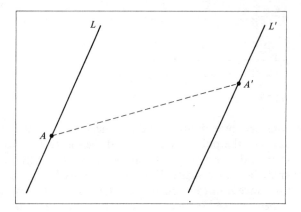

Figure 11.18

Suppose that L and L' intersect at some point P' on L'; then P' is the image of some point P of L. But P' is also on L, so P and P' are corresponding points under translation. Thus, the direction of the translation is along L, and every point of L will be carried into another point of L and not L'. See Figure 11.19. Since it is not the case that every point of L is carried into another point of L, the only alternative to our conclusion is impossible. Hence, the lines L and L' do not intersect. If L and L' do not intersect, we say that they are

■ D.10

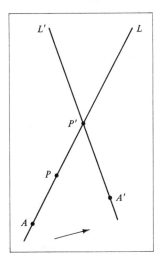

Figure 11.19

parallel. Hence, a straight line is carried into a parallel line under a translation.

■

Example 2

If two sides of a quadrilateral are equal and parallel, show that the other pair of sides is equal and parallel, that is, that the figure is a parallelogram.

SOLUTION AB and DC are the two equal and parallel sides of the quadrilateral $ABCD$. Let T be the translation in the direction DC, where the distance a is the length of the segment CD. Then, $D' = T(D) = C$, and $A' = T(A) = B$. Thus, $A'D' = BC$, but the length of AD is equal to the length of $A'D'$, so $AD = A'D' = BC$; that is, the other two sides have equal lengths. By Example 1, we know that AD is parallel to $A'D'$, so the sides AD and BC are both equal and parallel. Thus, the figure is a parallelogram. See Figure 11.20. To illustrate this example yourself, draw two lines AB and DC. Make the lines two inches long (measure carefully), and be sure that they are parallel. Draw DA as in Figure 11.20. Cut a strip of paper,

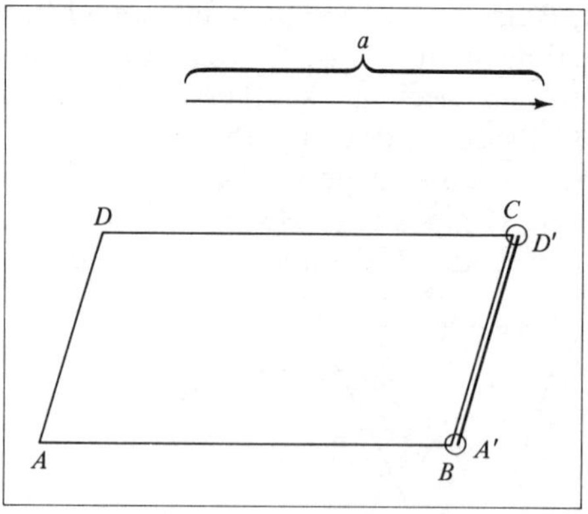

Figure 11.20

and mark off the distance from *D* to *A* on the paper. Slide the paper so that *D* moves along the line *DC* and *A* moves along *AB*. When you reach the point *C* on *DC*, what point coincides with *B* on *AB*?

EXERCISES

1. Is the set of proper rigid motions closed? (*Hint*: Will the composition of two rigid motions produce a rigid motion?)

2. Argue that the set of improper rigid motions is not closed under the operation of performing one after the other. (*Hint*: Look at the results of two reflections (flips).

3. a. Is there a proper rigid motion that acts as the identity?
   b. Is the inverse of a proper rigid motion again a proper rigid motion?
   c. Is the set of proper rigid motions a group?

4. Show that the set of translations that carries a straight line into itself forms a group. (See the exercise set on page 425.)

5. Can you show that the set of translations under the operation of composition is a group? a commutative group?

ROTATIONS

The second kind of proper rigid motion is a *rotation*. All our rotations will be counterclockwise unless otherwise stated. If we have a figure on a sheet of paper and over this we place a sheet of glass upon which the figure is marked, then imagine a pin through the glass at some point *O*; using the pin as an axle, we rotate or turn the sheet of glass. The points of the plane under the design on the glass are called the *image of F under the rotation*. See Figure 11.21.

To gain a feeling for rotations, make an equilateral triangle out of cardboard. Punch a hole in the center and rotate the triangle around its center. Trace the triangle in several positions. Now cut a rectangular slip of paper and place one of its sides against the side of the triangle. Rotate the triangle

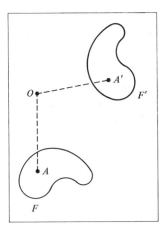

Figure 11.21

and rectangle as a fixed figure through 90°, 180°, 270°, tracing the rectangle
■ after each rotation. The result should clarify Figure 11.21.

There are an infinite number of rotations about the point *O*, depending
on how much we turn the sheet of glass. In order to be more specific, we
describe the rotation by the angle through which we rotate. If *A'* is the
image of *A* under a rotation (Figure 11.21), we say that we have performed
a rotation about *O* through the angle *A'OA*.

Example 1

■ **D.11**   Rotate the figure **F** through half a complete turn about the point *O*.
The angle *A'OA* is 180°, or half a complete turn, as seen in Figure 11.22.

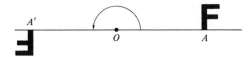

Figure 11.22

Example 2

Rotate the figure **F** about the point *O* through 90°. See Figure 11.23.

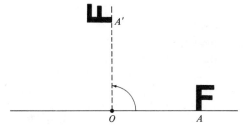

■

Figure 11.23

We suggest that you verify these examples by drawing a figure on a rectangular piece of paper. Then, using a pencil to hold down one end, rotate the paper around the pencil point.

There are several properties of rotations that we should mention.

■ D.12 1. A translation leaves no point of the plane fixed, whereas under a rotation a single point of the plane is left fixed. To say this using the notation of this chapter, we simply say that there is a point A such that $A' = A$ under a rotation. Do you know what point remains fixed?

2. For a rotation about the point O, the distance from O to A is equal to the distance from O to A'. That is, A and its image, A', both lie on a circle with its center at O.

3. If an angle is rotated about some point O, as in Figure 11.24, the size or measure of the image under the rotation remains the same. For

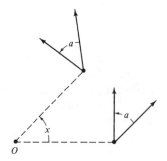

Figure 11.24

example, if a right angle at A is rotated through a half-turn about O (see Figure 11.25), the angle at A' will be a right angle. Notice that the point around which we rotated has remained fixed.

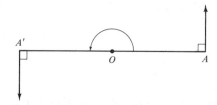

Figure 11.25

From the above we see that a rotation through an angle of measure x rotates every line through an angle of measure x. This is seen in Figure 11.26 by looking at some line L that is rotated through x. Choose some point P on L; then both P and P' are on the same circle with center O. Since the angle that L makes with a radius to P is the same as the angle that L makes with a radius to P', it should be clear that the line L turns through an angle of measure x, as does the radius in going from P to P'.

■ D.13 Let us use the above discussion to examine the question of whether or

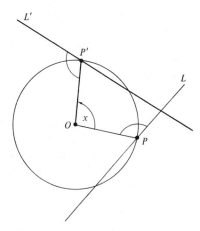

Figure 11.26

not the composition of two rotations is a rotation. If the first rotation is a rotation through x degrees, then every line in the plane is rotated through x degrees. If the second rotation is through y degrees, then every line is rotated under the composition through x and y degrees and is thus a rotation. This argument holds except for one case, which is shown in the following examples.

Example 1

Rotate the plane first about the point corresponding to 1 and then about the point corresponding to 2. Let the first rotation be through 90° and the second be through 270°; that is, **F′** is the rotation of **F** through

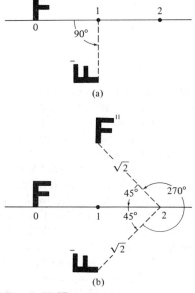

Figure 11.27

90° about the point 1 as in Figure 11.27a. **F″** is the rotation of **F′** through 270° about the point 2 as in Figure 11.27b.

As we see from the last figure, the result of composing these two rotations is that **F** corresponds to **F″**. This is equivalent to a translation; namely, ■ a translation of the plane at an angle of 45° to our base line.

Example 2

Mark a sheet of tracing paper with an **F**. Place the tracing paper over a sheet of paper that has a number line drawn on it, so that the *A* is the 0 point. Put a pin or a pencil point on the *A* point (see Figure 11.28) and rotate the tracing paper counterclockwise through 270°. Then put the pin or pencil on the *B* point and rotate 90° counterclockwise. Do you get ■ a vertical translation downward? See Figure 11.28.

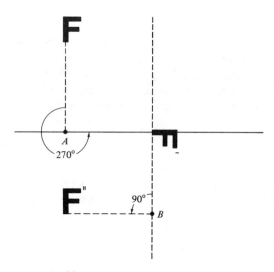

Figure 11.28

If we try a number of cases where the sum of the rotations is 360°, we see that the result is a translation. Thus, we formulate the answer to ■ **D.13** our original question this way: The composition of two rotations is a rotation except if the sum of the measures of the two rotations is 360°; ■ then the composition of these rotations is a translation.

■ **D.14** What happens when a translation is composed with a rotation? Suppose we perform the translation first. What happens to every line in the plane? It is carried into a parallel line. When the rotation is performed, what happens to each translated line? It is rotated. So the result is that every line is rotated through some angle. That is, the result of translating and then rotating is a rotation. For a given translation and rotation, can you find the center of the resulting rotation? (*Hint:* Find the point that is car-■ ried into itself by composition of the translation and rotation.)

Consider a triangle *ABC* which is translated to *A′B′C′* first and then rotated a half-turn around a point P', the midpoint of the side $B'C'$. See Figure 11.29.

■ D.14 ABC is mapped under T_a so that A maps to A', B to B', and C to C'. We include the point P, which maps to P' and lies on the side BC. Notice the position of $A'B'C'$ with respect to the original figure, as shown in Figure 11.30.

Notice in Figure 11.30 how the position of $A'B'C'$ relates to $A''B''C''$. In Figure 11.31 you can see the composition (translation to the right a distance of "a" composed with the rotation around the point P').

Try connecting the points A to A'', B to B'', C to C'', and P to P'. This may help you to answer the question of what rotation is equivalent to the composition of a translation and rotation. Notice that we have con-
■ nected some of the points in Figure 11.31.

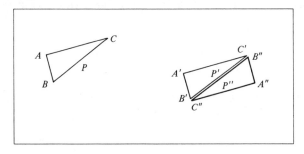

Figure 11.29

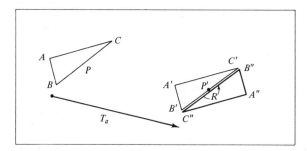

Figure 11.30

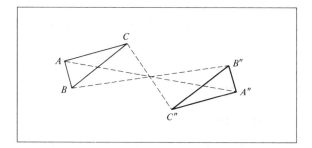

Figure 11.31

Suppose that we rotate first and then translate. What is the result? You might consider doing this by rotating the figure *ABC* as shown and then translating to see what results.

We would like to again point out that exercises like the ones mentioned above are very exciting for children. All one needs in terms of materials is some cutout squares, triangles, rectangles, or circles which can be moved over a piece of paper and traced after several operations are done. It will be of great help to you to experiment in this way as you read. Keep a pack of matches or some cutout squares handy.

EXERCISES

1. a. Rotate the plane 180° about the point corresponding to 1 on the number line.
  b. Rotate the result obtained in a above 180° about the point corresponding to 2 on the number line.
  c. What is the result of performing a and then b?

2. Show that the composition of two half-turns is a translation.

3. The following questions refer to the set of rotations about a fixed point.
  a. Is it closed?
  b. Is there any rotation that acts as an identity under composition? which?
  c. Can you find inverses in the set?
  d. Is the operation of composition commutative in the set?

4. Draw a number line on a sheet of paper, and place a pencil perpendicular to the number line. Now let R_a and R_b, with measures a and b respectively, be rotations about the points 1 and 2. Suppose $a + b = 360°$.
  a. After performing R_a and R_b on the pencil, is the pencil parallel to its original position?
  b. Suppose a point P remained fixed after performing R_a and R_b. Place the pencil through P and perform R_a and R_b. What is the terminal position of the pencil?
  c. From b, argue that no point remains fixed under R_a followed by R_b.
  d. Why is the composition of R_a by R_b a translation?

5. Use the conditions of Exercise 4, but assume that $a + b \neq 360°$.
  a. Show that the composition of two rotations is again a rotation.
  b. Why is it not a translation?
  c. Can you find the center of the rotation that corresponds to the composition of two rotations?

6. Using a figure drawn on tracing paper, place it over the 0 point of the number line on another sheet of paper. Translate the figure to the right two units, then rotate through 90° about the point corresponding to 1. Find the center of the rotation that corresponds to the above composition.

7. Argue that the composition of a rotation and a translation is a rotation. Use examples for your argument.

SECTION E **REFLECTIONS**

BEHAVIORAL OBJECTIVES

E.1 Be able to illustrate an improper rigid motion by a physical model.

E.2 Be able to state the properties of a reflection.

E.3 Be able to use the properties of a reflection to show that a reflection maps a line parallel to the line of reflection into another parallel line.

E.4 Be able to use the properties of a reflection to show that the angles opposite equal sides of a triangle are equal.

IMPROPER RIGID MOTIONS

We will now discuss improper rigid motions. Before we formalize this idea, consider how a plane mirror works. When you look at yourself in a mirror, the image of your right ear is really on the left in the mirror. If you touch your right ear you will see your hand in the mirror, and although it is facing out from the mirror, it is on the left side of the image.

Make a regular figure—triangle, square, or rectangle—on a piece of paper. Draw a line *L* and place a mirror on the line. How would you trace the image of the figure as it appears in the mirror?

Again, we illustrate this type of rigid motion with our sheet of glass placed over some design, **F**, in the plane. Draw a line *L* on the glass, then pick up the glass, turn it over, and put it back so that line *L* falls exactly where it was before (see Figure 11.32). The set of points, **F′**, under the design on the glass is called the image of **F** under a *reflection* in line *L*.

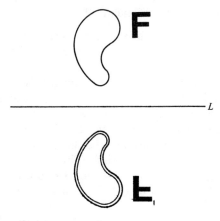

Figure 11.32

Can you see why this is called a reflection? Suppose line *L* represented a mirror. Would **F′** represent the image of **F** in that mirror? What do you notice about the distance of figure **F** from *L* and figure **F′** from *L*? What happens to the image of a figure if you move the figure itself closer and closer to the mirror? The motion that carries **F** onto **F′** is called a reflection in the line *L*. Let us observe some of the properties of reflections in a line by beginning with a point *P*.

■ **E.2** 1. If *P′* is the image of *P* under a reflection in *L*, then *P* is the image of *P′* under a reflection in *L*. See Figure 11.33.

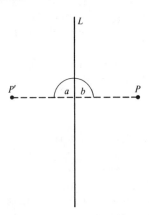

Figure 11.33

2. Because our motion of the sheet of glass preserves distance, the distance from *P* to the line *L* is equal to the distance from *P'* to the line *L*.

3. Since the angles are preserved by the motion of the piece of glass, the two angles, *a* and *b*, formed by the line segment *PP'* and *L* are equal. Since the measure of *a* and *b* add up to 180°, *a* and *b* are each right angles.

4. Line *L* is the perpendicular bisector of the segment *PP'*.

We should also note that the line *L* is carried into itself by the reflection. Thus, the fixed points of a reflection in the line *L* is the line itself. Are there any other fixed points?

■ E.1   It might be helpful again to consider an example in which we use a more familiar geometric figure to illustrate the concept of a reflection. Consider the quadrilateral *ABCD* and the line *L*. You can label a piece of paper *ABCD* on the front in each corner and then label the corners on the other side of the paper *A'B'C'D'* and consider what happens when you flip the piece of paper over a pencil laid on a desk. This is again a good way to deal with this idea in the elementary school. Be sure that *A* and *A'* are the same corner. Then label *B* and *B'*, *C* and *C'*, and *D* and *D'*. See Figure 11.34.

If you replace the pencil in Figure 11.34 with a mirror held in a vertical position and mark the corners of the quadrilateral with symbols such as @, $, #, and %, and if you hold the paper close to the vertical mirror, you will see quickly why this is called a reflection. Using mirrors in this manner is also a good way to teach children more about the physical and mathematical world and how they often coincide. Try moving the paper farther and farther away from the mirror. See Figure 11.35 for clarification, and try this idea with any square mirror and plane figure.

A third method of teaching this idea is by creasing the piece of paper along *L* (see Figure 11.33). Make a pinhole at *P* through the double thickness of paper to produce *P'*. This can be done with the Figure 11.34 by

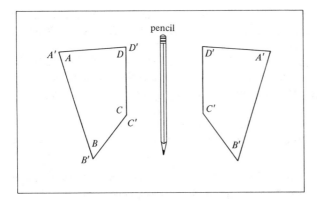

Figure 11.34

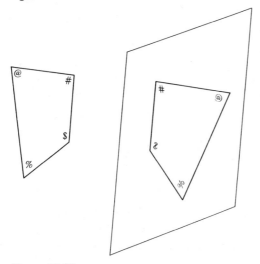

Figure 11.35

making pinholes at *A*, *B*, *C*, and *D*. In this way we would produce *A'*,
■ *B'*, *C'*, and *D'*.

■ **E.3** Suppose that the line *M* is parallel to the line of reflection, *L*, as in Figure
11.36; what can then be said about *M'*, the image of *M*? Suppose that
M' intersected *L* at some point *P'*. Then since *P'* is on *L* and all the points
of *L* are carried into themselves by the reflection, the point *P* must be on

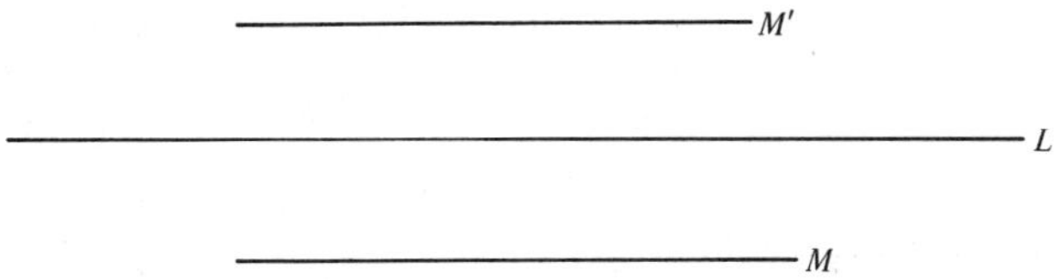

——————————————————————— *M'*

——————————————————————————— *L*

————————————————— *M*

Figure 11.36

both *M* and *L*. That is, *M* intersects *L*. But we are working under the condition that *M* is parallel to *L*. Hence, *P* is not on *L*, which tells us that *P'* is not on *L* and that *M'* must be parallel to *L*. Thus, the image of a line parallel to the line of reflection is a line parallel to the line of reflection. Let us demonstrate how reflections can be used to do a problem that you might recognize as a familiar theorem from high school geometry.

Theorem

E.4   *If two sides of a triangle have equal length, then the angles opposite them have equal measure.*

SOLUTION   Let *AC* and *AB* be the equal sides in triangle *ABC*. Let the line *L* bisect the angle at *A*, as in Figure 11.37.

Figure 11.37

Now reflect the triangle in line *L*. Since the angles formed by *L* at *A* are equal, the image of *AC* will fall on *AB*, and the image of *AB* will fall along *AC*. See Figure 11.38. Since *AC*

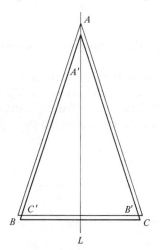

Figure 11.38

is equal to *AB*, the point *B'* will be *C* and the point *C'* will be *B*. Thus, triangle *ABC* is its own image under a reflection in *L*. It now should be obvious that the measure of the angle at *C* is equal to the measure of the angle at *C'*, which in turn is equal to the measure of the angle at *B*. Similarly, the measure of the angle at *B* is equal to the measure of the angle at *C*.

EXERCISES

1. Show that two reflections in the same line are inverse transformations, that is, that their composition is the identity transformation.

2. If *M* is a reflection in a line *L*, show that M^2; that is, *M* composed by itself, is the identity transformation.

3. Show that the set $\{M, M^2\}$ is a group under the operation of composition.

4. If *N* is a line that intersects the line of reflection *L* at *P*, does *N'* intersect *L*? If so, at what point?

5. What is the image of a circle under a reflection in a line?

6. Sketch the image of a circle under a reflection if
a. the line of reflection is tangent to the circle (by tangent we mean that the line touches the circle at one point only).
b. the line of reflection intersects the circle at two points.
c. the line of reflection contains a diameter.

SECTION F **CONGRUENCE**

BEHAVIORAL OBJECTIVES

F.1 Be able to use rigid motions to define congruent figures.

F.2 Be able to give examples of congruent figures.

F.3 Be able to use rigid motions to prove geometry theorems.

At this point we could begin to prove the theorems of high school geometry. In order to do this, we need to know what congruent figures are. So we make the following definition.

Definition

■ **F.1**   *Two figures are congruent if and only if one of them is the image of the*
■   *other under a rigid motion.*

We have already seen an example of congruent triangles in the previous section, where we showed that the angles opposite equal sides in a triangle
■ **F.2** had equal measure. Recall that under a reflection in *L*, *A'B'* was *AC*. That is, the image of *AB* was *AC*. Thus, we can say that *AB* is *congruent*

to *AC*. Since the angle at *C* is mapped by the rigid motion of reflection in *L* onto the angle at *B*, we say that angle *C* is congruent to angle *B*.

We also showed that triangle *ABC* was congruent to itself where *A* = *A'*, *B* = *C'*, and *C* = *B'*. That is, the triangle was congruent to itself "flipped over." In this case we say that triangle *ABC* is congruent to triangle *ACB*.

As usual, mathematicians use symbols for words that are used fairly often. The usual symbol for congruent is ≃, the symbol for angle is ∠, and for triangle it is △. Historically, the symbol ≃ is a combination of the symbols ~, meaning "the same shape as," and =, meaning "having equal measure."

Thus, the three examples of congruence given above can be summarized as follows.

1. $AB \simeq AC$
2. $\angle B \simeq \angle C$
3. $\triangle ABC \simeq \triangle ACB$

■ **F.2**
▲ **F.3**

If two squares are to be congruent, what conditions must hold? To answer this, let us consider two squares *ABCD* and *EFGH*, as in Figure 11.39. It should be obvious that there is a translation *T* such that $T(A) = E$.

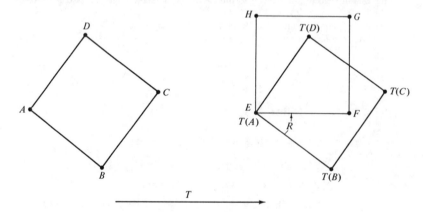

Figure 11.39

Then rotate, by *R*, the image of the first square so that $RT(A) = E$ and $RT(AB)$ falls above *EF*. If the squares are to be congruent, $RT(B)$ must be *F*. This requires that $AB \simeq EF$, and we have the answer to our question:
■ The sides must be congruent.

Using Figure 11.39 as a model, it is relatively easy to prove the familiar theorem of geometry: If two sides and the included angle of one triangle are congruent to two sides and the included angle of another, then the triangles are congruent. The proof of this theorem is usually assumed in most high school geometry courses; it uses rigid motions and is easy and,
▲ in fact, was used by Euclid himself.

EXERCISES

1. Show that for any angle A, $\angle A \simeq \angle A$.

2. Show that if in $\triangle ABC$ and $\triangle DEF$, $\angle B \simeq \angle E$ and $\overline{AB} \simeq \overline{DE}$ and $\overline{BC} \simeq \overline{EF}$, then $\triangle ABC \simeq \triangle DEF$.

3. Show that the opposite angles formed by two intersecting lines are congruent.

4. What condition is necessary for two circles to be congruent?

5. Show that all the sides of a square are congruent to each other.

Abelian group. See Commutative group
Abundant numbers, 165
Addition table
  base ten, 101
  base twelve, 91
Addition
  associative property of, 260
  associative property on I, 183–184
  associative property on naturals, 115–116
  associative property on whole numbers, 115
  base five, 75
  base twelve, 91
  cancellation laws, 188–190
    for rationals, 276–278
  closure
    on I, 181
    for rationals, 258
    for whole numbers, 110
  closure property for naturals, 110
  commutative property of, 112, 259
  commutative property on I, 182–183
  of decimal fractions, 329–330
  of fractions, 308–310
  identity element in I, 185–186
  identity of, for rationals, 260–261
  identity property on N and W, 118
  independence of choice of ordered pairs, 256
  of integers, 178–180
    definition of, 179, 181
  of inverses on rationals, 261–262
  mod 3, 228–231
    properties of, 229
  of natural numbers, 99–101
  properties of
    on integers, 188
    on natural numbers, 109–119
    table of, 122
    on whole numbers, 109–119
  of rationals, 252–254
  of standard sets, 100
  table of properties, 122
  of whole numbers, 99–101
    definition of, 100
Additive inverse, 186–188
  definition for I, 187
Angle, 413–414

measure of, 414
Antecedent, 388
Arabic system. See Hindu-Arabic system
Associative property, 41, 114
  of addition on I, 183–184
  of addition on naturals, 115
  of addition on rationals, 260
  of intersection, 43
  mod 3, 229
  of multiplication in base five, 78–79
  of multiplication on I, 202
  of multiplication on naturals, 116
  of rationals under addition, 260
  of rationals under multiplication, 265–267
  of reals, 350
  of union, 41–42

Babylonian numeration system, 64–65
Base eight, 86
Base five, 74–81
  addition in, 75–78
  counting in, 81
  multiplication in, 77–81
Base four, 85
Base system, example of in numeration system, 55–58
Base systems
  base five, 74–81
  base twelve, 88–92
  powers of, 84–87
Base twelve, 88–92
  addition table for, 91
  multiplication table for, 91
Base two. See Binary system
Basic fraction, 249
Basic pair, 171, 175
Between operation, 97
Bibliography
  discovery, 137
  games, 408
  mathematics laboratory, 341
  structure in school mathematics, 52–53
Biconditional, 388–391
Binary operation, 95–97
  between operation, 97
  definition of, 97
  first number operation, 96

Binary operation (*Continued*)
  greater number operation, 97
Binary system, 81–84
  counting in a, 81
Braids, 242–247

Cancellation laws
  for addition in *I*, 188–190
  applications of, 297–300
  intuitively defined, 296
  on rationals, 276–279
Cancellation property
  of addition, 126
  for "less than," 130–131, 295–300
  of multiplication, 127, 211
Cardinal number. *See* Cardinality
Cardinality, definition of, 95
Cartesian product. *See* Cartesian set
Cartesian set, 48–50
  definition of, 49
  geometric model of a, 49–50
  model for multiplication, 103
Chain law. *See* Extended transitive law
Closed curve, 416
Closure, 37–38, 109
  for addition of whole numbers, 182
  for addition on *I*, 181
  for addition on rationals, 258
  for multiplication on *I*, 199
  for multiplication on rationals, 263
  for naturals under addition, 110
  for naturals under multiplication, 110
  for reals, 350
  for subtraction on *I*, 193
  for subtraction on rationals, 280
  for whole numbers, 110
Closure property
  for intersection, 38
  for union, 38
Common divisor, 155
Common fractions, 301
  definition of, 305
Common multiple, 153
Commutative group, 188, 230, 274
Commutative property, 39–40, 110–113
  for addition, 112
  for addition on *I*, 182–183
  for addition on rationals, 259
  of intersection, 40
  for multiplication, 112
  for multiplication on *I*, 200
  for multiplication on rationals, 264–265
  of reals, 350
  of union, 40
Complement, of a set, 14
Composite number, 140
Composition of mappings, 423–425
Composition of rotations, 435–438

Compound sentences, solving, 371–374
Compound statement, definition of, 377
Conditional, 387–388
  equivalent statements, 397–398
Congruence, 443–445
Congruent figures, definition of, 443
Conjunction, 377–378, 380–382
  truth table for, 380–382
Consequent, 388
Contradiction, law of, 405
Contrapositive, 397
  law of, 405
Converse, 397
Correspondence, integers and subset of rationals, 272–274
Counting
  base nine, 89
  base ten, 89
  base three, 89
  base twelve, 89, 228
  history of, 54
Counting numbers, development of, 94–95. *See also* Natural numbers
Curve, 416–418
  closed, 416
  simple closed, 416
Cyclic group, 235–238

Decimal fractions, 324–326
  addition and subtraction of, 329–331
  multiplication of, 332–335
  repeating and terminating, 326–338
Deductive proof, 401–406
Deficient numbers, 165
Definition, characteristics of, 106
Denominator, 252, 302
Digit operation, 96
Discovery sequences, 215–218
  "Find My Secret," 217–218
Discovery teaching, 133–136, 215–218
Disjoint sets, 18, 31
Disjunction, 377–378, 380–382
  truth table for, 380–382
Distributive property, 46–47, 119–121, 125, 206
  for intersection over union, 48
  mod 3, 233
  multiplying in base five, 78–80
  for *N* and *W*, 120–121
  for rationals, 275
  of reals, 351
  for union over intersection, 48
Divisibility, 145–152
  powers of five, 150–151
  powers of three, 148–149
  powers of two, 147–148
  Sieve of Eratosthenes, 145

Division
  closure for, 123
  of decimal fractions, 337–339
  definition of, 106
  definition of, for integers, 211
  definition of, for rationals, 282–285
  of fractions, 320–322
  identity property of, 124
  on integers, 209–211
  mod 3, 232
  of natural numbers, 105–107
  properties of, on naturals, 123–125
  property of one, 287
  right identity for, 287
  of whole numbers, 105–107
Division on naturals, properties of, 123–125
Double negation, law of, 400
Duodecimal system. *See* Base twelve

Egyptian numeration system, 59–60
Elements, of a set, 2
Empty set, 7–9
Equal fractions, 306
Equal mapping, definition of, 424
Equality
  as an equivalence relation, 126
  of rational numbers, 251
  of sets, 20
Equivalence, logical, 393–396
Equivalence class, 171–177, 228
  mod 3, 231
  rationals, 249
Equivalence relation
  definition of, 24
  of integers, 171–177
Equivalence statement, definition of, 390
Equivalent pairs, 177, 212
  in addition, 180
  definition of, 175
  for integers, 174
Equivalent sets, 22–25
Eratosthenes, Sieve of, 145
Euclid's algorithm, 159
Excluded middle, law of, 399
Exclusive "or," 365
Expanded form, 70–71
Expanded notation. *See* Expanded form
Exponential notation, 144
Extended transitive law, 401

Factor
  definition of, 138
  exponential notation, 144
Factorization, 138–139
  prime, 141–142
Field
  definition of, 233

of rational numbers, 274–275
Field properties of rationals, 275
Finite set, 3
First number operation, 40, 96
Fraction
  basic, 249
  improper, 311
  mixed number, 312
  notation used with, 311–314
Fractions, 301–307
  addition of, 308–310
  addition of decimal, 329–330
  common, 301, 305
  decimal fractions, 324–326
  definition of, 305
  division of, 320–322
  division of decimal, 337–339
  equality of, 305–307
  geometric argument for addition of, 308–309
  least common denominator, 312
  multiplication by power of ten, 333–336
  multiplication of, 316–319
  multiplication of decimal, 332–335
  multiplying mixed numbers, 319–320
  physical examples of, 302–304
  physical model for multiplication of, 316–317
  properties of, 307
  properties of addition and subtraction, 314–315
  properties of multiplication and division, 323
  quotient of, 321
  repeating and terminating decimal, 326–328
  represented on number line, 304
  sets of equal, 306
  subtraction of, 310–311
  subtraction of decimal, 330–331
Fundamental theorem of arithmetic, 143–144

Games, 407–408
  suggested readings for, 408
Geoboard, 242
Graphs, using to solve open sentences, 354–355
Greater number operation, 97
"Greater than," 128–132
Greatest common divisor, 155–157
  Euclid's algorithm, 159
  factor method, 157–158
Group, 188, 220–225
  commutative, 188, 230, 274
  cyclic, 235–238
  definition of, 221
  order of a, 223

Group (*Continued*)
   rotation, 236, 238
   symmetry, 240
Groupoid, 219–220

Half-turn, 436–438
Hexagon, 417
Hindu-Arabic numeration system, 69–71
   characteristics of, 69–71
Hypothesis, 401

Identity, 43–44, 224
   for addition on I, 185–186
   of addition on naturals, 118
   of addition on rationals, 260–261
   for division, 124
   left, 118
   in mod 3, 117–118
   for multiplication, 203
   of multiplication on naturals, 118
   of multiplication on rationals,
      267–268
   of reals under addition, 350
   right, 118
   for subtraction, 124
Identity element, 117–118
   for intersection, 44
   for union, 44
Identity mapping, 424
"If and only if," 123, 189, 295
Image, under rotation, 432
Implication, 393–396
   definition of, 390
   transitive law of. *See Syllogism*
Improper fraction, 311
Improper rigid motions, 439–443
   physical models of, 440–442
Inclusive "or," 365
Inequality
   cancellation on rationals, 295–300
   solution sets involving, 365–371
   using addition to define, 293
Infinite set, 3
Infinitude of primes, 151–152
Integer, definition of, 172
Integers
   addition of, 178–180
   associativity of multiplication, 202
   closure of multiplication, 199
   commutativity of multiplication, 200
   correspondence of, with subset of
      rationals, 272–274
   definition of addition, 181
   definition of multiplication, 198
   division of, 209–212
   equal, 176
   multiplication of, 195–207
   multiplicative identity of, 203
   multiplicative inverses of, 24
   operations on, summary of, 207

order on, 212–214
   as sets, 171–177
   subtraction of, 191–193
   summary of properties, 288–289
Intersection
   attributes of, 33
   combined with union, 44–48
   definition of, 33
   distributive property over union, 47
   identity for, 44
   of sets, 30–31
Invariant, 422
Inverse, 186, 224
   addition on rationals, 261–262
   additive, 186–188
   left, 186
   logical, 397
   multiplication on I, 204
   for multiplication on rationals, 268–
      271
   in reals, 351
   right, 186
   used in solving open sentences,
      356–358
Irrational number, 333
Irrational numbers
   approximating, 345–346
   existence of, 343–345
Irrational points, 333
Isomorphic, 182, 237–238
   definition of, 237
Isomorphism, 237–238
Isomorphs. *See* Isomorphic

Laboratory approach, 340–341
Lattice points, 50
Law
   of contradiction, 400
   of contrapositive, 405
   of double negation, 400
   of excluded middle, 399
Least common multiple, 153–155
Left identity. *See* Identity
Lemma, 343
"Less than," 128–132
Line, 410
Line segment, 410
   measure of a, 418
Logic, laws of, 399–401
Logical equivalence. *See* Equivalence

Madison project, 215–216
Mapping
   composition of, 423–425
   definition of, 422
   definition of equal, 424
   equal, 424
   identity, 424
   rigid, 422. *See also* Rigid motion
Mathematical sentence, 352

Mathematics laboratory, 340–341
  suggested readings for, 341
Mayan numeration system, 66–68
Measure
  of an angle, 414
  of a line segment, 418
Members, of a set, 2
Mixed number, 312
Mixed numbers, multiplication of,
    319–320
Mod 3, 117, 228–234, 238
  addition, 228–231
  distributive property, 233
  equivalence classes, 231
  multiplication, 232
  subtraction, 228–231
Modular addition, with braids, 244
Modular system, 227–228
  compared with base twelve, 227
Modus ponens, 402
Modus tollens, 403
Monoid, 219–220
Multiplication
  associative property of, for naturals,
    116
  associativity on *I*, 202
  base five, 77–81
  base twelve, 91
    table for base twelve, 91
  cancellation law for, 211
  cancellation law for rationals,
    278–279
  closure
    on *I*, 199
    on rationals, 263
    for whole numbers, 110
  closure property for naturals, 110
  commutative property of, 112
  commutativity
    on *I*, 200
    of rationals, 264–267
  of decimal fractions, 332–335
  defined in terms of addition, 105
  definition of, 105
    for integers, 198
    for rationals, 263
  examples of, with decimal fractions,
    333–334
  of fractions, 316–319
  identity for, 203
    on rationals, 267–268
  identity property on *N* and *W*, 118
  of integers, 195–199
    definition of, 198
  inverses
    in *I*, 204
    in rationals, 268–271
  mod 3, 232
  of naturals, 103–105
    properties of, 109–119

  physical model for, 103
  physical model for fractions,
    316–317
  properties of
    on naturals, 109–119
    on whole numbers, 109–119
  of rationals, 255–257
  table of properties, 122
  with powers of reals, 364–349
  of whole numbers, 103–105
    properties of, 109–119
    table of, 104
  zero property of, 204, 271
Multiplication inverse. *See* Inverse

Natural numbers
  addition of, 99–101
  attributes of, 94, 98
  development of, 94–95
  division on, 105–107
  multiplication of, 103–105
  order of, 128
  Peano postulates for, 98
  properties of addition, 109–112
  properties of division, 123–125
  properties of subtraction, 123–125
  subtraction of, 101–102
  summary of properties, 288–289
  table of properties, 122
Notation, exponential, 144
Null set, 8
Number
  composite, 140
  irrational, 333
  on the number line, 167
  prime, 140
Number line, 187–189, 333
  order on a, 291
Number property of sets, 10–12, 34–35
  definition of, 13
Number property of union of sets,
    34–35
Number ray, 168
Numbers
  abundant, 165
  approximating irrational, 345–346
  deficient, 165
  existence of, 343–345
  perfect, 165
  polygonal. *See* Polygonal numbers
  prime twins, 165
  Pythagorean triples, 164–165
  real, 333
Numeration systems, 54–59
  Babylonian, 64–65
  base system, 55–58
  Egyptian, 59–60
  Mayan, 66–68
  positional
    base twelve, 88–92

Numeration systems (*Continued*)
　binary, 81–84
　principles used in, 55–59
　　addition, 55
　　base, 56
　　multiplication, 56
　　position, 56–58
　　repetition, 55
　　subtraction, 55
　Roman, 60–63
Numerator, 252, 302

One-to-one correspondence, 21–22,
　　237, 333
　between integers and rationals,
　　272–274
Open sentence, 131, 352
Open sentences
　effect of universe on solution set,
　　357
　involving multiple operations,
　　358–364
　solution sets involving inequality,
　　365–371
　solving, 354–374
　solving compound, 371–374
　using graphs to solve, 354–355
　using inverses to solve, 356–358
Operation
　commutative property of, 40
　first number, 40
Operations
　associative, 114
　closed, 37, 109
　commutative, 110–113
　distributive, 121
　on integers, summary of, 207
　primary, 101
　secondary, 101
　on sets, 27–34
　subtraction on *I*, 191–193
　union, 27
　union combined with intersection,
　　44–48
Or
　exclusive, 365
　inclusive, 367
Order
　based on number line, 291
　definition of, on rationals, 289–295
　finite group, 223
　generalizations for, 299
　of integers, 212–214
　properties of, 128
　trichotomy property of, 128
Ordered pair, 48

Parallelogram, 417, 431
Partition, 228
　definition of, 169

Peano postulates, 98
Pegboard, 242
Pentagon, 417
Perfect numbers, 165
Plane, 409
Point, 409
Points, irrational, 333
Polygon, 417
Polygonal numbers
　square, 163
　triangular, 161–162
Positional numeration system, 58
Positional numeration systems
　base five, 74–76
　base twelve, 88–92
　bases other than ten, 74–84
　binary, 81–84
　Hindu-Arabic, 69–71
　principles of, 73–74
Power set, 12–13
　intersection of a, 35–36
　union of a, 35–36
Primary operations, 101, 105
Prime
　infinitude of, 151–152
　relative, 158
Prime factor method for G.C.D.,
　　157–158
Prime factorization, 141–142
Prime number, 140
Principles, in numeration systems
　addition, 55
　base, 56
　multiplication, 56
　positional notation, 57–59, 70
　repetition, 56
　subtraction, 55
Prior knowledge, used in proof, 200
Proof, deductive, 401–406
Proper rigid motion, 426–438
Proper subset, 9–10
Pythagorean theorem, 164
Pythagorean triples, 164–165

Quadrilateral, 417
Quotient, 105, 209, 321
　of fractions, 321

Rational number
　definition of, 251
　equivalence class of a, 250–251
Rational numbers
　addition of, 252–254
　associative property of addition on,
　　260
　cancellation laws for inequality,
　　295–300
　closure
　　of addition on, 259
　　of multiplication, 263

commutative property
  of addition on, 259
  under multiplication, 264–267
definition of order on, 289–295
development of, 248–252
distributive property for, 275
division of, 282–285
equality of, 251
field of, 274–275
field properties of, 275
historical development of, 248
identity
  for addition, 260–261
  for multiplication, 267–268
independence of ordered pairs for
  addition, 256
inverses
  for addition, 261–262
  for multiplication, 268–271
multiplication of, 255–257
multiplying by powers of, 346–349
summary of properties of, 288–289
Ray, 411
Real number, definition of, 333
Real number line, 333
Real numbers
  distributive property of, 351
  inverses under addition and
    multiplication, 351
  properties of, 350–352
Reciprocal, 268
Rectangle, 417
Reflexive property, 24
  of equivalent pairs, 177
Relation, equivalence, 24–25
Relative complement, 14
Relatively prime, 158
Rhombus, 417
Right identity, 124, 287. *See also*
  Identity
Rigid mapping, 422
Rigid motion, 420–424. *See also*
  Mapping
  improper, 426, 439–443
  physical model of, 421
  proper, 426–438
  rotation, 432–438
  translation of, 428–432
  types of, 426–427
Roman numeration system, 60–63
Rotation, 432–438
  image under, 432
Rotation group, 236–238
Rotations
  composition of, 435–438
  half-turn, 436–438
  properties of, 433–438

School Mathematics Study Group,
  216–217

Secondary operations, 101, 105
Semigroup, 219–220
Sentence, mathematical, 352
  open, 352
Sentences
  solving compound, 371–374
  translating symbolic statements to,
    378
Set, 1
  complement of a, 14–18
  described in words, 2
  elements of a, 2
  empty, 7
  equality of a, 20
  finite, 3
  infinite, 3
  members of a, 2
  membership in a, 3
  number property of a, 10, 34
  power set, 12
  relative complement of a, 14–18
  solution, 131, 354
  standard, 99
  subset, 8–10
  tabulation of a, 2
  universal, 7
  well-defined, 2
Set builder notation, 4
Sets
  as integers, 171
  disjoint, 18, 31
  of equal fractions, 306
  equivalent, 22–25
  one-to-one correspondence of,
    21–22
  operations on, 27–34
  power set of intersection, 35–36
  power set of union, 35
  union of, 27–29, 34
Sieve of Eratosthenes, 145
Simple closed curve, 416
Simplication for ∧, 404
Solution. *See* Solution set
Solution set, 131
  for compound sentences, 371–374
  definition of, 353
  effect of universe on a, 357
  equations involving multiple
    operations, 358–364
  sentences involving inequality,
    365–371
  summary for finding a, 364
Solving open sentences, 354–374
Square, 417
Square numbers, 163
  standard set, 94–95, 99
Statement, definition of, 377
Statements, 376–377
  involving negation, 383–386
Straight line, translation of a 430–431

Structure, role of, in school
    mathematics, 51–53
Subgroup, 225–226
    definition of, 226
Subset, 8–10
    definition of, 9
    proper, 9–10
Subtraction
    associative properties of, 123–124
        closure, 123
        commutation, 123–124
        on *I*, 193–194
    closure of, 280
        on *I*, 193
    of decimal fractions, 330–331
    definition of, for rationals, 279–280
    of fractions, 310–311
    identity property of, 124
    of integers, 191–193
    mod, 3, 230–231
    of natural numbers, 101–102
    properties of, on naturals, 123–125
    of whole numbers, 101–102
Successor operation, 96
Syllogism, law of, 400
Symbolic statements, translating
    sentences to, 378
Symmetric property, 24
Symmetry, 422
Symmetry group, 240

Tautological statements. *See* Tautology
Tautology, 397–399
    definition of, 397
Transformations, 420–426. *See also*
    Rigid motions; Mapping
Transitive law, extended, 401
Transitive law of implication. *See*
    Syllogism
Transitive property, 24
    of equivalent pairs, 177
Translations, 428–432
Trapezoid, 417
Triangle, 417
Triangular numbers, 161–162
Trichotomy, 128, 212
Trichotomy property for rationals, 290
Truth set. *See* Solution set

Truth table, 379
    conjunction, 380–382
    disjunction, 380–382
Truth value, 376
Truth values, 378–380

Undefined term, 1
Union
    combined with intersection, 44–48
    definition of, 29
    distributive property over
        intersection, 47
    identity for, 44
    number property of, 34–35
    power set of, 35
    properties of, 29
    of sets, 27–29, 34
Unit distance, 167
Unitary operation, 96
    digit operation, 96
    successor operation, 96
Universal set, 7
Universe, effect on solution set, 357

Vertex, 414

Whole numbers, 98
    addition of, 99–101
    definition of, 100
    division of, 105–107
    multiplication of, 103–105
    order of, 128–129
    properties of
        addition, 109–119
        division, 123–128
        subtraction, 123–128
    subtraction of, 101–102
    summary of properties, 288–289
    table of properties, 122

Zero
    as a divisor, 286
    identity for reals, 350
    properties of, 286
Zero property
    division, 285
    multiplication, 118–119, 204, 271,
        351
    subtraction, 194, 281

73 74 75 76   9 8 7 6 5 4 3 2 1